Textiles for Functional Applications

Edited by Bipin Kumar

Published in London, United Kingdom

IntechOpen

Supporting open minds since 2005

Textiles for Functional Applications
http://dx.doi.org/10.5772/intechopen.91596
Edited by Bipin Kumar

Assistant to the Editor: Viraj Somkuwar

Contributors
Zeynep Omerogullari Basyigit, Mallika Datta, Devarun Nath, Srijan Das, Mukesh Kumar Singh, Ion Razvan Radulescu, Lilioara Surdu, Emilia Visileanu, Cristian Morari, Marian Costea, Thilagavathi Govindharajan, Viju Subramoniapllai, Harozila Ramli, Tajul Shuhaizam Said, Ramratan Guru, Anupam Kumar, Rohit Kumar, Justine Decaens, Ahmad Ibrahim, Edison Omollo Oduor, Lucy Wanjiru Ciera, Edwin Kamalha, Yamini Jhanji, Jamiu Mosebolatan Jabar, Ikra Iftekhar Iftekhar Shuvo, Patricia I. Dolez, Dominic Lachapelle, Bipin Kumar, Viraj Somkuwar, Joseph-Émile Laquerre, Patricia Forcier, Vincent Deregnaucourt, Olivier Vermeersch

First published in London, United Kingdom, 2021 by IntechOpen
IntechOpen is the global imprint of INTECHOPEN LIMITED, registered in England and Wales, registration number: 11086078, 5 Princes Gate Court, London, SW7 2QJ, United Kingdom
Printed in Croatia

British Library Cataloguing-in-Publication Data
A catalogue record for this book is available from the British Library

Additional hard and PDF copies can be obtained from orders@intechopen.com

Textiles for Functional Applications
Edited by Bipin Kumar
p. cm.
Print ISBN 978-1-83968-629-0
Online ISBN 978-1-83968-630-6
eBook (PDF) ISBN 978-1-83968-631-3

For EU product safety concerns:
IN TECH d.o.o., Prolaz Marije Krucifikse Kozulić 3, 51000 Rijeka, Croatia, info@intechopen.com or visit our website at intechopen.com.

Meet the editor

Dr. Bipin Kumar graduated from the Indian Institute of Technology (IIT) Delhi, India, with a Ph.D. in Textile Engineering in 2013. Currently, he is an assistant professor in the Department of Textile Technology at IIT Delhi. Prior to joining IIT Delhi, he was a research assistant professor (2016–2017) at the Hong Kong Polytechnic University. He also served as a cultural ambassador (2014–2016) to the United States at the University of California Davis via the prestigious Fulbright Fellowship program. His main research focuses on smart fibrous/polymeric materials and related fabric structures. He has more than forty publications in leading refereed SCI journals of materials, textiles, and medical fields, seven patents, three authored books, and many other academic publications. He has been awarded several prestigious awards including the Entrepreneur Faculty Award (IITDAA, 2020), IEI Young Engineer Award (Institute of Engineers, 2019), Teaching Excellence Award (IIT Delhi, 2018), DST INSPIRE Faculty Award 2016 (DST, 2016), Award for Excellence in Postdoctoral Research (UC Davis, 2016), and many more. He holds editorial membership at several international refereed journals including *Journal of Engineered Fibers and Fabrics and Fabrics, American Association of Textile Chemists and Colorists Journal of Research, Fibres & Textiles in Eastern Europe, Advance Research in Textile Engineering, Current Trends in Fashion Technology & Textile Engineering and Journal of Textile Engineering & Fashion Technology.* Dr. Kumar's research team launched their first commercial product, KAWACH Mask, in May 2020 at the time of the COVID-19 emergency. He is leading two successful start-ups at IIT Delhi, ETEX and SWATRIC, to commercialize several textile innovations for engineering and medical applications, including KAWACH masks, antiviral T-shirts, anti-radiation garments, Indian flags, E-textiles, anxiety garments, and more.

Contents

Preface **XI**

Section 1
Introduction **1**

Chapter 1 **3**
Introductory Chapter: Functional Textiles
by Bipin Kumar and Viraj Somkuwar

Section 2
Textile Material, Design and Technology **11**

Chapter 2 **13**
Antimicrobial Agents for Textiles: Types, Mechanisms and Analysis Standards
by Ahmad Ibrahim, Joseph-Émile Laquerre, Patricia Forcier, Vincent Deregnaucourt, Justine Decaens and Olivier Vermeersch

Chapter 3 **43**
Textiles Functionalization - A Review of Materials, Processes, and Assessment
by Mukesh Kumar Singh

Chapter 4 **75**
Applications of Silk in Biomedical and Healthcare Textiles
by Edison Omollo Oduor, Lucy Wanjiru Ciera and Edwin Kamalha

Chapter 5 **101**
Application Technologies for Functional Finishing of Textile Materials
by Zeynep Omerogullari Basyigit

Chapter 6 **123**
Surface Design Technique through Tradition Technique
by Harozila Ramli and Tajul Shuhaizam Said

Section 3
Testing and Analysis **139**

Chapter 7 **141**
Smart Textiles Testing: A Roadmap to Standardized Test Methods for Safety and Quality-Control
by Ikra Iftekhar Shuvo, Justine Decaens, Dominic Lachapelle and Patricia I. Dolez

Chapter 8 **171**
Modeling and Validating Analytic Relations for Electromagnetic Shielding Effectiveness of Fabrics with Conductive Yarns
by Ion Razvan Radulescu, Lilioara Surdu, Emilia Visileanu, Cristian Morari and Marian Costea

Section 4
Application Areas **193**

Chapter 9 **195**
Face Mask: A Novel Material for Protection against Bacteria/Virus
by Thilagavathi Govindharajan and Viju Subramoniapllai

Chapter 10 **209**
Antimicrobial Functional Textiles
by Jamiu Mosebolatan Jabar

Chapter 11 **219**
Functional Textile for Active Wear Clothing
by Ramratan Guru, Anupam Kumar and Rohit Kumar

Chapter 12 **237**
Textiles for Noise Control
by Mallika Datta, Srijan Das and Devarun Nath

Chapter 13 **261**
Sportswear: Acumen of Raw Materials, Designing, Innovative and Sustainable Concepts
by Yamini Jhanji

Preface

Textiles have been an integral part of human life for centuries. They were initially thought to be used for coverage, storage, protection, ropes, fishing, and more. A textile material and its structures offer many unique characteristics such as flexibility, drape, breathability, and strength, which are otherwise not easily achieved in other engineering materials and structures. Further, their surfaces can be functionalized to achieve new properties such as super-hydrophobicity, shape memory, phase change, anti-microbial characteristics, and so on. Today, their applications have widened in multiple areas including medical, civil, defense, energy, transport, and protection industries due to the development of smart materials and their fibers, processing methods, and innovative technologies. This book emphasizes textile science, engineering, technology, and the design processes involved in developing textile products for medicine, sports, protection, and filtration. The content of this book is divided into four sections. The first and Introduction section is followed by the second which discusses the material and technology used for fabricating different types of functional textiles. The third section focuses on other characterization techniques and analysis of various results published in the scientific literature. The final section describes the different application areas where functional textiles have been utilized to enhance the efficiency of processes.

Following the Introductory chapter, Chapter 2 discusses the enhancement of healthcare textiles using different antimicrobial agents. It examines different types of antimicrobial agents such as metallic, non-metallic, and non-leaching agents, along with their application methodologies. Chapter 3 provides a critical review of different types of intelligent functionality imparted to textiles such as stimuli-responsive textiles, phase-change material, enzymes used for functionalization, and more. Chapter 4 discusses the use of silk in functional textiles, as silk fiber shows excellent mechanical properties that are beneficial in functional textiles. Chapters 5 and 6 describe the application techniques to transfer various materials and treatments over a textile fabric and the compatibility of these materials and textiles. Chapters 7 and 8 cover the standard methodology for testing and assessing the properties of functional textiles. Due to the COVID-19 pandemic, face masks have become an essential healthcare product, thus it is necessary to examine the functional aspects of these masks. Chapter 9 critically reviews different face masks, their manufacturing techniques, and testing parameters. Chapter 10 discusses the application of antimicrobial agents on various textile surfaces, as every type of fiber has different compatibility with the various antimicrobial agents. The chapter examines the advantages and disadvantages of using different materials as substrates for antimicrobial applications. Functional textiles are a significant contribution to activewear. As such, Chapter 11 discusses different types of material, fabric, and fabric structures used to design activewear clothing. It presents various studies on the material used for activewear clothing and the effect of different parameters on permeability properties such as water vapor permeability, air permeability, and moisture management. Chapter 12 discusses textiles used for noise control and noise barriers for acoustic management. It covers the mechanisms and textile materials used for sound absorption while Chapter 13 discusses concepts employed in sportswear and sports textile industry.

I would like to acknowledge the contributing authors for their valuable time and efforts in writing their chapters. I would also like to thank Viraj Somkuwar (Ph.D. scholar, IIT Delhi) for helping with the review of multiple chapters and helping with the final editing of the book.

Bipin Kumar
Indian Institute of Technology Delhi,
Department of Textile and Fiber Engineering,
New Delhi, India

Section 1

Introduction

Chapter 1

Introductory Chapter: Functional Textiles

Bipin Kumar and Viraj Somkuwar

1. Introduction

Recent developments in high-performance fibers, fabrics, and manufacturing technologies become the main driving force behind the emergence of functional textiles. The majority of functional textiles were originally used in defense applications, but due to their popularity, they are now available to the general population. The field of functional clothing is vast and varied, with each function having its own set of specifications, material needs, and corresponding technologies and methods. Hence, the designing and from the prospect of manufacturing, the functional clothing becomes a challenge due to limited set of standards and varying requirements according to the needs. Functional textiles have a wide range of applications, including saving lives, adapting to hostile environments, and improving performance and quality of life [1, 2].

The recent COVID-19 epidemic had a significant impact on the world, posing difficulties to the health care infrastructure, society, culture, and economic system. The outbreak has failed all the advance medical treatments, but the crucial role was played by the non-pharmaceutical measures in reducing the transmission of viruses. The functional and smart textile has played a crucial role in designing the personal protective equipment (PPE) and telemedicine for strengthening the healthcare system. The breakthrough research in the field of nanomaterial, surface treatments, and finishing technology has given an edge to the functional textiles to successfully prevent this spread of viruses and disease. The personal protective equipment such as masks, surgical gowns, and gloves are the examples of textile functional clothing used for COVID19 protection. The fabrics were treated with antiviral and antimicrobial treatment to enhance its protection efficiency [3–5]. Another major factors that help the people during the lockdown situation are the telemedicine technology, which again mainly depends upon the smart and wearable textile which contains some electronic sensing functionality. The functional clothing used for telemedicine contains, electronic-embedded sensor, or the textile material itself converted into the sensor, which monitor the human vital parameters, such as heart rate, breathing pattern, blood oxygen level [6].

The focus of this book is to review key materials, manufacturing technologies, and the application methodologies for the designing and development of functional and technical textiles. This chapter will provide an overview of the principles of different types of functional textiles and their applications in various industries.

1.1 Functional textile: definition

All the textile clothing is meant to perform several objectives in our daily life from esthetics to provide basic protection from the various external factors. "Functional clothing" can therefore be defined as "a generic term that includes all

IntechOpen

such types of clothing or assemblies that are specifically engineered to deliver a pre-defined performance or functionality to the user, over and above its normal functions" [7]. This type of clothing can be produced by using high-performance fiber, novelty finishes, or intrinsic modification of conventional material. The clothing is expected to perform some specific functions, which can be protecting the person from a hazardous working environment, facilitating the movement during sporting activity, assisting a physically challenged individual or enhancing the endurance of a sports person. Functional clothing can be used for protection against the life-threatening viruses and diseases in medical treatment. It can also have the electronic functionality embedded inside the clothing, which can be used for transmitting the signals wirelessly and monitor the human vital to provide telemedicine facilities [1].

1.2 Classification of functional textiles

The standard technical textile is categorized according to the application, such as Sportech (sports textile), Protech (protective textile), and MedTech (medical textile). Designing a product for a specific end user opens up a new classification system that includes current technology. The classification for the functional textile can be made according to the functionality and requirements. As the material, manufacturing technology may be the same or varied for the specific application. For example, material selection is based on the user's physiological and psychological needs, whereas technology is chosen based on the required functionality, ergonomics, comfort, and fit (**Table 1**).

1.2.1 Protective clothing

It is one of the largest areas of functional clothing. The designing of the fabric varies for every function and required special attention. Environmental factors such as heat, cold, snow, and wind demand different types of fiber, fabric construction, and treatment. The challenge in designing the protective clothing is to offer maximum protection without affecting the metabolic heat transmission [8]. A fire protective clothing assembly schematic is shown in **Figure 1**, depicting the fabric layer arrangement to protect the skin from different hazards and simultaneously maintaining the body's thermal balance [9, 10]. Some primary requirements are

Class	Description
Protective clothing	Environmental protection—extreme heat or cold, rain, snow, dust, UV rays Biological, chemical, and radiation—hazardous chemicals, toxic gases, germs, radioactive substances Injury protection—cut, ballistic, impact protection
Medical functional clothing	Surgical protection—viruses, germs, bacterial protection Therapeutic—pressure management in lymphatic and venous disorder, scar management
Wearable clothing	Sensing—biological and physiological monitoring, telemedicine Communication—wireless monitoring, remotely tracking
Sports clothing	Training—performance enhancement, fatigue management, body shaping Activewear—moisture, sweat management, heat stress management
Special needs clothing	Clothing for the elderly, pregnant women, infants, and disabled

Table 1.
Classification of functional textile clothing.

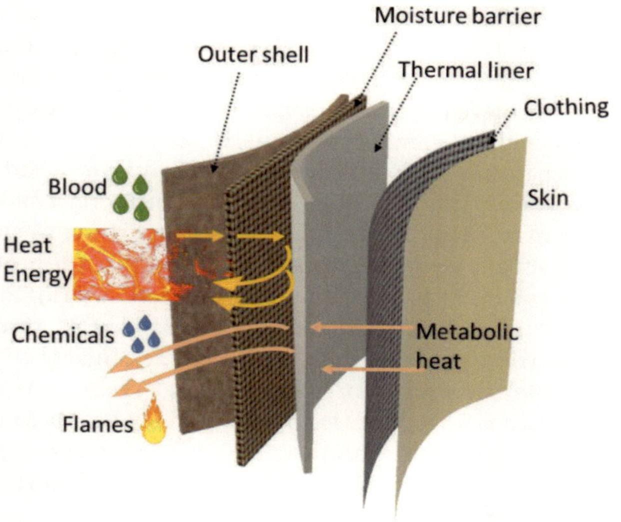

Figure 1.
Fire-protective functional clothing assembly.

lighter weight and volume, ergonomic design, and better moisture management. The biological, chemical, and radiation protection demands a barrier between the source of radiation and the human skin; hence, the clothing should contain a reflective material or a coating to prevent the penetration of these substances through the clothing [11, 12].

1.2.2 Medical functional clothing

Textiles have been utilized for medical treatment from a long time for surgery, wound protection, etc. The major function of this clothing is to protect from the bacteria, viruses, and body fluid infection. The clothing uses intrinsic antimicrobial textile material or substances coated onto the fabric. The fabric used for therapeutic treatments uses elastic garment to impart a compression on the infected area. The pressure exerted by the compression garments helps in the movement of the blood and lymphatic fluid from the affected area to reduce the venous disease. The pressure garments use elastic yarn and fabric construction, such as knitting to develop compression garments [13–15]. The functional clothing used for COVID19 protection uses a nonwoven fabric as a filter medium in the mask and PPE kit. Nonwoven fabric's porosity can be modified to obtain appropriate virus protection, and they can be coated with antiviral and antibacterial compounds to improve their protective performance [3–5].

1.2.3 Wearable clothing

Textile-based monitoring and treatment devices are becoming more preferable due to comfortability and portability, and use discretely and carefully among children and elderly people. Furthermore, the interactive textiles can be utilized for real-time monitoring and ultra-personalization, such as data measurement and storage for individual customers to have accurate and precise diagnoses [16–18]. There are several textiles-based TENGs applications reported in real-time monitoring, such as heart rate detection (ECG), neurobiological rehabilitation, gait recognition, pulse detection, motion sensors, respiration detection, and thermotherapy [17, 19–21]. E-textiles constitute two domains—textiles and electronics, which vary in type, material, and behavior from each other. Textiles are soft, flexible, porous, and susceptible to different conditions, while in general, electronics are rigid, precise, and guarded. It is always challenging to achieve material-specific properties for the wearable application while complying with the contrasting properties of electronics and textiles. Triboelectric nanogenerator (TENG) devices constitute one such platform for merging textile with electronics. The TENGs use the triboelectric effect and electrostatic induction to transform mechanical energy into electricity. The self-generation of an electrical signal without any power supply and the response to the health stimulus creates the TENG as a viable alternative for the wearable monitoring applications. The triboelectric effect is well known for almost a thousand years, through which material becomes electrically charged due to friction. The contact between two materials of opposite polarity creates an electrochemical interaction between the surface molecules, which is responsible for the generation of triboelectric charges on their surfaces. However, upon the separation, these triboelectric charges become the driving force for the electron to flow through the electrode to equalize the potential difference created (**Figure 2**) [20, 22].

1.2.4 Sports functional clothing

The growing popularity of textiles in the sports industry has increased the demand for functional clothing, which has become a critical aspect in improving a

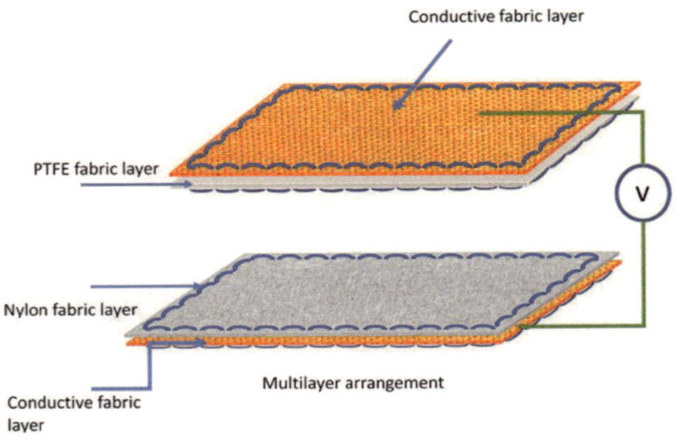

Figure 2.
Schematic of triboelectric nanogenerator arrangement [22].

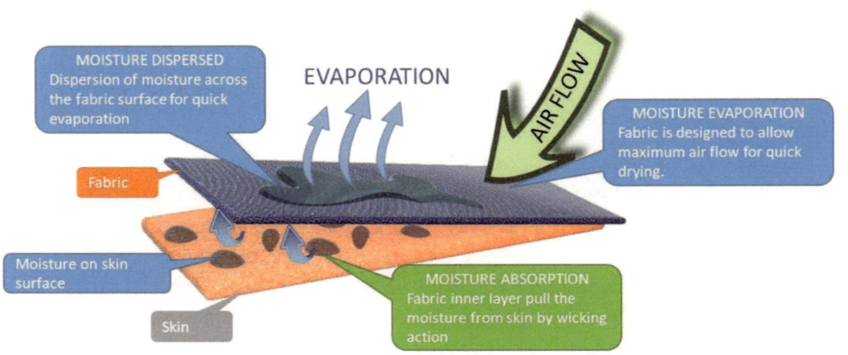

Figure 3.
Mechanism of moisture transport through sports clothing.

sportsperson's overall performance. The main characteristics of sports functional clothing are moisture management, quick moisture transport, temperature management, odor control, lightweight, and fit. The textile fiber is specially modified to enhance its surface area to facilitate moisture wicking helps in maintaining thermal balance during sports activity (**Figure 3**) [23]. Sports functional clothing is also used for enhancing the performance of the sportsperson by the mechanism of compression and aerodynamic design [23, 24]. The compression property of the garment helps in regulating the blood flow to a specific muscle groups providing enhanced energy and oxygen supply, which also helps in faster muscle recovery [14]. The aerodynamic design helps to reduce the wind and air resistance by systematically controlling the morphology of the fiber, shaping the garment, and structural arrangement of the fabric components [16].

2. Market and future scope

The need for functional fibers, processes, and technologies is growing in the 21st century, and such products are not being only used in apparel or garment, but also in other different applications, such as medical, automotive, agriculture, sports, geotextile, and others. The recent outbreak of COVID19 has resulted in more demand for medical textile products. The acceptance of smart healthcare products in daily life will be the next significant technological change going to happen globally. Smart Wearable E-Textile Medical Technology is one of the top-trending technology topics across the globe. It includes the integration of smart sensor/actuator materials in garments for non-invasive health monitoring. Such e-textiles help to detect the individual's vital signs and retransmit them *via* wireless sensor technology to provide continuous feedback on the health status. For the development of any functional or smart textile product, it combines knowledge from interdisciplinary backgrounds including material science, interfacial physics, biomechanics, textile engineering and design, and other engineering stream. Additionally, the need for new technologies is expected to be developed, which could bring down the manufacturing cost and make such products a commercially successful. It has been projected a massive growth rate of over 6% CAGR for the technical and functional textiles from 2020 to 2025, expected to reach over $222.4 billion global markets by 2025.

Author details

Bipin Kumar* and Viraj Somkuwar
Department of Textile and Fibre Engineering, IIT Delhi, New Delhi, Delhi, India

*Address all correspondence to: bipiniitd18@gmail.com

References

[1] Pan N, Sun G. Functional Textiles for Improved Performance, Protection and Health. Sawston, United Kingdom: Elsevier; 2011

[2] Majumdar A, Singh SP, Ghosh A. Modelling, optimization and decision making techniques in designing of functional clothing. Indian Journal of Fibre and Textile Research. 2011;**36**(4):398-409

[3] Beesoon S, Behary N, Perwuelz A. Universal masking during COVID-19 pandemic: Can textile engineering help public health? Preventive Medicine. 2020;**139**:106236

[4] Darby S et al. COVID-19: Mask efficacy is dependent on both fabric and fit. Future Microbiology. 2021;**16**(1):5-11

[5] Ha JF. The COVID-19 pandemic, personal protective equipment and respirator: A narrative review. International Journal of Clinical Practice. 2020;**74**(10):e13578

[6] Hollander JE, Carr BG. Virtually perfect? Telemedicine for COVID-19. The New England Journal of Medicine. 2020;**382**(18):1679-1681

[7] Gupta D. Functional clothing—Definition and classification. Indian Journal of Fibre and Textile Research. 2011;**36**(4):321-326

[8] Scataglini S, Andreoni G, Gallant J. Smart clothing design issues in military applications. In: International Conference on Applied Human Factors and Ergonomics. Switzerland AG: Springer; 2018

[9] Song G, Mandal S, Rossi R. Thermal Protective Clothing for Firefighters. United Kingdom: Woodhead publishing, Elsevier; 2016.

[10] Lawson JR. Fire fighters' protective clothing and thermal environments of structural fire fighting. In: Performance of Protective Clothing: Sixth Volume. West Conshohocken, PA, USA: ASTM International; 1997

[11] Nurfaizey A et al. Functional nanofibers in clothing for protection against chemical and biological hazards. In: Functional Nanofibers and their Applications. Sawston, United Kingdom: Woodhead Publishers; 2012. pp. 236-261

[12] Bhuiyan MR et al. Advances and applications of chemical protective clothing system. Journal of Industrial Textiles. 2019;**49**(1):97-138

[13] Garg H et al. Polyethylenimine-based shape memory polyurethane with low transition temperature and excellent memory performance. Macromolecular Materials and Engineering. 2020;**305**(8):2000215

[14] Kumar B, Hu J, Pan N. Smart medical stocking using memory polymer for chronic venous disorders. Biomaterials. 2016;**75**:174-181

[15] Kumar B, Das A, Alagirusamy R. Study of the effect of composition and construction of material on sub-bandage pressure during dynamic loading of a limb in vitro. Biorheology. 2013;**50**(1-2):83-94

[16] Leutheuser H et al. Textile integrated wearable technologies for sports and medical applications. In: Smart Textiles. Switzerland AG: Springer; 2017. pp. 359-382

[17] Coyle S et al. Textile-based wearable sensors for assisting sports performance. In: 2009 Sixth International Workshop on Wearable and Implantable Body Sensor Networks. Berkeley, CA, USA: IEEE; 2009

[18] Islam GN, Ali A, Collie S. Textile sensors for wearable applications: A comprehensive review. Cellulose. 2020;**27**(11):6103-6131

[19] Yi F et al. Recent advances in triboelectric nanogenerator-based health monitoring. Advanced Functional Materials. 2019;**29**(41):1808849

[20] Mathew AA, Chandrasekhar A, Vivekanandan S. A review on real-time implantable and wearable health monitoring sensors based on triboelectric nanogenerator approach. Nano Energy. 2020;**80**:105566

[21] Pragya A et al. Designing and investigation of braided-cum-woven structure for wearable heating textile. Engineering Research Express. 2020;**2**(1):015003

[22] Somkuwar VU, Pragya A, Kumar B. Structurally engineered textile-based triboelectric nanogenerator for energy harvesting application. Journal of Materials Science. 2020;**55**(12):5177-5189

[23] Dong Y et al. Materials design towards sport textiles with low-friction and moisture-wicking dual functions. Materials and Design. 2015;**88**:82-87

[24] Raj AN, Yamunadevi S. Application of textile fibres for technical and performance enhancements in sports. International Journal of Multidisciplinary Research and Development. 2016;**3**(12):40-45

Section 2

Textile Material, Design and Technology

Chapter 2

Antimicrobial Agents for Textiles: Types, Mechanisms and Analysis Standards

Ahmad Ibrahim, Joseph-Émile Laquerre, Patricia Forcier, Vincent Deregnaucourt, Justine Decaens and Olivier Vermeersch

Abstract

The large surface area, and ability to retain moisture of textile structures enable microorganisms' growth, which causes a range of undesirable effects, not only on the textile itself, but also on the user. Moreover, textiles used in health care environments are required to possess antimicrobial property to minimize spread of pathogenic infection. Anti-microbial property can be imparted via chemical finishing with an antimicrobial agent. Currently the use of antimicrobial agents includes metal compounds (notably copper and silver particle), chitosan, halogenated phenols "triclosan", quaternary ammonium compounds, antibiotics (a class of antimicrobials produced from microorganisms that act against one another), and N-halamines. The possibility of bacterial resistance limits antibiotic use to specific medical applications, and triclosan is known for being dangerous to the environment and is currently under scrutiny for possible endocrine disrupting to human being. Although quaternary ammonium compounds are stable and easily manufactured, microbial resistance is also a concern. Quaternary ammonium compounds (QACs), Polyhexamethylene Biguanide (PHMB), chitosan and N-halamines are listed under bound or non-leaching type antimicrobials. The bulk of current chapter focuses on the different family of antimicrobial agents used for textiles and their mechanisms.

Keywords: Finishing textiles, nanoparticles, silver particles, quaternary ammonium, antibacterial effect, ecological antibacterial

1. Introduction

Microorganisms play both beneficial and harmful roles in our lives. Some of the beneficial roles include production of oxygen via photosynthesis, nitrogen fixation, circulation of carbon by decomposition of dead organic matter, formation of crude oil, and helping animals such as cows digest their food. They are used by humans in making bread, beer, cheese, and antibiotics. Some of the harmful effects are caused by the virulence of pathogenic microorganisms, i.e., infection causing bacteria such as *Staphylococcus aureus* (*S. aureus*), *Escherichia coli* (*E. coli*), and *Enterococcus faecalis* (*E. faecalis*). Health care associated infections can be controlled by inhibiting the various routes of transmission that causes an infection to spread from an infected person to healthy person. One of the various routes through which an infection can spread is the direct contact with infected individuals; infected water and food; contact with inanimate objects such as textiles used in scrubs, doctor's coats, surgical gowns, bed-sheets, pillow covers, and curtains. The control of the spread of infections via infected individuals, water and food can be achieved by developing hygienic practices.

Textiles have been recognized as a media for the growth of microorganisms such as fungi and bacteria. The growth of these microorganisms on textiles inflicts unwanted effects not only on the textile material, but also on the consumer. These effects can include the generation of unwanted odor, discoloration in the fabric, an increased probability of contamination, and an overall reduction in the fabric mechanical strength [1, 2]. The spread of infections through textile materials can be controlled by the use of antimicrobial textiles that kill pathogens on contact or hinder their ability to reproduce prior to being transferred on to another material or person. Antimicrobial textiles are made by treating textile substrates with antimicrobial agents or by using textile fiber with inherent antimicrobial efficiency. Antimicrobial agents are bound to textiles by different methods depending on the chemistry between the antimicrobial agent and the textile [3]. Consumers' attitudes toward hygiene and their desire for comfort and well-being have created a rapidly increasing market for antibacterial materials. Therefore, there has been extensive research in recent years in this area. Estimations have shown that there was approximately a production of 30,000 tons of antimicrobial textiles in Western Europe and 100,000 tones worldwide in year 2000. It was also estimated that the production increased by over 15% annually in Western Europe between 2001 and 2005, making antimicrobial textiles a rapidly growing sector of the textile market [1]. While synthetic fibers have been known to be more resistant to microorganisms due to hydrophobicity, natural fibers are more vulnerable to microorganism attack. In addition, soil, sweat, and dust can be nutrient sources for microorganisms [2]. Socks, active-wear, shoe linings, and lingerie account for approximately 85% of the total antimicrobial textile production. In addition, there has recently been a large market for antimicrobial fibers in air filters, outdoor textiles, upholstery, and medical textiles [1].

Other than the antimicrobial ability, there are certain basic requirements to be satisfied by an antimicrobial agent for its successful application on textiles rendering them to be used commercially. The basic requirements of a good antimicrobial agent for textile substrates are summarized below [3, 4]:

- Should possess affinity for specific fabric and fiber types.
- Be easy to apply on textile substrates.
- Be able to inactivate undesirable microbes while simultaneously not affect desired microbes.
- Inert to chemicals to which the textile might be exposed during processing.
- Durable to repeated laundering, dry cleaning, ironing and prolonged storage including resistance to detergents used to care for the textiles.
- Stable during usage without degrading into hazardous secondary products.
- Not adversely affect the user or the environment.

2. Antibacterial family

Many antibacterial product and chemistry are available in the current market using different technologies. Most antibacterial agents applied on textiles have been used for many years in food preservatives, disinfectants, wound dressings, and pool sanitizers. The attachment of these compounds to textile surfaces or their binding with the fiber can reduce their activity largely and limits the antibacterial agents' availability. In addition, the antibacterial agent can gradually be lost during the washing and use of the textile material. The most widely used antimicrobial agents for textile applications are based on metal salts (for e.g., silver), quaternary ammonium compounds (QAC), halogenated phenols (for e.g., triclosan), polybiguanide (for e.g., PHMB), chitosan, and N-halamines [5]. The aim of this section is to present the general family of antibacterial textile finishing.

In general, antibacterial agents can either kill the microorganisms (−cidal) or inhibit their growth (−static). Almost all the commercial antimicrobial agents used in textiles (silver, polyhexamethylene biguanide (PHMB), quaternary ammonium compounds, and triclosan) are biocides. They can damage the cell wall or disrupt the cell membrane permeability, and inhibit the activity of enzymes or synthesis of lipids, while all these functions are essential for microorganism's survival [3].

The antibacterial material can be separated in two categories: antimicrobials with controlled release or 'leaching' mechanism and bound or non-leaching type antimicrobials. The mechanism of the leaching type will act upon contact of the cell. On the other hand, the non-leaching types will diffuse a disruptive chemical to the cell. This type is preferred for an environment supporting the diffusion of the chemical, such as water.

2.1 Antimicrobials with controlled release or 'leaching' mechanism

The antimicrobial agents that belong to this category do not form strong bonds with the textile substrate. The chemical species responsible for biocidal activity are released slowly from the treated fabric surface, thus killing all the microbes surrounding the agent. An advantage of leaching antimicrobials effect are their superior antimicrobial activity than compounds based on other modes of action on the same fabric under similar environmental conditions [6]. The flip side is that the antimicrobial agent in the textile substrate is depleted eventually and loses its effectiveness. Metal salts (e.g., silver) and halogenated phenols (e.g., triclosans) are examples of antimicrobial agents that utilize the leaching mechanism [7].

2.1.1 Metal and metal salts

The interest of metal and metal oxide particle reside in the high antibacterial activity against microorganism, durability and stability, while having a low mammalian cell toxicity, meaning they are safe for close to the skin application. Even at very low concentrations, many heavy metals are toxic to microorganisms. Metal particles are synthesized from different precursor and reducing agent to obtain different end material, morphology or to lower the impact of on cost or environment. Plethora of synthesis reaction are available from the scientific literature. However, the reaction principle is similar for each technique, using a sol–gel. The precursor is usually a water-soluble salt such as silver nitrate, copper chloride, and zinc acetate. The metal ion is reacted with a reducing agent, such as conventional reducer like sodium borohydride, citric acid, citrate, and ascorbic acid, or with bio-based reducer such as glucose, polysaccharide, cellulosic fiber and plant or microorganism extract. The precursor is mixed with a reducing agent under different conditions such as heat, mixing, sonication to surpass the activation energy of the reaction. Strong reducing agent will require milder reaction condition, while weak reducing agent will require stronger reaction condition. During the reaction, the particle could be stabilized using a capping agent in order to control the shape, size and stability of the final product. In some reaction, the reducing agent will also be used as a capping agent. While metals such as zinc, cobalt and copper have had some applications in past years as antibacterial agents for textiles, silver, having an MIC value of 0.05–0.1 mg/L against *E. coli* is still the most widely used metal in textile applications and wound dressings [3, 7–9]. Moreover, this metal is less toxic in the human body than other heavy elements with a smaller risk for exposure through inhalation, ingestion, or dermatological exposure [10]. It was found that AgNP was much less toxic to human cells than silver ion [11]. A concentration of silver ion higher than 1 μg/ml is toxic to human mesenchymal stem cells while the concentration of AgNP can be higher than 2.5 μg/mL. AgNP can destroy bacteria even at a nano-molar level while silver ion needs to be at a micro-molar level.

Silver can be applied in other forms: silver ion exchangers, silver salts, and silver metals. Silver zirconium phosphate and silver zeolites are examples of ion exchangers. Silver chloride (AgCl), nanosilver chloride, and AgCl microcomposites (AgCl nanoparticles attached to titanium dioxide as a carrier material) are types of silver salts. Silver metal can be used in the form of filaments and silver metal composites [12]. With concerns regarding bacterial resistance to silver [3], there is efforts to increase the efficiency of metal-based antimicrobial. Other metal based antimicrobial agents found to exhibit good antimicrobial properties are based on copper and zinc compounds, in the form of their sulfides and sulfates [13]. Many studies on metal salts have focused on preparation of nano sized metal particles, which has led to the development of new generation of biocides [5]. Above all, AgNP (Silver Nanoparticles), a nanometric form of silver element without an ionic charge, can be used as a catalyst, an optical sensor and an antibacterial agent [14–16]. The antibacterial activities of the silver ion and salts are well studied, but research about antibacterial mechanism of AgNP is relatively recent [14]. Different methods have been developed to synthesize and incorporate AgNP in some biomedical applications, and some reports have proven AgNP to be a potent antibacterial agent, that is effective against both Gram-positive and Gram-negative bacteria [17–19].

2.1.1.1 Mechanisms of metal and metal salts antimicrobial action

All silver-based antimicrobials generate and release different amounts of silver ions, with silver metals releasing the least, silver ion exchangers releasing the most, and silver salts somewhere in between [20]. In the presence of moisture, silver releases ions that bind the bacterial cell's surface with proteins. On binding, the following action occurs [21].

- Denaturing effect of the silver causes DNA to get condensed and lose its replication abilities.
- Induces inactivation of bacterial proteins by reacting with thiol group [21, 61].

The form of the silver used impacts its antibacterial effectiveness. For example, a concentration of $AgNO_3$ should be higher than 1 mM to kill silver resistant E.coli. While only 80 nM of AgNP is necessary for the same result [17]. The antibacterial efficacy of silver is directly proportional to the amount of bioactive silver ions released in the presence of moisture, as well as its ability to penetrate bacterial cell membranes [10]. Silver is effective at low concentrations and promotes wound healing without appreciable toxic risk. However, there is a small risk of developing allergies to silver [22, 23]. In fact, silver and copper ions can disrupt or kill the microbes via different mechanism path. First, the ions can diffuse trough the cell membrane and bond to the enzyme of the cell. The enzymatic activity of the cell is decreased, which inhibit the growth of the cell until the death of the cell. Second, Silver ion can kill microbes by binding to intracellular proteins and inactivating them, can inhibit the synthesis of ATP (Adenosine triphosphate) and lead to DNA (Deoxyribonucleic acid) denaturation [24]. To observe the killing mechanism of silver ion more directly, TEM (Transmission electron microscopy) and X-ray techniques were used to facilitate the investigation. When *Escherichia coli* (E.coli) and *Staphylococcus aureus* (*S. aureus*) were treated with $AgNO_3$, the cytoplasm membrane detached from the cell wall. Subsequently, DNA and protein failed to function and finally the cell wall was damaged [21]. Third, the silver cation can damage bacterial cell walls, proton leakage and result in cellular structural changes. It can induce proton leakage through the membrane of the cell and cell death. The silver cation is highly reactive in a concentration between 5 and 40 mg/L [25]. Regarding the AgNP, their exact antibacterial mechanism has not been clearly revealed to date [11, 26]. The reduced nano-silver did not show antibacterial activity toward E.coli, but when it was mixed with partially oxidized nano-silver, the mixture showed significant inhibition to the growth of E.coli. Thus, the antibacterial activity of AgNP is a result of surface oxidization as AgNP is sensitive to oxygen [17].

Others metals oxides of interest are titanium dioxide (TiO_2) and zinc oxide (ZnO). The mechanism of those compounds is believed to be mostly from the generation of reactive oxygen species (ROS). Those compounds prevent the antioxidant defense system and damage the cell membrane of the microbe. This mechanism is catalyzed by ultraviolet light. It is of particular interest as an adjuvant to the UV disinfection, which is of growing usage for disinfection against COVID-19. However, this also means the efficiency of those metal oxides is largely influenced by the environment in which they are used. The efficiency of the metal oxide could be reduced in the presence of antioxidant or pigment, often used in synthetic textile. The morphology of the particle will have a great impact on the stability of the product as well as the antibacterial activity. In general, the greater surface area will provide a greater activity, but decrease the durability for the leaching type.

Currently, silver is used in a large number of antimicrobial commercial textile products at a relatively low cost. The silver is in the form of ultra-fine metallic particles and is mainly applied to polyesters, in the finishing stage. Ruco-BAC®, SilverClear®, UltraFresh®, Silpure®, AlphaSan®, Microfresh®, Solefresh®, GuardYarn®, and SmartSilver® are some of the commercially available antibacterial agents applied on textiles [3, 27, 28]. In the case of synthetic fibers, metal and metal salts particles can be incorporated into the polymer prior to extrusion (or before electrospinning for nanofibers). For example, silver can diffuse into the fiber surface and in the presence of moisture it can form silver ions. Gradual release can lead to an extended period of antibacterial activity. In addition, silver nanoparticles can be padded onto cellulosic and synthetic fabrics, resulting in a durable antibacterial finish [29]. While metals and metal salts has excellent antimicrobial activity, leaching from treated textiles into laundering effluent is problematic. Ionic silver is highly toxic to aquatic organisms, with the EPA setting water quality criteria at 1.9 ppb in salt water and 3.4 ppb in fresh [30]. Effluent from both home laundering and industrial application can transfer silver into sewage treatment facilities, depleting necessary bacterial communities. Research conducted by Geranio et al. found that fabric treated with AgCl released only 2.7 ppb (2.4 ppb for AgCl plus a binder) of total silver per gram of textile after the first wash cycle [31]. As the effectiveness of silver depends on the release of silver ions, too few ions result in a lack of antimicrobial action, and too many yield an excess leading to pollution and waste. Success depends on finding the balance between minimum antimicrobial concentration and effectiveness.

2.1.2 Halogenated phenols (Triclosan)

2,4,4′-trichloro-2′-hydroxydiphenyl ether, commonly known as Triclosan is a broad-spectrum antibacterial agent, having a Minimum Inhibitory Concentration (MIC) of less than 10 ppm against most kinds of bacterial species. Triclosan has been used since 1960 in a wide variety of consumer products including toothpastes, hand soaps, deodorants, mouthwashes, shower gels, etc. Its mode of action is inhibiting bacterial growth by blocking biosynthesis of lipids. As a relatively small molecule, triclosan can be used by exhaustion, combined with dyes, or applied after dyeing. Through melt-mixing or suspension polymerization, triclosan can be incorporated directly into synthetic polymers (**Figure 1**) [5, 32].

Triclosan inhibits the growth of microbes by using an electrochemical mode of action to penetrate and disrupt the cell wall of microbes. When incorporated within a polymer, it migrates to the surface and protects the material [3, 33]. When embedded in β-cyclodextrin triclosan forms a complex and can exhibit antimicrobial action with minimum quantities [34]. Some researchers claim that triclosan inhibits a specific function i.e., lipids synthesis in a bacteria [35]. Others claim that lower levels of triclosan resistance by strains of bacteria shows that triclosan inhibits bacterial cell function in multiple ways. A decrease in the antimicrobial efficiency of triclosan treated material when the material is subjected to repeated home wash cycles has been reported by [36]. One of the greatest concerns regarding triclosan is that when exposed to sunlight, it breaks down into 2, 8-dichlorodibenzo-p-dioxin, a chemical related to other harmful polychlorinated dioxins. Therefore, it has raised a lot of concern in European governments, and its application in consumer products is banned in some countries [37, 38].

2.2 Bound or non-leaching type antimicrobials

The antimicrobial agents that belong to this category are chemically bound to the textile substrate. Hence, the antimicrobial can act only on the microbe that are in contact with the treated textile's surface. By virtue of its binding nature, these antimicrobials are not depleted and therefore potentially may have higher durability than [39]. However, compounds on a treated fabric might get abraded or deactivated with long-term usage and lose their durability [40]. The antimicrobial agents listed under this category are Quaternary Ammonium Compounds (QACs), Polyhexamethylene Biguanide (PHMB), chitosan and N-halamines.

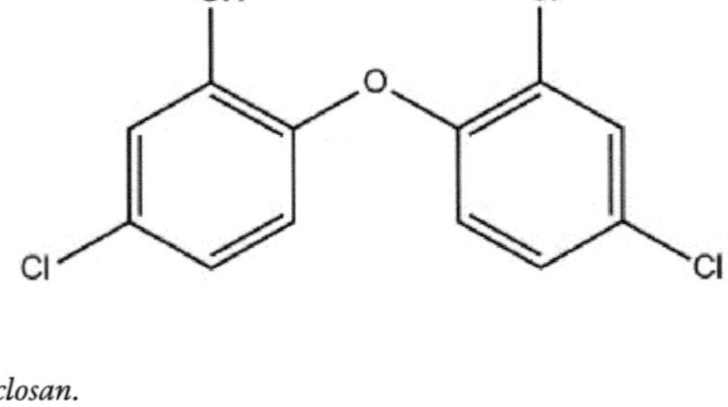

Figure 1.
Molecular structure of Triclosan.

DOI: http://dx.doi.org/10.5772/intechopen.98397

2.2.1 Quaterny ammonium (QAC)

Surface-active agents (surfactants) contain two distinct regions in their molecules: a long chain hydrophobic hydrocarbon tail and a hydrophilic head. Based on the charge of the hydrophilic group, they are classified into cationic, anionic, nonionic, and amphoteric compounds. Among the wide range of these surfactants, the cationic agents (Quaternary Ammonium Compounds or QACs) are known to be the most effective (**Figure 2**). QACs have significant antimicrobial properties and are excellent for deodorization and hard surface cleaning. They are used as biocides in a variety of consumer products, including toothpaste, mouthwash, shampoo, soap, deodorant, etc. The application of QACs as disinfectants goes back to 1936, where Dunn investigated the antibacterial properties of alkyldimethyl-benzylammonium chloride and found the phenol-coefficients against *S. aureus* and S. Typhosa (**Table 1**). The most widely used QACs are monoquaternary ammonium such as alkyltrimethylammonium bromide, and diquaternary ammonium salts such as alkanediyl-α,ω-bis (dimethylalkylammonium bromide). Murugan et al. studied the antibacterial behavior of five novel insoluble bead-shaped, polymer-supported multi-quaternary ammonium salts containing two to six quaternary ammonium groups [41]. The QACs showed excellent antibacterial activity against S.aureus, Klebsiella pneumonia and *Pseudomonas aeruginosa* (P.aeruginosa). Murugan et al. also found that the antibacterial activity increased as the number of QACs in the structure increased [41]. However, there are also reports of bacterial resistance to QACs [42, 43]. Following alkyldimethyl-benzylammonium chloride, other QACs such as Cetyldimethyl-benzylammonium halide and N-(acylcolaminoformyl methyl)pyridinium chloride were studied and were found to have high phenol-coefficients. These solutions were known to be both bacteriostatic and bactericidal, according to the period of exposure and concentration [44].

The attachment of QACs to a textile material is known to be predominantly by ionic interactions between the anionic fiber and the cationic QAC. Therefore, in the case of fabrics that contain sulfonate or carboxylic groups, QACs can be attached to fibers by using an exhaustion dyeing process [45–47]. In the case of synthetic fibers, which contain fewer reactive sites and are quite resistant to antibacterial finishing modifications (such as Nylon 66); dye molecules can act as bridges to bind the functional molecules to fibers [48]. For example, acid dyes can be used to dye the fabric and then QACs can be applied under alkaline conditions. This ionic bonding between the QAC and the dye is relatively strong and provides a semi-durable antibacterial finishing [47, 48]. Hence a dyed fabric can achieve higher add on levels of QACs and antimicrobial efficacy as compared with undyed fabrics [48]. One commercial QAC- based antibacterial textile is Bioguard®. The active antimicrobial agent is 3-trimethoxysilylpropyldimethyloctadecyl ammonium chloride, also known as AEM 5700 or Dow Corning 5700 Antimicrobial agent, which has an MIC = 10–100 mg/L against Gram-negative and Gram-positive bacteria. This

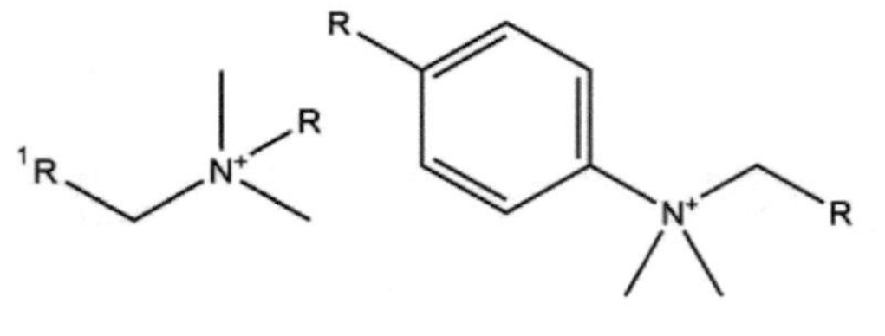

Figure 2.
General molecular structure of Quaterny ammonium (QAC).

Generation	Compound example	Description
1st	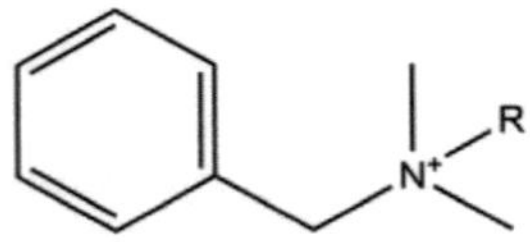	Benzalkonium, alkyl chains, C12 to C18
2nd	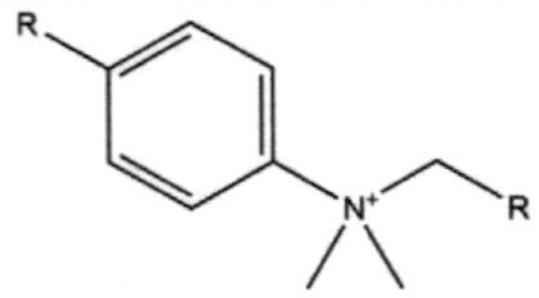	Aromatic rings with hydrogen and chlorine, methyl and ethyl groups
3rd	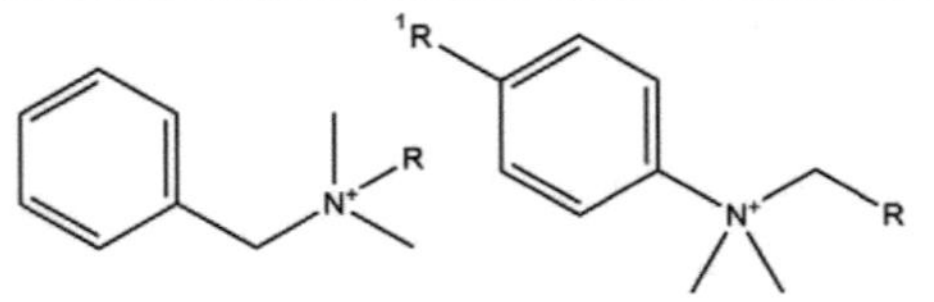	Dual QACs; mixture of alkyl dimethyl benzyl ammonium chloride (lower toxicity)
4th	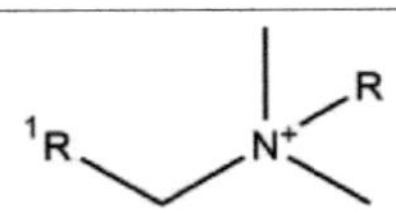	Dialkylmethyl aminos with twin chains
5th	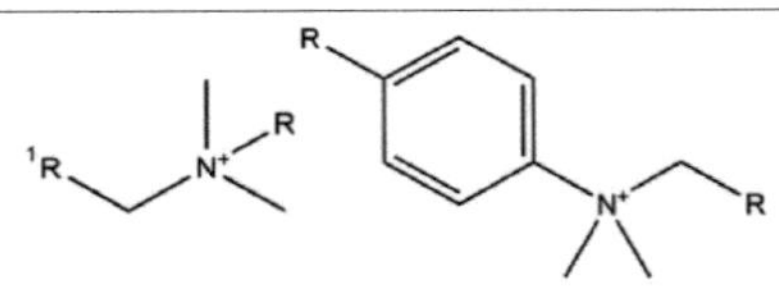	Synergistic combinations of dual QACs
6th	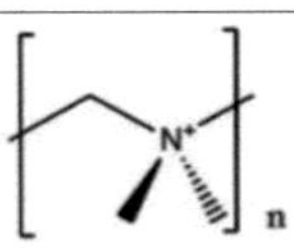	Polymeric QACs
7th	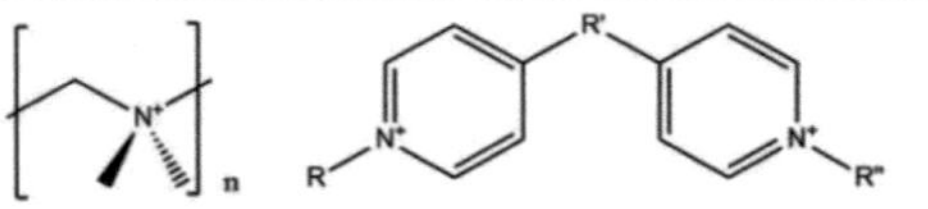	Bis-QACs with polymeric QACs

Table 1.
Molecular structures and description of the different generations of quaternary ammonium (QAC).

compound is made into aqueous solution and applied by spraying, padding, and foam finishing. During drying, silane forms covalent bonds with the textile, resulting in excellent durability. This compound has been commercially used on nylon, cotton, and polyester. Recently, novel quaternary ammonium functional dyes have been applied on textiles in order to combine antimicrobial finishing and dyeing of textiles in a single step [49–51].

2.2.1.1 Mechanisms of Quaterny ammonium (QAC) antimicrobial action

QACs are active against a broad range of microorganisms such as fungi, Gram-positive and Gram-negative bacteria, and some viruses. QACs have a positive charge on the N atom and inflict a variety of effects on microorganisms, including the disruption of cell membrane, denaturation of proteins, and damage to cell structures. It has been proposed that during the inactivation of bacteria, the quaternary ammonium group remains intact and can retain its antibacterial ability as long as the QAC is attached to the fibers [48]. When a microbe approaches a QAC treated fabric, the free end of the agent's molecule reacts with the cell wall and causes a leakage of the negatively charged species in the microbe cell. It eventually causes the cell's death [39, 52]. The cationic ammonium group and the negatively charged bacteria membrane are attracted to each other. Consequently, the interactions result in the formation of a surfactant-microbe complex that interrupts all the normal functions of the membrane [53]. QACs affect bacterial DNA, causing a loss of multiplication ability [5]. If the long hydrocarbon chain is bonded to the cationic ammonium in the structure of the QAC, two types of interactions between the agent and the microorganism can occur: a polar interaction with the cationic nitrogen of the ammonium group and a non-polar interaction with the hydrophobic chain. Penetration of the hydrophobic group into the microorganism consequently occurs, enabling the alkylammonium group to physically interrupt all key cell functions [5]. The efficiency of the quaternary ammonium depends on the generation and chain substitution. It is known that the germicidal power increases with an increase in carbon chain length, while the surface activity also increases in the same way [44]. The QACs with 12–18 carbons have been used extensively as disinfectants. The typical dosage is under 1%, and even under 0,1% for some application.

To resume, the quaternary ammonium compounds are membrane active agents, their target site is at the inner (cytoplasmic) membrane in the bacteria (or plasma membrane in yeasts) [8, 44]. One of the mechanisms proposed for the antimicrobial action of QACs is in this sequence:

- Adsorption and penetration of QAC in the microorganism's cell wall
- Reaction with the lipid or protein cytoplasmic membrane, which will disorganize the membrane
- The leakage of low molecular weight intracellular material
- Degradation of nucleic acids and proteins
- Wall lysis which is caused by the autolytic enzymes.

Without detailing the studies carried out on the toxicity of quaternary ammoniums, different experiences were carried out on their ocular toxicity [54, 55], contact dermatitis [56], their skin sensitizer (human contact allergen) and asthma [57, 58]. Quaternary ammonium compounds are known to cause occupational asthma. It was found that nurses exposed to a class of QAC and all exhibited early or delayed asthma symptoms when handling disinfectant solutions containing QAC. The same study was done with products lacking in QAC and the results were negative [59]. These results have been confirmed by a multitude of studies [57–59]. In parallel, it has been reported that repeated occupational exposure after handling QACs as powders or solutions could cause sensitization [60]. In conclusion, the studies above all confirm the link between prolonged exposure to quaternary ammonium compounds and asthma. However, regarding ocular and dermal irritation, it seems that the quaternary ammonium compounds allergenicity is likely to be related to the compound's solubility. Apparently, no quaternary ammonium compounds can be regarded as allergens. In most of the studies that classify these compounds as irritants/allergens, the lipid or water-soluble compounds have been studied, while the non-soluble QACs certainly do not have the same properties.

2.2.2 Polyhexamethylene biguanide (PHMB)

PHMB is a hetero disperse mixture of polyhexamethylene biguanide (**Figure 3**). Polyhexamethylene biguanide (PHMB, commercially known as Vantocil) is a broad-spectrum antibacterial agent with low toxicity, having an MIC = 0.5–10 ppm. It has been previously used as a disinfectant in pool sanitizers, mouthwashes, wound dressings, and in the food industry. PHMB can disrupt the integrity of cell membranes [61].

The halide form of PHMB i.e., polyhexamethylene biguanide hydrochloride is applied on cellulosic materials [62]. PHMB is found to form hydrogen bonds with cellulosic fibers. With the increase in the concentration of PHMB there is a dominant increase in hydrogen bond formation between PHMB and fibers [63]. When the fabric treated with PHMB is exposed to a bacterium, the biocide interacts with the surface of the bacteria and is transferred to the cytoplasm and cytoplasmic phospholipids in the bacterial membrane. This biocide is positively charged, and therefore it mainly reacts with negatively charged species and includes aggregation, leading to increased fluidity and permeability. This results in the leakage of inner material from the outer membrane and eventually causes death of an organism [52].

2.2.3 Regenerable N-halamines and Peroxyacids

N-halamines are heterocyclic compounds containing one or two covalent bonds formed between nitrogen and halogen [64]. N-halamines contain one or more nitrogen-halogen covalent bonds formed by the chlorination or bromation of imide, amide or amine groups. The halogen, which is usually chloride, is replaced with hydrogen in presence of water or chloroform and acts as biocide (**Figure 4**) [65]. By using chlorine-containing N-halamine compounds, durable antibacterial finishing can be achieved on textiles. N-halamines are broad-spectrum disinfectants, which have been used previously for water treatment. Their antibacterial activity is known to be due to the oxidative properties of halamine bond (N-Cl). In order to kill the bacteria, N-Cl will be transformed to N-H, which can be recharged with chlorine (during laundering, by using bleach). The product of the reaction is reversible, meaning the N-halamide can be regenerated with the presence of chlorine compound. This function is found in hypochlorite, commonly found in bleach solution. The regeneration with bleach can be done during the washing process. This novel regenerable method was first proposed by Sun and Xu for the treatment of cotton fabric [66]. Since then, many different heterocyclic N-halamines have been applied on polyester, nylon, keratinous fibers, and cotton through covalent bonding. In all these studies, it was demonstrated that regenerable and durable antibacterial activity can be achieved by recharging the fabric in aqueous chlorine solutions.

N-halamine compounds, of which N-chloramine is one form, can provide instant and complete kill of a broad-spectrum of microorganisms. The antibacterial property is based on active chlorine, Cl^+. Two mechanisms can be used to explain

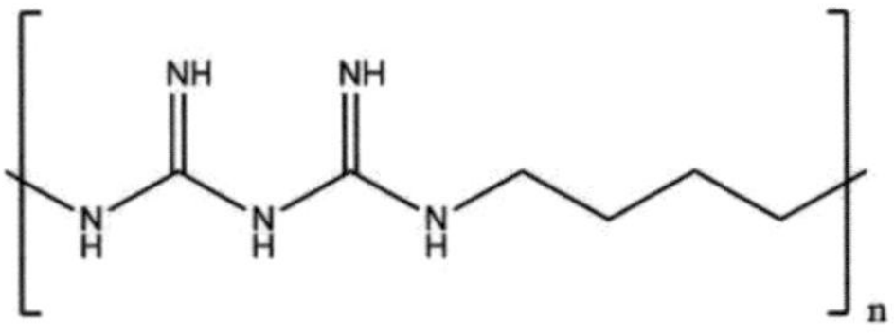

Figure 3.
Molecular structure of Polyhexamethylene biguanide (PHMB).

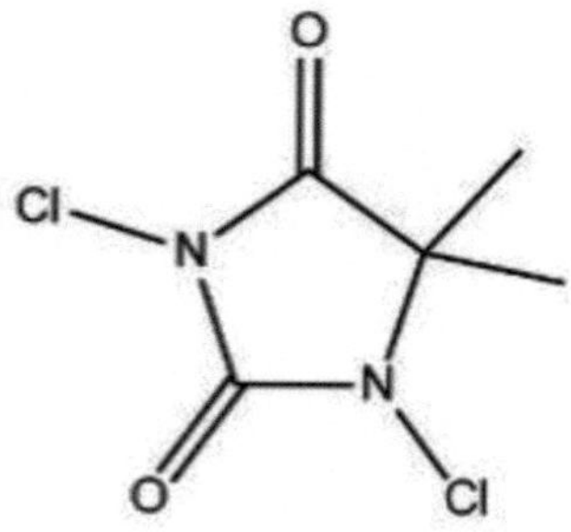

Figure 4.
Molecular structure of N-halamines.

the antibacterial activity of N-chloramine. One mechanism is that free chlorine is released into water and then forms HClO or ClO^-. The other is that chlorine binds directly to acceptor regions in bacteria and greatly influences their enzymatic and metabolic processes [64]. It was found that the antibacterial activity mainly attributed to the second mechanism because the dissociated chlorine is limited [67]. N-halamines possess stability that is suitable for long-term use, storage and regeneration. N-chloramine can be achieved by the reaction between sodium hypochlorite solution and imide, amide or amine groups. N-halamines have been used in water treatment and incorporated into cellulose-containing fabrics, polyester fibers and polyamide [68–70]. Although no research has directly addressed N-halamine in wound dressing, it has been grafted onto fibers or fabrics so it may be used in wound dressing [69, 70].

Halamine can be applied on different textile including cellulose, polyamide and polyester fibers [71–74]. It has also been found to have extraordinarily durable biocidal functions in a series of laundering tests [75]. However, N-halamine materials are found to be decomposed upon exposure to ultraviolet irradiation as in direct sunlight [76]. The main problem with N-halamines was that they result in a significant amount of absorbed chlorine (or maybe other halogens), which can remain on the fabric surface, resulting in unpleasant odor and fabric discoloration. The use of bleach and the presence of strong oxidizing degrade the dye on the textile, which leads to discoloration of the textile. This antimicrobial technology is best used on bleach resistant textile. One method known to resolve this problem is using a reduction step to remove the residual unbounded halogen from the surface of fabric [75–79]. An alternative antibacterial finishing agent is known to be peroxyacids (such as peroxy acetic acid, which is extensively used in hospitals.) Peroxyacids should convert to carboxylic acid in order to deactivate bacteria, but can be regenerated by reacting with an oxidant (such as hydrogen peroxide). Despite the stability of the peroxyacids on the fabric during prolonged periods, the antibacterial activity reduces largely after a number of washing and recharging cycles [73, 74].

2.2.4 Chitosan

Chitosan is derivatized by the deacetylation of Chitin, the main component of shrimp, crab, and lobster shells. Chitin, a poly (β-(1–4)-N-acetyl-D-glucosamine) is a natural polysaccharide. Chitin is synthesized by many living organisms. Chitin is the second most abundant polysaccharide in nature after cellulose [80]. When chitin is acetylated to at least about 50%, then it is called chitosan [28]. Chitosan (**Figure 5**) contains three reactive sites including a primary amine and two primary or secondary hydroxyl groups per glucosamine unit. As a result, it is readily

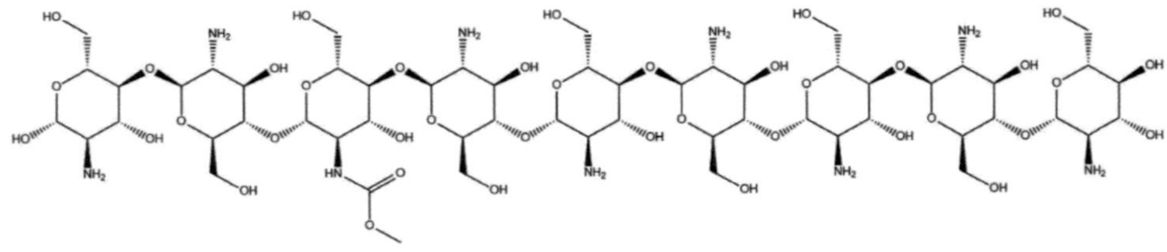

Figure 5.
Molecular structure of chitosan.

subject to chemical modification. The structural characteristics of chitosan mimic glycosaminoglycan components of the extracellular matrix, so the biocompatibility, biodegradability, antibacterial, hemostatic and antioxidant activities and muco-adhesive properties impart versatility [81]. Chitosan's good antibacterial activity along with its biodegradability, biocompatibility, and most importantly nontoxicity makes it an ideal biocidal agent in food science, pharmaceuticals, medicine, and textile applications. Despite all these advantages, chitosan lacks good solubility above pH 6.5. Its applications in a commercial context are not as wide as might be expected [82]. One of the potential problems with an effective chitosan based anti-microbial agent is that chitosan is insoluble in water and possesses high molecular weight. The high molecular weight increases the viscosity of the medium and causes detrimental effect on the hand and feel of the fabric [83]. Chitosan can be used to spin antimicrobial fibers or as a finishing agent for surface modification. Therefore, researchers are exploring chitosan derivatives that are soluble in water over a wide pH range for expanding the chitosan applications.

2.2.4.1 Chitosan derivatives

Given that chitosan does not dissolve in aqueous media at neutral and alkaline pH's and its antimicrobial activity likewise is not particularly good in neutral or alkaline solutions, so there is many causes to chemically modify chitosan. These modifications were made with the aim of proposing more soluble chitosan derivatives better suited for textiles. Recent researchers reported that chitosan derivatives have better water solubility, antibacterial and antioxidant properties [84]. Chitosan can be modified to include quaternary ammonium groups, alkyl and aromatic groups, substituents having free amino or hydroxyl groups, carboxy-alkyl groups and amino acids and peptides [85]. And different applications have been found for these chitosan derivatives. Among the derivatives of chitosan we cite: Carboxymethyl Chitosan, N,N,N,-trimethyl chitosan (TMC) and Chitosan nanoparticles (CSNP).

The modification of the structure of chitosan by the addition of carboxymethyl in the structure of chitosan allows the manufacture of carboxymethyl chitosan (CMC). Compared to chitosan, CMC is characterized by high solubility at neutral and alkaline pHs. This modification does not affect its characteristic properties [86]. In addition, CMC has superior antimicrobial activity, biocompatibility and safety for humans. Usually, there are O-CMC, N-CMC, N, singlet O and N, N-dicarboxymethyl chitosan that have different chemical structures (**Figure 6**). For antimicrobial properties, the antimicrobial activity of different types of CMCs against *E. coli* has been shown to increase by converting NO-CMC < Chitosan <O-CMC due to the reduced number of protonated amino groups in NO-CMC [86]. And against *S. aureus*, O-CMC and N-CMC also have improved antimicrobial properties [86–88].

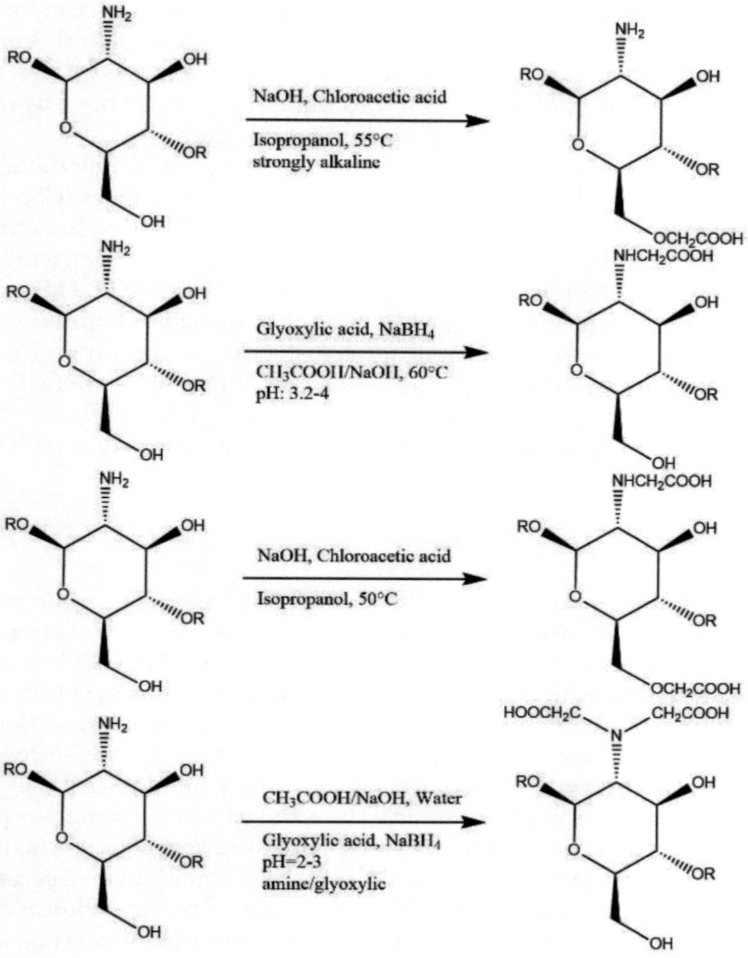

Figure 6.
As example, some carboxymethyl derivatives of chitosan [86].

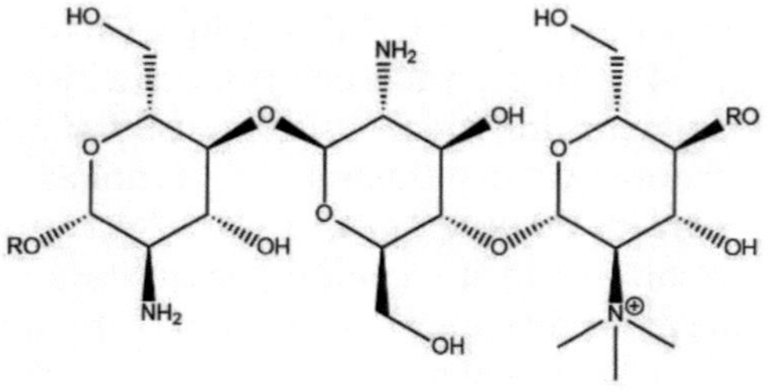

Figure 7.
Molecular structure of N, N, N-trimethyl chitosan [89].

The second chitosan derivative fairly presented in the literature is TMC (**Figure 7**). It is a partially quaternized derivative [89]. It is obtained by nucleophilic substitution of the primary amino group in position C-2 by a quaternary amino group [89]. This modification facilitates the aqueous solubility of TMS at neutral and basic pH due to the presence of a permanent positive charge independent of pH (the quaternary amino amino groups) [90, 91]. This high positive charge density is responsible for the high antibacterial performance reported for TMC compared to Chitosan [92–94]. It is essential to point out that this modification takes place under alkaline conditions using sodium iodide as catalyst and N-methyl-2-pyrrolidone as solvent [95]. In addition, it can take place by reaction with dimethyl sulfate [96, 97] or dimethylformamide [98].

Among the derivatives of Chitosan, which have a higher antimicrobial activity than the chitosan solution, there are the chitosan nanoparticles (CSNP). For the moment, there is no clear explanation for this high efficiency, but one of the hypotheses given is based on the high specific surface area of nanoparticles as well as their better affinity for the microbial cell wall [99]. Unlike TMC, CNSPs are prepared with simple methods, without organic solvent or high shear force. Among these methods there are: emulsion crosslinking, coacervation (precipitation), ionic gelation, spray drying, microemulsion, diffusion of emulsifying solvent and polyelectrolyte complex [100]. The degree of deacetylation of chitosan and its molecular weight are factors that affect the formation and size of CNSPs [101].

2.2.4.2 Mechanisms of chitosan's antimicrobial action

Chitin is a film-forming polymer with antibacterial and fungi-static property. It triggers the defensive mechanism in host inducing certain enzymes like phytoalexins, chitinases, pectinases, glucanases, and lignin in plants [28]. Chitosan and its derivatives have been studied as antimicrobial agents against bacteria, fungi and viruses, in experiments involving in vivo and in vitro interactions with chitosan in different forms (solutions, nanoparticles, films, fibers and composites). Chitosan can react in two different mechanisms, killing or inhibiting the growth of microorganisms (biocide or Biostatic). However, its action often takes place without distinction between activities [102]. The antimicrobial performance of Chitosan or one of its derivatives depends largely on its molecular structure and its properties such as molecular weight [103], the degree of deacetylation and its water solubility [104–108]. In addition, its pH and its concentration in solution affect its effectiveness against microorganisms [109]. Chitosan has MIC of 0.05–0.1% (w/v) against most common kinds of bacteria. The mode of its antibacterial action is not yet fully understood [110–112], but it is possible that the amine groups provide positive charges which can react with the negatively charged surface of microbes; therefore, they can change the cell permeability, which finally leads to intracellular substance leakage [113, 114]. Chitosan acts primarily as a disruptor of the outer membrane and not as a penetrating agent. Using transmission electron microscopy, impaired membrane function was demonstrated by shrinkage, implying that water and ion leakage had occurred. However, other studies have proposed a mechanism by encapsulation where chitosan forms a polymeric substance around the bacterial cell preventing nutrients from entering the cell and its subsequent death [115]. In addition, a third mechanism based on metal chelation, removal of spore elements and binding to nutrients essential for microbial growth is proposed. This mechanism is based on the strong binding capacities known for chitosan to metals. This absorption of cations occurs by metal chelation favored by the amino groups present in the chitosan molecules [115]. The efficiency of this mechanism depends largely on the pH of the medium. A high pH favors this absorption mechanism by the fact that the amino groups of the chitosan will not be protonated under these conditions. This will allow the pair of electrons on the nitrogen atom to be available for donation to metal ions. Some studies have assumed that the metal can behave as an electron acceptor which connects via $-NH_2$ functions to one or more chitosan chains, forming bridges with hydroxyl groups [116].

2.2.4.3 Chitosan and chitosan nanoparticles (CSNP) on textiles applications

Even though, chitosan has already been utilized for the treatment of fibrous materials, a comprehensive research on their use for antimicrobial functionalization of viscose fibers for development of modern medical textiles for applications in medical devices is still missing. There is a lack of information regarding the behavior of different chitosans in contact with the cellulose materials. In addition, in-depth knowledge of their physical, chemical, and biological properties is missing [117, 118].

For textiles, chitosan and its derivatives could constitute one of the products that can be used for the finishing processes [116]. In addition, they could be used for the production of raw materials such as chitin and chitosan fibers [116]. The latter are widely used, alone or in admixture with other products such as viscose, in the medical field for the manufacture of non-woven, which can be used for dressings, or as a carrier for medicaments in the form of hollow fibers [119, 120]. Microbiological tests showed antimicrobial activity, no cytotoxicity was detected for a chitosan-polypropylene nonwoven [119]. Viscose tampons treated with chitosan were utilized for maintaining the physiological pH of vagina and acting as moisturizing agent, while simultaneously providing antimycotic and antibacterial activity [121]. Textiles treated in such a way were effective against gram-negative and gram-positive bacteria [122]. Some studies have shown that chitosan-based products provide rapid healing and less dense skin lesions compared to standard products [122–124]. By way of example, it has been found that the treatment of cotton with a carboxymethyl chitosan derivative at a concentration of 0.1% provides antibacterial protection against *E. coli* and *S. aureus*, as well as an improvement in the wrinkle recovery. Alternatively, core-shell particles based on chitosan (shell) and poly (n-butyl acrylate) (PBA) (core) have been designed as a novel antibacterial coating for textiles [125]. These particles applied to a cotton fabric showed an antibacterial performance of over 90% against *S. aureus*. The application of chitosan on cotton takes place without the need for a crosslinking agent. Studies have shown that an intermolecular hydrogen bond between the hydroxyl groups of cotton and the amino groups of chitosan [125]. Antibacterial efficacy against *S. aureus* was examined for chitosan nanoparticles loaded with silver (CSNP) applied to polyester. An efficiency of 100% was obtained. This efficacy is attributed to the synergistic effect of silver and CSNP [126].

2.2.5 Natural antibacterial

In addition to chitin and chitosan, antibacterial performance has been identified for other naturally occurring products such as honey. The latter and because of its antibacterial activity allows the treatment of certain burns [127–129]. This antibacterial performance, against certain large bacteria-infecting wounds such as *S. aureus* known for its resistance to methicillin, is attributed to the presence of certain molecules such as hydrogen peroxide and bioflavonoids. The latter have the capacity to inhibit the synthesis of nucleic acids [130].

At the same time, we must not forget the products of plant origin (*Aloe Vera*, tea tree oil and eucalyptus, extracts of neem, grapefruit seeds and tulsi leaves, etc.) which have a high antibacterial performance. This efficiency is attributed to their content in certain phenol type products (simple phenols, phenolic acids, quinines, flavonoids, flavones, tannins and cumarins), terpenoids, essential oils, alkaloids, lectins, polypeptides and polyacetylenes [131]. In addition to their antibacterial performance, these components are known for their antioxidant properties. The combination of these two functionalities constitutes an important advantage for biomaterials that can be used for medical applications such as dressings. It ensures a reduction in the formation of reactive oxygen species, which are strongly involved in the pathogenesis of injury. Usually these species cause the formation of biomolecules (lipids, proteins and nucleic acids) at the level of injury as well as the depletion of mitochondrial DNA in human skin [132]. Among the powerful anti-oxidants are the flavonoids, which are used as anti-inflammatory, antimicrobial and anticancer agents [132]. But, durability and resistance to washing remains the weak point for the antimicrobial finish based on natural agents. This weakness results from their difficulty in forming bonds with textile materials [133]. Certain methods have been developed to increase this durability. Among these methods, mention may be made of microencapsulation [133–135], the use of a crosslinking agent [131] and the immobilization of bioactive liquids in sol–gel matrices are also described in the literature [135]. For example, antimicrobial activity against *S. aureus* has been recorded for cotton fabrics treated with *Aloe Vera* extract by a dry drying process. This effectiveness was sustainable even after 50 wash cycles [136].

3. Standard tests for antimicrobial activity

The antimicrobial efficacy of textiles could be characterized by different methods of analysis. These methods are standardized and divided into two categories: 1) qualitative, such as AATCC TM147, AATCC TM30 (American Association of Textile Chemists and Colorists Test Method), ISO 20645, ISO 11721 (International Organization for Standardization) and SN 195920, SN 195921 (Swiss standard) and 2) quantitative, such as AATCC TM100, ISO 20743, SN 195924, JIS L 1902 (Japanese industry standards) and ASTM E 2149 (or its modification).

Qualitative methods are characterized by their speed and simplicity. They are mainly based on the agar diffusion test. As diffusion through agar occurs at different rates depending on the textiles and the nature of the antimicrobial agents used, this category of methods is not suitable for all types of textiles. Some differences could be identified between the different qualitative methods. As an indication, we can mention that the textile is laid on an inoculated agar plate for AATCC TM147. While it is place between two agars plates, with one side inoculated for ISO 20645. Usually the qualitative method has an incubation period. After this period (24–48 h) depending on the type of microorganisms tested, the plates are examined for bacterial growth directly underneath the fabric and around its edges (zone of inhibition). The appearance of the zone of inhibition depends on the ability of the antimicrobial agent to diffuse into the agar and it is binding to the textiles. The appearance of a zone of inhibition and its size are indicative of the rate of release of the active agent and its antimicrobial efficacy. It is important to specify that zone of inhibition does not necessarily imply that microorganisms have been killed; they might have only been prevented from growing. By qualitative methods, the efficacy of different agents cannot be compared [137].

On the other hand, quantitative methods can be used for the majority of antimicrobial agents and textile supports. However, they require a longer time compared to qualitative methods. In addition, they are more expensive because they involve a real count of the microbes to measure the antimicrobial effectiveness. This method of measurement makes it easier to compare the effectiveness of different antimicrobial agents on the same textile support, for example [138–141]. Quantitative methods are much more specific depending on the mechanism of action of the antibacterial agent. For example, ISO 20645 can be tested with only leaching types because the configuration does not allow observation under the textile [141]. AATCC TM100 and JIS L 1902 can be tested with leaching and non-leaching type's antimicrobial. These methods (AATCC TM100, ISO20743 and JIS L 1902) are based on similar principle: specified amount (weight, size, and surface area) of sample swatches or substrate are inoculated with a specified number of microorganisms [142]. The inoculum is put in contact with the treated surface via three different methods: absorption, transfer and printing (ISO 20743). The absorption method uses an inoculated broth with a standardized species and concentration. The broth is absorbed by the textile sample. The sample is incubated in different condition depending on the method, to promote the bacterial growth. This method allows testing a leaching, non-leaching or a combination of antibacterial textile as well as bacteriostatic and bactericidal. However, this method is not recommended for textile with hydrophobic treatment or low absorption capacities [143]. While, the transfer method uses an agar plate who is inoculated with the tested bacteria. The contaminated plate is put in contact with the textile for 60 seconds, and after the sample is incubated. This method is used to replicate the contact of the antimicrobial textile with a contaminated surface. Whereas, the printing method apply the bacteria via a printer. This method allows faster incubation time (1 to 4 h, ISO 20743) and faster sample preparation with the automated printer (ISO20743). Finally, the dynamic shake flask method (ASTM E 2149 (or its modification) is particularly appropriate for non-leaching antimicrobials whilst the dynamic contact conditions are applied to the samples [140]. It can be used to assess the activity of the antimicrobial textile as a qualitative test. This method has been used for testing antimicrobial activity of cotton fabrics (or cellulose fibers) treated with the nanoparticles [144, 145], as well as functionalized wool [146], cotton and viscose fibers coated with chitosan [147, 148], and some other fabrics.

Once the microorganism and incubation application protocol are applied according to the desired method, the microorganism count will take place via two different techniques: the plate count method and the luminescence method. The count plate method consists of recovering the microorganism from the broth by re-plating and the number of surviving organisms. The number of colonies forming unit (CFU) is counted and the bacteria concentration is obtained by multiplying the dilution rate. The ATP concentration is quantified via a spectrophotometer according to the luminescence method. This measurement will be compared with a calibration curve prepared according to the ATP standard. The quantification of the ATP of the inoculum is carried out before and after exposure to the antimicrobial treatment. The number of surviving organisms is counted as CFU and results are usually presented as percentages or log10 reduction in contamination relative to the initial inoculum of microorganisms or the untreated control.

It should be noted that antimicrobial analysis methods are quite sensitive to contaminate. For this reason, tests are usually done under tightly controlled conditions to ensure reproducibility of results. However, carrying out tests in such a standardized environment does not reflect the reality of using textiles treated with antimicrobial agents [137]. Another factor that affects these tests is the efficiency of microbial extractions from the sample tissues. In addition, the absence of an absolute standard of effectiveness facilitates changes in the protocols applied creating inconsistencies between laboratories at national and international levels. Taking into account all the factors affecting the effectiveness of antimicrobial tests, certain additional methods are applied in complementarity. These methods include colorimetric analyzes [149], viability test [150], viability staining and microscopy [151], and fluorescent staining coupled with flow cytometry [152]. Despite the advancements made to date, the poor reproducibility of test results is the Achilles point of these tests. Over time, some attempts to establish a correlation between the different analytical techniques have taken place. For example, AATCC TM147 and JIS L 1902 were found to give the same result for a textile sample with a non-leaching antimicrobial [137]. Nevertheless, the strong differences are always an obstacle.

4. Conclusion

In conclusion, the microorganism presence on textile can be eliminated or the growth can be slowed by treating with a variety of antimicrobial agent. Multiple antimicrobial families were presented in this chapter, including synthetic and natural chemicals (**Table 2**). Textiles are susceptible to microorganism growth because of the structure and the ability to retain moisture of the textile. Therefore, the microorganism growth can generate multiple undesired consequence, such as hosting and transmitting harmful microbe, creating odor, mold, degradation, discoloration and biofouling for example. Textile can be treated with antimicrobial agent to reduce, slow or eliminate the microbial growth and spread. The antimicrobial was categorized in two types in this chapter, leaching and non-leaching. The non-leaching types are bound to the textile and react with the microbes upon direct contact. On the other side, the leaching types release antimicrobial in the environment at a controlled rate to disrupt the microorganism at proximity of the textile. A summary of the common reagents discussed in this chapter is gathered in the **Table 3**.

The antimicrobial efficiency, the durability and skin compatibility of the treated textile must be assessing during the development of an antimicrobial treatment to minimize the risk. The antimicrobial activity testing can be categorized by quantitative and qualitative method. The qualitative method is useful for routine quality control and for the screening of multiples iterations during the development of a product, such as determining the wash durability. However, this method can lead to subjective determination. Instead, the quantitative method eliminates the possible subjectivity with plate count and luminescence technique. In addition, the skin should not be harmed be the treated textile. The safety for the skin can be evaluated with cytotoxicity test to human cell and irritation test in-vitro and in-vivo. The properties of the antimicrobial should be assessed before commercialization of an antimicrobial treated product.

Mechanism	Reagent	Fiber	Remarks
Leaching	**Metals (silver)**	Nylon, Wool, Polyester and Regenerated- cellulose	Slow release, durable, depletion of Ag might occur
	Triclosan	Polyester, Nylon, Acrylic, Polypropylene and Cellulose- acetate	Breaks down into toxic dioxine, large amount needed, bacterial resistance, banned in some European countries
Non-Leaching	**QACs**	Cotton, wool, Nylon, Polyester and Acrylic	Very durable, covalent bonding, possible bacterial resistance
	PHMB	Nylon, Polyester and Cotton	Large amount needed, bacterial resistance
	N-halamine	Wool, Nylon, Cotton and Polyester	Requires regeneration, unpleasant odor from residual Cl
	Peroxyacids	Cotton and Polyester	Poor durability, requires regeneration
	Chitosan	Wool, Polyester and Cotton	Low durability, adverse effect on fabric handle

Table 2.
Conventional reagents used in the antimicrobial finishing of textiles.

Test	Method	Title	Principle	Antimicrobial type	Uses
ASTM E2149	20	Standard Test Method for Determining the Antimicrobial Activity of Antimicrobial Agents Under Dynamic Contact Conditions	Dynamic shake flask test	Non-leaching type	Qualitative: Screening, routine quality-control
AATCC 147	Parallel Streak	Antibacterial Activity of Textile Materials: Parallel Streak	Zone diffusion assay: Agar	Leaching-types	Qualitative: Screening, routine quality-control
AATCC 100	—	Test Method for Antibacterial Finishes on Textile Materials: Assessment of	Cell suspension intimate contact test	Leaching and non-leaching	Quantitative
XP G 39–010	—	Propriétés des étoffes - Étoffes et surfaces polymériques à propriétés antibactériennes - Caractérisation et mesure de l'activité antibactérienne	Cell suspension intimate contact test	Leaching and non-leaching	Quantitative
JIS L 1902	Absorption method	Testing Method for Antibacterial Activity of Textiles Quantitative Test	Cell suspension intimate contact test	Leaching and non-leaching	Quantitative
JIS L 1902	Transfer method	Testing Method for Antibacterial Activity of Textiles Quantitative Test	Transferred Agar plate contact test	Leaching-types	Quantitative
JIS L 1902	Printing method	Testing Method for Antibacterial Activity of Textiles Qualitative Test	'Dry' inoculum intimate contact test	Non-leaching	Quantitative
JIS L 1902	Halo Method	Testing Method for Antibacterial Activity of Textiles Qualitative Test	Zone diffusion assay: Agar	Leaching-types	Qualitative: Screening, routine quality-control
ISO 20645	Agar diffusion plate test	Textile fabrics — Determination of antibacterial activity — Agar diffusion plate test	Zone diffusion assay: Agar	Leaching-types only	Qualitative: Screening, routine quality-control
ISO 20743	Absorption Method	Textiles - Determination of antibacterial activity of antibacterial finished products: Absorption Method	Cell suspension intimate contact test	Leaching and non-leaching	Quantitative

Test	Method	Title	Principle	Antimicrobial type	Uses
ISO 20743	Transfer Method	Textiles - Determination of antibacterial activity of antibacterial finished products: Transfer Method	Cell suspension intimate contact test	Leaching-types	Quantitative
ISO 20743	Printing Method	Textiles - Determination of antibacterial activity of antibacterial finished products: Printing method	'Dry' inoculum intimate contact test	Non-leaching	Quantitative

Table 3.
Comparison of antimicrobial test method for textile.

Antimicrobial resistance should also be a concern when developing an antimicrobial treatment for a textile because of the large quantity of antimicrobial agent required achieving the antimicrobial activity and durability. The risk/reward should always be considered before applying antimicrobial product to a textile. The risk of antimicrobial resistance can be minimized. First off, the antimicrobial should not come close to the minimal inhibitory concentration (MIC) of the treatment during the useful life of the product to guarantee an effective product. The MIC of the antimicrobial product can be reached because of poor wash durability of the antimicrobial product. Second of, the synergy, mechanism of different antimicrobial product can be combined to reduce the resistance of a gene to a pathway. A complex antimicrobial mechanism is believed to be more efficient and more complex for the microbe to develop a set of successful mutated gene against the antimicrobials [153].

Acknowledgements

This work was supported by the Natural Sciences and Engineering Research Council of Canada (NSERC) (COVPJ 553781 - 20), and the MEI-Québec (Ministère de l'Économie et de l'Innovation).

Author details

Ahmad Ibrahim*, Joseph-Émile Laquerre, Patricia Forcier, Vincent Deregnaucourt, Justine Decaens and Olivier Vermeersch
Groupe CTT, Saint-Hyacinthe, QC, Canada

*Address all correspondence to: aibrahim@gcttg.com

References

[1] Majeti, K.; Ravi, N.V. A review of chitin and chitosan applications. Reactive and Functional Polymers, 2000, 46,. 1, 1-27.

[2] Deepti, G.; Adane, H. Multifunctional properties of cotton fabric treated with chitosan and carboxymethyl chitosan. Carbohydrate Polymers, 2007, 69, 1, 164-171.

[3] Gao, Y.; Cranston, R. Recent advances in antimicrobial treatments of textiles. Textile Research Journal, 2008, 78, 60-72.

[4] Kramer, A.; Guggenbichler, P.; Heldt, P.; Junger, M.; Ladwig, A.; Thierbach, H.; Weber, U.; Daeschlein, G. Hygienic relevance and risk assessment of antimicrobialimpregnated textiles. Current Problems in Dermatology, 2006, 33, 78-109.

[5] Simoncic, B.; Tomsic, B. Structures of novel antimicrobial agents for textiles-A review. Textile Res. J. 2010, 80, 1721-1731.

[6] Kut D.; Orhan, M.; Gunesoglu, C.; Ozakin, C. Effects of environmental conditions on the antimicrobial activity of treated cotton knits. AATCC Review, 2005, 5, 25-28.

[7] Wolfgang, S.D.; Peter, H. J. Antimicrobial finishes. in Chemical Finishing of Textiles. Woodhead Publishing, 2004, 213.

[8] Mcdonnell, G.; Russel, A. Antiseptics and disinfectants: activity, action and resistance. American Society of Microbiology, 1999. 12, 147-179.

[9] Purwar, R.; Joshi, M. Recent Developments in Antimicrobial Finishing of Textiles-A Review. AATCC Review, 2004. 4: 22-26.

[10] Lansdown, A.; Hipler, U.; ed, Elsner, P., ed. "Silver in Health Care: Antimicrobial Effects and Safety in Use." Biofunctional Textiles and the Skin. Current Problems in Dermatology, 2006, 33, 17-34

[11] Kittler, S.; Greulich,C.; Koller. M.; Epple, M. Synthesis of PVP-coated silver nanoparticles and their biological activity towards human mesenchymal stem cells. Materialwissenschaft und Werkstofftechnik 2009, 40, 258-264.

[12] Windler, L.; Height, M.; Nowack, B. "Comparative Evaluation of Antimicrobials for Textile Applications." Environment International, 2013, 53, 62-73.

[13] Nakashima, T.; Sakagami,Y.; Ito, H.; Matsuo, M. Antibacterial activity of cellulose fabrics modified with metallic salts. Textile Research Journal, 2001, 71, 688-694.

[14] Kim, J.S.; Kuk, E.; Yu, K.N.; Kim, J.H.; Park, S.J.; Lee, H.J. Antimicrobial effects of silver nanoparticles. Nanomed-Nanotechnol. 2007, 3, 95-101.

[15] Pradhan, N.; Pal, A.; Pal, T. Silver nanoparticle catalyzed reduction of aromatic nitro compound. Colloid Surface A. 2002, 196, 247-257.

[16] McFarland, A.D.; Van Duyne, R.P. Single silver nanoparticles as real-time optical sensors with zeptomole sensitivity. Nano Letters 2003, 3, 1057-1062.

[17] Lok, C.N.; Ho, C.M.; Chen, R.; He, Q.Y.; Yu, W.Y.; et al. Proteomic analysis of the mode of antibacterial action of silver nanoparticles. J. Proteome Res. 2006, 5, 916-924.

[18] Baker, C.; Pradhan, A.; Pakstis, L.; Pochan, D. J.; Shah, S.I. Synthesis and antibacterial properties of silver nanoparticles. J. Nanosci. Nanotechnol. 2005, 5, 244-249.

[19] Aymonier, C.; Schlotterbeck, U.; Antonietti, L.; Zacharias, P.; Thomann, R.; Tiller, J. C.; Mecking, S. Hybrids of silver nanoparticles with amphiphilic hyperbranched macromolecules exhibiting antimicrobial properties. Chem. Commun. 2002, 3018-3019.

[20] Height, Murray. "Silver Use in Textiles." VA. Proc. of Nano Release Steering Committee Workshop, U.S. EPA Potomac Yard Conference Center, 2777 S. Crystal Drive, Arlington. http://205.251.124.92/Research Foundation/Documents/Nano Release%202011%20Workshop%20 Presentation%20PDF%20Files/ HeightNanoreleaseWorkshop_ SilverTextiles_10May2011_print.pdf.

[21] Feng, Q.L.; Wu, J., Chen; G.Q., Cui; F.Z., Kim; T. N.; Kim, J.O. A mechanistic study of the antibacterial effect of silver ions on Escherichia coli and Staphylococcus aureus. Biomedical Material Research, 2000, 52, 662-668.

[22] Lansdown, A. B. G. Silver I: Its antimicrobial properties and mechanism of action. Journal of Wound Care, 2002, 11, 125-130.

[23] Lansdown, A. B. G. A review of the use of silver in wound care: facts and fallacies. British Journal of Nursing 2004, 13, S6-S19.

[24] Leaper, J.D. Silver dressings: their role in wound management. Int. Wound J. 2006, 3,282-294.

[25] Klasen, H.J. Historical review of the use of silver in the treatment of burns. I. Early uses. Burns 2000, 26, 117-130.

[26] Vimala, K.; Mohan, Y.M.; Sivudu, K.S.; Varaprasad, K.; Ravindra, S.; Reddy, N.N.; Padma, Y.; Sreedhar, B.; MohanaRaju, K. Fabrication of porous chitosan films impregnated with silver nanoparticles: A facile approach for superior antibacterial application. Colloid Surface B. 2010, 76, 248-258.

[27] Thomson Research Associates. Available from: http://www.ultra-fresh.com/tra/.

[28] Rinaudo, M.; Chitin and Chitosan: Properties and Applications. Progr. Polymer Sci., 2006, 31, 603-632.

[29] Williams, J.F.; HaloSource, V.; and Cho, U.; Antimicrobial Functions for Synthetic Fibers: Recent Developments. AATCC Review, 2005, 5, 17-21.

[30] Troy M.B.; Westerhoff, P. "Nanoparticle Silver Released into Water from Commercially Available Sock Fabrics." Environmental Science & Technology 2008, 42, 4133-4139.

[31] Geranio, L.; Heuberger, M.; Nowack, B. The Behavior of Silver Nanotextiles during Washing." Environmental Science & Technology, 2009, 43, 8113-8118.

[32] Kalyon, B.D; Olgun, U., Antibacterial Efficacy of Triclosan-incorporated Polymers. Am. J. Infect. Contr, 2001, 29, 124-125.

[33] Mansfield, R. G. Keeping it fresh. Textile World, 2002, 152, 42-45.

[34] Lu, J.; Hill, M. A.; Hood, M.; Greeson, D. F.; JR.; Horton, J. H.; Orndorff, P. E., Herndon, A. S.; Tonelli, A. E. Formation of antibiotic, biodegradable polymers by processing with Irgasan® DP300R (triclosan) and its inclusion compound with β-cyclodextrin. Journal of Applied Polymer Science, 2001, 82, 300-309.

[35] Mcmurry, L.M.; Oethinger, M.; Levy, S.B. Triclosan targets lipid synthesis.Nature. 1998, 94, 531-532.

[36] Orhan, M.; Kut, D.; Gunesegulu, C. Use of triclosan as antimicrobial agent in textiles. Indian Journal of Fibre and Textile Research, 2007, 32.114-32.118.

[37] Glaser, A. The Ubiquitous Triclosan- A common antibacterial

agent exposed Available from: http://www.beyondpesticides.org/pesticides/factsheets/Triclosan%20cited.pdf.

[38] Williams, R.M. Triclosan - A Controversial Antibacterial. Available from: http://www.townsendletter.com/May2006/healthrisk0506.htm.

[39] Malek, J. R.; Speier, J. L. Development of organosilicone antimicrobial agent for the treatment of surfaces. Journal of Coated Fabrics, 1982, 12, 38-46.

[40] Shindler, W. D.; Hauser, P.J. Antimicrobial finishes. Chemical finishing of textiles. Cambridge England: Woodhead Publishing Limited, 2004, 165-174.

[41] Murugan, E.; Gopinath, P.; Shanmugayya, V.; Mathivanan, N. Antibacterial activity of novel insoluble bead-shaped polymer-supported multiquaternary ammonium salts. J. Appl. Polym. Sci. 2010, 117, 3673-3678.

[42] Townsend, D.E.; Greed, I.; Ashdown, N.; Grubb, W.B. Plasmid-mediated resistance to quaternary ammonium compounds in methicillin-resistant staphylococcus aureus. Med. J. Australia 1983, 2, 310.

[43] Tennent, J.M.; Lyon, B.R.; Gillespie, M.T. Cloning and expression of Staphylococcus-aureus plasmid-medicated quaternary ammonium resistance in Escherichia-coli. Antimicrob. Agents Chemother. 1985, 27, 1, 79-83.

[44] Resuggan, J.C.L.; The antibacterial activity of quaternary ammonium compounds. Journal of Applied Microbiology, 1952, 15, 166-171.

[45] Cai, Z.S.; Sun, G. Antimicrobial Finishing of Acrilan Fabrics with Cetylpyridinium Chloride: Affected Properties and Structures. Journal of Applied Polymer Science, 2005, 97, 1227-1236.

[46] Kim, Y.H.; Sun, G. Functional Finishing of Acrylic and Cationic Dyeable Fabrics: Intermolecular Interactions. Textile Research Journal, 2002, 72, 1052-1056.

[47] Son, Y.A.; Sun, G. Durable Antimicrobial Nylon 66 Fabrics:Ionic Interactions with Quaternary Ammonium Salts. Journal of Applied Polymer Science, 2003. 90, 2194-2199.

[48] Kim, Y.H.; Sun, G. Dye Molecules as Bridges for Functional Modifications of Nylon: Antimicrobial Functions. Textile Research Journal, 2000, 70, 728-733.

[49] Liu, J.; Sun, G. The synthesis of novel cationic anthraquinone dyes with high potent antimicrobial activity. Dyes and Pigments, 2008, 77, 380-386.

[50] Ma, M.; Sun, G. Antimicrobial cationic dyes. Part 3: simultaneous dyeing and antimicrobial finishing of acrylic fabrics. Dyes and Pigments, 2005, 66, 33-41.

[51] Ma, M.; Sun, Y.; Sun, G. Antimicrobial cationic dyes: part 1: synthesis and characterization. Dyes and Pigments, 2003, 58, 27-35.

[52] [52]Mulder, G. D.; Cavorsi, J. P.; Lee, D. K. Polyhexamethylenebiguanide (PHMB): An addendum to current topical antimicrobials. Wounds, 2007, 19, 173-182.

[53] Tiller, J.C.; Liao, C.J.; Lewis, J.; Kilbanov, A.M. Designing surfaces that kill bacteria on contact. Proc. Natl. Acad. Sci. U.S.A. 2001, 98, 5981-5985.

[54] Cserhati, T., Alkyl Ethoxylated and Alkylphenol Ethoxylated Nonionic Surfactants: Interaction with Bioactive Compounds and Biological Effects. Journal of Environmental Health Perspectives, 1995, 103, 358-364.

[55] Yanni, J.M. Ophthalmic, anti-allergy compositions suitable for use with contact lenses. 2002, Alcon Universal Ltd.: U.S. Patent 5641805.

[56] Schallreuter, K.U.; Schulz, K.H.; Wood, J.M. Induction of Contact Dermatitis in Guinea Pigs by Quaternary Ammonium Compounds: The Mechanism of Antigen Formation. Environmental Health Perspectives, 1986, 70, 229-237.

[57] Reynolds, J. Martindale: the extra pharmacopoeia, ed. 31. 1996, London: Pharmaceutical Press.

[58] Bernstein, J.A.; Stauder, T.; Bernstein, D.I.; Bernstein, I.L. A combined respiratory and cutaneous hypersensitivity syndrome induced by work exposure to quaternary amines. J Allergy Clin Immunol, 1994, 94, 257-259.

[59] Purohit, A.; quaternary ammonium compounds and occupational asthma. International Archives of Occupational and Environmental Health, 2000, 73, 423-427.

[60] Marple, B.; Roland, P.; Benninger, M. Safety review of benzalkonium chloride used as a preservative in intranasal solutions: an overview of conflicting data and opinions. Otolaryngol Head Neck Surg 2004, 130.

[61] Feng, Q.L.; Wu, J.; Chen, G.Q.; Cui, F.Z.; Kim, T.N.; Kim, J.O. A mechanistic study of the antibacterial effect of silver ions on Escherichia coli and Staphylococcus aureus. J.Biomed. mater. Res. 2000, 52, 662-668.

[62] Payne, J. D.; Yates, J. E. U.S. Patent No. A1 0271,707. Washington, DC: U.S. Patent and Trademark Office, 2007.

[63] Blackburn, R. S.; Harvey, A. L.; Kettle, L.; Payne, J. D.; Russell, S. J. Sorption of poly (hexamethy lenebiguanide) on cellulose: mechanism of binding and molecular recognition, Langmuir, 2006, 22, 5636-5644.

[64] Lin, J.; Winkelman, C.; Worley, S. D.; Broughton, R. M.; Williams, J. F. Antimicrobial treatment of nylon. Journal of Applied Polymer Science, 2001, 81, 942-947.

[65] Qian, L.; Gang,S. Durable and regenerable antimicrobial textiles: Chlorine transfer among halamine structures. Industrial and Engineering Chemistry Research. 2005, 44, 4, 852-856

[66] Sun, G.; Xu, X.J., Durable and Regenerable Antibacterial Finishing of Fabrics: Biocidal Properties. Textile Chemist and Colorist, 1998, 30, 26-30.

[67] Williams, D.E.; Elder, E.D.; Worley, S.D. Is free halogen necessary for disinfection? Appl. Environ. Microbiol. 1988, 54, 2583-2585.

[68] Sun, Y.Y.; Sun, G. Novel regenerable N-halamine polymeric biocides. I. Sythesis, characterization, and antibacterial activity of hydantoin-containing polymers. J. Appl. Poly. Sci. 2000, 80, 2460-2467.

[69] Sun, Y.Y.; Sun, G. Novel regenerable N-halamine polymeric biocides. II. Grafting hydantoin-containing monomers onto cotton cellulose. J. Appl. Poly. Sci. 2001, 81, 617-624.

[70] Lin, J.; Cammarata, V.; Worley, S.D.; Broughton, R.M.; Tzou,Y. M.; Huang, T.S. Infrared charactrization of biocidal Nylon. Polymer 2001, 42, 7903-7906.

[71] Sun, Y.Y. and Sun, G., Regenerable N-halamine Polymeric Biocides. III. Grafting Hydantoin-containing Monomersonto Synthetic Fabrics. J. Appl. Polymer Sci., 2001. 81: 1517-1525.

[72] Sun, Y.Y.; Sun, G., Novel Refreshable N-halamine Polymeric Biocides: Grafting Hydantoin-containing

Monomers onto High Performance Fibers by a Continuous Process. J.Appl. Polymer Sci., 2003. 88, 1032-1039.

[73] Huang, L.K.; Sun, G., Durable and Oxygen Bleach Rechargeable Antimicrobial Cellulose: Sodium Perborate as an Activating and Recharging Agent. Ind. Eng. Chem. Res, 2003. 42, 5417-5422.

[74] Huang, L.K.; Sun, G., Durable and Regenerable Antimicrobial Cellulose with Oxygen Bleach: Concept Proofing. AATCC Review, 2003. 3, 17-21.

[75] Sun, G.; Xu, X.; Bickett, J. R.; Williams, J. F. Durable and regenerable antibacterial finishing of fabrics with a new hydantoin derivative. Industrial Engineering and Chemical Research, 2001, 40, 1016-1021.

[76] Kocer, H.B.; Akdag, A.; Worley, S.D.; Acevedo, O.; Broughton, R.M.; Wu, Y. Mechanism of photolytic decomposition of N-halamine antimicrobial siloxane coatings. ACS Applied Material Interfaces. 2010, 2, 2456-2464

[77] Liu, S.; Sun, G., Durable and Regenerable Biocidal Polymers: Acyclic Nhalamine Cotton Cellulose. Ind. Eng. Chem.Res, 2006, 45, 6477-6482.

[78] Qian, L.; Sun, G., Durable and Regenerable Antimicrobial Textiles: Improving Efficacy and Durability of Biocidal Functions. J. Appl. Polymer Sci. 91, 2588-2593.

[79] Sun, G.; Allen, L.C.; Luckie, E.P.; Wheatley, W.B.; Worley, S.D. Disinfection of Water by N-Halamine Biocidal Polymers Ind. Eng. Chem. Res, 1995. 34, 4106-4109.

[80] Choi, C.; Nam, J.; Nah, J. Application of chitosan and chitosan derivatives as biomaterials. Journal of Industrial and Engineering Chemistry, 2016, 01, 33, 1-10.

[81] Dragostin, O. M.; Samal, S. K.; Dash, M., Lupascu, F.; Pânzariu, A.; Tuchilus, C.; Profire, L. New antimicrobial chitosan derivatives for wound dressing applications. Carbohydrate Polymers, 2016, 05, 141, 28-40.

[82] Wang, X.; Du, Y.; Luo, J.; Yang, J.; Wang, W.; Kennedy, J. .. A novel biopolymer/rectorite nanocomposite with antimicrobial activity. Carbohydrate Polymers, 2009, 07, 77, 449-456.

[83] El-tahlawy, K.F.; El-bendary, M.A.; Elhendawy, A. G.; Hudson, S. M. The antimicrobial activity of cotton fabrics treated with different crosslinking agents and chitosan, Carbohydrate Polymers,2005, 60, 421-430.

[84] Alves, N.; Mano, J. Chitosan derivatives obtained by chemical modifications for biomedical and environmental applications. International Journal of Biological Macromolecules, 2008, 12, 43, 401-414.

[85] Sahariah, P.; Másson, M. Antimicrobial Chitosan and Chitosan Derivatives: A Review of the Structure–Activity Relationship. Biomacro molecules, 2017, 18, 3846-3868.

[86] Upadhyaya, L.; Singh, J.; Agarwal, V.; Tewari, R. P. Biomedical applications of carboxymethyl chitosans. Carbohydrate Polymers, 2013, 91, 452-466.

[87] Anitha, A.; Rani, V. D.; Krishna, R.; Sreeja, V.; Selvamurugan, N.; Nair, S.; Jayakumar, R. Synthesis, characterization, cytotoxicity and antibacterial studies of chitosan, O-carboxymethyl and N,O-carboxymethyl chitosan nanoparticles. Carbohydrate Polymers, 2009, 11, 78, 672-677.

[88] Mishra, D.; Bhunia, B.; Banerjee, I.; Datta, P.; Dhara, S.; Maiti, T. K.

Enzymatically crosslinked carboxymethyl–chitosan/gelatin/ nano-hydroxyapatite injectable gels for in situ bone tissue engineering application. Materials Science and Engineering: C, 2011, 10, 31, 1295-1304.

[89] Dehousse V.; Garbacki N.; Colige A.; Evrard B. Development of pH– responsive nanocarriers using trimethylchitosans and methacrylic acid copolymer for siRNA delivery. Biomaterials, 2010, 31, 7, 1839-1849.

[90] Jintapattanakit A.; Mao S.; Kissel T.; Junyaprasert V.B. Physicochemical properties and biocompatibility of N-trimethyl chitosan: effect of quaternization and dimethylation. European Journal of Pharmaceutics and Biopharmaceutics, 2008, 70, 2, 563.

[91] Prezottoda, S.L.; Douglas, B.; Regali, S.M.H.; Odilio B G. In vitro activity of water-soluble quaternary chitosan chloride salt against *E. coli*. World Journal of Microbiology and Biotechnology, 2010, 26, 11, 2089-2092.

[92] Tao, X.; Meihua, X.; Mingchun, L.; Huili, H.; Shengquan. Z. Synthesis, characteristic and antibacterial activity of N,N,N-trimethyl chitosan and its carboxymethyl derivatives. Carbohydrate Polymers, 2010, 81, 4, 931-936.

[93] Sadeghi A.M.M.; Dorkoosh F.A.; Avadi M.R.; Saadat P.; Rafiee-Tehrani M.; Junginger H.E. Preparation, characterization and antibacterial activities of chitosan, Ntrimethyl chitosan (TMC) and N-diethylmethyl chitosan (DEMC) nanoparticles loaded with insulin using both the ionotropic gelation and polyelectrolyte complexation methods. International Journal of Pharmaceutics, 2008, 355, 1-2, 299-306.

[94] Vidar, R.O.; Jukka H.;, Tapio, N.; Martha, H.; Tomi, J.; Thorsteinn, L.; Einarsson Jón M.; Sigrídur, J.; Margrét, V.; Már. M. Antibacterial activity of methylated chitosan and chitooligomer derivatives: Synthesis and structure activity relationships. European Polymer Journal, 2007, 43, 6, 2660-2671.

[95] Sieval A.B.; Thanou M.; Kotze´ A.F.; Verhoef J.C.; Brussee J.; Junginger H.E. Preparation and NMR characterization of highly substituted N-trimethyl chitosan chloride. Carbohydrate Polymers, 1998, 36, 2-3, 157-165.

[96] Douglas, D.B.; Odílio B.G. A novel method for obtaining a quaternary salt of chitosan. Carbohydrate Polymers, 2007, 69, 2, 305-310.

[97] Pattarapond, G.; Warayuth, S.; Uracha, R.; Nuttaporn, R.P.; Issara, S.; Onanong, N.; Somsak, S.; Saowaluk, C.; Satit, P. Novel quaternized chitosan containing β-cyclodextrin moiety: Synthesis, characterization and antimicrobial activity. Carbohydrate Polymers, 2011, 83, 2, 905-913.

[98] Vidar, RO.; Jukka, H.; Sigrídur, J.; Hákon, S.; Már, M. N-selective 'one pot' synthesis of highly N-substituted trimethyl chitosan (TMC). Carbohydrate Polymers, 2008, 74, 3, 740-744.

[99] Lifeng, Q.; Zirong, X.; Xia, J.; Caihong, H.; Xiangfei, Z. Preparation and antibacterial activity of chitosan nanoparticles. Carbohydrate Research, 2004, 339, 16, 2693-2700.

[100] Sunil A.A; Nadagouda N.M.; Tejraj M.A.; Recent advances on chitosan-based micro- and nanoparticles in drug delivery. Journal of Controlled Release, 2004, 100, 1, 5-28.

[101] Quan, G.; Tao, W.; Colette, C.; Paul, M. Modulation of surface charge, particle size and morphological properties of chitosan–TPP nanoparticles intended for gene

delivery. Colloids and Surfaces B: Biointerfaces, 2005, 44, 2-3, 65-73.

[102] Rejane C.G.; Douglas, B.;, Assis Odilio B. G. A review of the antimicrobial activity of chitosan. Polímeros, 2009, 19, 241-247.

[103] Nan, L.; Xi-Guang, C.; Hyun-Jin, P.; Chen-Guang, L.; Cheng-Sheng, L.; Xiang-Hong, M.; Le-Jun, Y. Effect of MW and concentration of chitosan on antibacterial activity of *Escherichia coli*. Carbohydrate Polymers, 2006, 64, 1, 60-65.

[104] Ming, K.; Xi-Guang, C.; Yu-ping, X.; Cheng-Sheng, L.; Le-Jun, Y.; Qiu-Xia, J.; Su, C.D.; Jin, P.H. Preparation and antibacterial activity of chitosan microshperes in a solid dispersing system. Frontiers of Materials Science in China, 2008, 2, 2, 214-220.

[105] Tomoki, T.; Masanao, I.; Isao, S.; Jun, S. Growth inhibitory effect on bacteria of chitosan membranes regulated with deacetylation degree. Biochemical Engineering Journal, 2008, 40, 3, 485-491.

[106] Avadi M.R.; Sadeghi A.M.M.; Tahzibi A.; Bayati Kh; Pouladzadeh M.; Zohuriaan- Mehr M.J.; Rafiee-Tehrani M. Diethylmethyl chitosan as an antimicrobial agent: Synthesis, characterization and antibacterial effects. European Polymer Journal, 2004, 40, 7, 1355-1361.

[107] Caiqin, Q.; Huirong, L.; Qi, X.; Yi, L.; Juncheng, Z.; Yumin, D. Water solubility of chitosan and its antimicrobial activity. Carbohydrate Polymers, 2006, 63, 367-374.

[108] Dina, R.; Kristine, B.; Albert, H.; Hans-Georg, S. Insights into the Mode of Action of Chitosan as an Antibacterial Compound. Applied and Environmental Microbiology, 2008, 74, 12, 3764-3773.

[109] Xiaohui, W.; Yumin, D.; Hui, L. Preparation, characterization and antimicrobial activity of chitosan–Zn complex. Carbohydrate Polymers, 2004, 56, 1, 21-26.

[110] Rabea E.I.; Badawy M.E.T.; Stevens C.V.; Guy, S.; Walter, S. Chitosan as Antimicrobial Agent: Applications and Mode of Action. Biomacromolecules, 2003, 4, 6, 1457-1465.

[111] Liu X.F.; Guan Y.L.; Yang D.Z.; Li Z.; Yao K.D. Antimicrobial action of chitosan and carboxymethylated chitosan. Journal of Applied Polymer Science, 2001, 79, 1324-1335.

[112] Lim, S., H.; Hudson, M. Review of chitosan and its derivatives as antimicrobial agents and their uses as textile chemicals. Journal of Macromolecular Science, 2004, C43, 223-269.

[113] Young, D.H.; Kohle, H.; Kauss, H. Effect of Chitosan on Membrane Permeability of Suspension-Cultured Glycine max and Phaseolus vulgaris Cells. Plant Physiology, 1982, 70, 1449-1454.

[114] Helander,I; M., Nurmiaho-Lassila,E. L.; Ahvenainen,R., Rhoades,J; Roller,S. Chitosan disrupts the barrier properties of the outer membrane of Gram-negative bacteria. International Journal of Food Microbiology, 2001, 71, 235-244.

[115] Simona, S.; Olivera, S.; Lidija, F.; Anita, J.; Karin, S.K. Chitosan-universally applicable biopolymer. Tekstilec, 2007, 50, 10, 243-261.

[116] Claudia, V. The use of mucoadhesive polymers in vaginal delivery. Advanced Drug Delivery Reviews, 2005, 57, 1692-1712.

[117] Elisabetta, G.; Vanna, S.; Claudia, J.; Cristina, B.M.; Paolo, G.

Mucoadhesive vaginal tablets as veterinary delivery system for the controlled release of an antimicrobial drug, acriflavine. AAPS PharmSciTech, 2002, 3, 3, 1-7.

[118] Antoni, N. Chitosan Medical Dressings. Fibres & Textiles in Eastern Europe, 2005, 13, 6, 16-18.

[119] Subhash, A. Medical Textiles and Biomaterials for Healthcare: Incorporating Proceedings of MEDTEX03 International Conference and Exhibition on Healthcare and Medical Textiles. Woodhead. Vol. 2006.

[120] Zemljič Lidija, F.; Olivera, Š.; Igor, B.; Andrej, Z.; Lusicky. M. Viscose Material Functionalised by Chitosan as a Potential Treatment in Gynaecology. Textile Research Journal, 2011, 81, 1183-1190.

[121] Hayano, S.; Fujieda, Y.; Yoshioka, S. Chitosan-Coating of Cellulosic Materials Using an Aqueous Chitosan-CO2 Solution. Polymer Journal, 2002, 34, 3, 144-148.

[122] Muzzarelli R.A.A.; Muzzarelli C. Chitosan Chemistry: Relevance to the Biomedical Sciences. in Polysaccharides I, Heinze Thomas, Editor. Springer Berlin Heidelberg, 2005, 151-209.

[123] Sang-Hoon, L.; Samuel M.H. Synthesis and antimicrobial activity of a watersoluble chitosan derivative with a fiber-reactive group. Carbohydrate Research, 2004, 339, 2, 313-319.

[124] Weijun, Y.; Fai, L.M.; John, X.; Leung, K.T.; Len, L.D.K.: Pei, L. Novel core-shell particles with poly(n-butyl acrylate) cores and chitosan shells as an antibacterial coating for textiles. Polymer, 2005, 46, 23, 10538-10543.

[125] Mingxi, W.; Yuanbin, S.; Zuobing, X.; Jing, H.; Rujun, Z.; Jia, Z. The green adsorption of chitosan tripolyphosphate nanoparticles on cotton fiber surfaces. Carbohydrate Polymers, 2014, 101, 812-818.

[126] Wazed, A.S.; Subbiyan, R.; Mangala, J. Synthesis and characterization of chitosan and silver loaded chitosan nanoparticles for bioactive polyester. Carbohydrate Polymers, 2011, 83, 2, 438-446.

[127] Efem, S.E.E. Clinical observations on the wound healing properties of honey. Br H Surg. 1988, 75,679-681.

[128] Wilkinson, J.M.; Cavanagh, H.M.A. Antibacterial activity of 13 honey against Escherichia coli and Pseudomonas aeruginosa. J. Med. Food 2005, 8, 100-103.

[129] Cooper, R.A.; Molan, P.C. Antibacterial activity of honey against strains of Staphylococcus aureus isolated from infected wound. J Soc Med. 1999, 92, 283-285.

[130] Tim Cushnie, T.P.; Lamb, A.J. Antimicrobial activity of flavonoids. Int. J. Antimicrob. agents 2005, 26, 343-356.

[131] Mangala, J.; Wazed, A.S.; Subbiyan, R. Antibacterial Finishing of Polyester/Cotton Blend Fabrics Using Neem (Azadirachta indica): A Natural Bioactive Agent. Journal of Applied Polymer Science, 2007, 106, 793-800.

[132] Lidija, F.Z.; Vanja, K.; Čakara. D. Antimicrobial and antioxidant properties of chitosan-based viscose fibres enzymatically functionalized with flavonoids. Textile Research Journal, 2011, 81, 1532-1540.

[133] Thilagavathi G.; Kannaian T. Combined antimicrobial and aroma finishing treatment for cotton, using micro encapsulated geranium (*Pelargonium graveolens* L' Herit. Ex. Ait.) leaves extract. Indian Journal of

Natural Products and Resources, 2010, 1, 348-352.

[134] Alonso D.; Gimeno M.; Sepulveda-Sanchez J.D.; Shirai K. Chitosan-based microcapsules containing grapefruit seed extract grafted onto cellulose fibres by a non-toxic procedure, Note. Carbohydrate Research, 2010, 345, 854-859.

[135] Haufe H.; Muschter K.; Siegert J.; Böttcher H. Bioactive Textiles by Sol-Gel Immobilized Natural Active Agents. Journal of Sol-Gel Science and Technology, 2008, 45, 97-101.

[136] Jothi D. Experimental study on antimicrobial activity of cotton fabric treated with aloe gel extract from *Aloe vera* plant for controlling the *Staphylococcus aureus* (bacterium). African Journal of Microbiology Research, 2009, 3, 5, 228-232.

[137] Tijana, R.; Lidija, F.Z.; Monika, N.; Marjetka, K.K.; Silva, S.; Nina, G.C.; Simona, S. Antimicrobial efficiency of functionalized cellulose fibres as potential medical textiles. in Science against microbial pathogens: communicating current research and technological advances, Méndez-Vilas A., Editor. Formatex: Badajoz, Spain, 2011, 36-51.

[138] Diana, C.; Simona, O.; Narcisa, V. Biofunctionalization of textile materials by antimicrobial treatments: a critical overview. Romanian Biotechnological Letters, 2010, 15, 4913-4921.

[139] Swofford H.W. An Overview of Antimicrobial Testing for Textile Applications: A look at commonly used antimicrobial testing protocols for textiles. AATCC Review, 2010, 10, 6, 5.

[140] Linda, T.; Bernhard, R. Improved methods for the investigation of the interaction between textiles and microorganisms. Lenzinger Berichte, 2006, 85, 54-60.

[141] Pinho, E.; Magalhães, L.; Henriques, M.; Oliveira, R. Antimicrobial activity assessment of textiles: standard methods comparison. Annals of microbiology, 2011, 61, 493-498.

[142] Askew, P. D. (2007). Analysis and assessment of current protocols to develop harmonised test methods and relevant performance standards for the efficacy testing of treated articles/ treated materials. ENV/JM/MONO.

[143] Microchem Laboratory, ISO 20743 - Assessment of Antibacterial Finished Products, Available from http://microchemlab.com/test/iso-20743-assessment-antibacterial-finished-products.

[144] Kim H.W.; Kim B.R.; Rhee Y.H. Imparting durable antimicrobial properties to cotton fabrics using alginate–quaternary ammonium complex nanoparticles. Carbohydrate Polymers, 2010, 79, 1057-1062.

[145] Daoud W.A.; Xin J.H.; Zhang Y-H. Surface functionalization of cellulose fibres with titanium dioxide nanoparticles and their combined bactericidal activities. Surface Science, 2005, 599, 69-75.

[146] Wang Q.; Jin G.; Fan X.; Zhao X.; Cui L.; Wang P. Antibacterial functionalization of wool via mTGase-catalyzed grafting of ε-poly-L-lysine. Applied Biochemistry and Biotechnology, 2010, 160, 2486-2497.

[147] Lidija, F.Z.; Tijana, R.; Tkavc, T. Adsorption and antibacterial activity of soluble and precipitated chitosan on cellulose viscose fibers. Journal of engineered fibers and fabrics, 2012, 7, 1, 50-57.

[148] Lidija, F.Z.; Simona, S.; Olivera, Š.; Karin, S.K. Characterization of Amino Groups for Cotton Fibers Coated with Chitosan. Textile Research Journal, 2009, 79, 3, 219-226.

[149] International ASTM. Standard test method for determining the antimicrobial activity of immobilized antimicrobial agents under dynamic contact conditions, 2001, ASTM.

[150] Ozer R.R.; Hill W.C.; Rogers M.E., Evans M. Development of colorimetric analytical methods to monitor quaternary amine grafted surfaces. Journal of Applied Polymer Science, 2010, 118, 2397-2407.

[151] Knittel D.; Schollmeyer E. Chitosans for permanent antimicrobial finish on textiles. Lenzinger Berichte, 2006, 85, 124-130.

[152] Hewitt C.J.; Franke R.; Marx A.; Kossmann B.; Ottersbach P. A study into the antimicrobial properties of an amino functionalised polymer using multi-parameter flow cytometry. Biotechnology Letters, 2004, 26, 549-557.

[153] Morais, D. S.; Guedes, R. M.; opes, M. A. Antimicrobiasl approaches for textiles: from research to market. Materials, 2016, 9, 498.

Chapter 3

Textiles Functionalization - A Review of Materials, Processes, and Assessment

Mukesh Kumar Singh

Abstract

Conventionally, textiles are known to cover up the human skin, but by scientific administration, clothing can be extended to serve other human skins' functions. Accepting the chemical and dermatological complexity of human skin, the effect of humidity, microbes, pH, temperature, and wind can be engineered by wrapping it by functional clothing. In this regard, the latest class of textile material has been added called functional textiles. Such clothing materials consist of the potential of delivering more than one functionality apart from its primary function to cover-ups the human body. This present chapter offers state-of-the-art viewpoints on the application of functional textiles, including assorted concerns. First, the skin responds to various environmental stimuli and then overviews various techniques to incorporate functionalities in textiles. Finally, the applications and future scope and possibilities of research in this field are included in this chapter. Miniaturisation to small micro to nanometre scale is registered as one of the most exciting meadows in engineering and science over the past few decades. This drift also grasps colossal potential to functionalise the textiles. Various techniques are available now to develop a thin uniform film of functional materials on clothing surface to offer extra functionalities hitherto unrevealed to textile processors. These technologies are based on layer-by-layer assembling, immobilisation of enzymes on textile surfaces, nanocoating of textile substances, plasma for nanoscale modifications, and loading of various functional biomaterials micro and nanoencapsulation by minimum influence on breathability, feel, handle, and strength. The manufacturing of functional textiles can be classified into two groups. One is to functionalise the fibre by adding dope additives, modifying the fibre forming polymer, and then converting it to clothing. The fibre surface is also functionalised by adding some resins on the fibre surface. The other is to modify the textile surfaces by functional biomaterials, resins, finishes.

Keywords: cosmetic textiles, skin care, skin moisturising, active textiles, microencapsulation, natural dyes, smart textile, enzyme immobilisation, wellness textiles

1. Introduction

The functionalisation of different surfaces is of curiosity since the chronicle emergence of various technologies in the antiquity. Different surface coatings and painting materials and techniques were developed to alter the aesthetic appearance, functional potential, and protection against the environment like resistance against oxidants [1, 2].

Initially, durability was prime criteria for selecting any fabric by the customer in the ancient era, and then aesthetic values and comfort index were primary attributes to decide the fabric choice. Now, customer's approach about clothing and textiles is shifting to search some additional functionality apart from traditional attributes in textiles. The functionality may come from protective clothing, cosmetotextiles, and temperature regulating textiles, industrial textiles, sports textiles, and automotive textiles. All the above textile materials must keep at least any one specific functionality to register as functional textiles.

The functional textiles market is growing with an excellent growth rate of 33.58% between 2015 to 2020. The global functional textile market was reached 4.72 billion US$ by 2020. India is a prime manufacturer in apparel and textile manufacturing and fourth-largest exporter in the international sector.

The functional textile sector has encountered a compound annual growth rate (CAGR) of 30% from 2015 to 2020 due to strong automotive, fitness, fashion, healthcare, military, and sports textiles.

The physical finishing process includes three methods: impregnation, padding, and coating, and its main drawback are that the bonding force is weak between the finishing agent and textiles. However, its strength is more durable, and functionality can be maintained for a long time. The chemical finishing method involves grafting a functional monomer onto a polymer substrate, to obtain a new functional textile. The advantage is that its functionality can be maintained for a long time. Biological, ecological finishing is a finishing method which has emerged in recent years, and it adopts biological enzymes with biological activity in the finishing of textiles.

Smart textiles do not necessarily imply a less sustainable option to ordinary textiles if they offer better user value, user attachment and longevity. This chapter discusses the difference between ordinary sustainable methods based on saving energy and resources and methods that tackle excessive consumption, such as user involved design to enhance product durability. It discusses the theoretical model of user involved design through a practical example of developing a smart, lightweight tracking tent and concludes with a set of general guidelines for developing sustainable smart- textile products.

The functional textiles can be defined as textiles consist of additional functions of adjusting and regulating various attributes like temperature, humidity, colour and controlled release of some additives from fibres. The most popular fibres used to manufacture various types of functional textiles are polyester and viscose. Other fibres are also used to manufacture functional textiles as the need for some specific functions. The significant demand for functional textiles arises from active and high-performance wear sectors.

Some leading manufacturer of functional textiles at the international horizon are Dyntex GmbH, Eclat Textile Corporation Ltd., Harvest SPF Textile Company Ltd., Kelheim Fibres GmbH, Sofileta, Trevira GmbH, Toung Loong Textile MFG.

2. Classification of functional textiles

Various authors and researchers have tried to classify the functional textiles on their own time, but it is not easy to propose an ideal classification [3].

Functional textiles have become instrumental for the advancement of the conventional technical textiles segment, representing a sector where traditional clothing crosses the usual frontiers and connects with the spheres of biotechnology, cosmetic science, computing potentials, flexible electronics, medicine, and

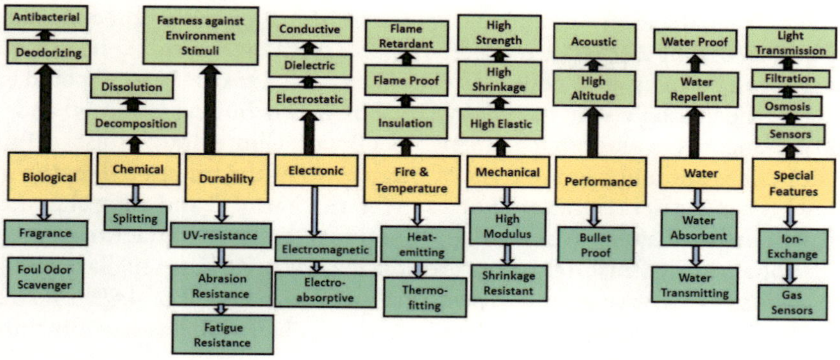

Figure 1.
Various stimuli and their corresponding functionalities for textiles.

nanotechnology among some more, to achieve the multidimensional and complex demands of the customers. By definition, Functional textiles are user-governed specified and customised or engineered products manufactured to fulfil the customer's performance needs under extreme conditions.

Gupta [3] classified the functional textiles logically in six categories. Now including three more, present classifications consist of nine different functional classes, as shown in **Figure 1**. It is essential to clarify that some more classes will introduce functional textiles classification as per the demand and availability of various functional textiles shortly. There is a slight difference between functional and technical textiles. All functional textiles may belong to technical textiles, but all technical textiles may not be functional. For example, protective functional textiles may belong to protective surgical masks for doctors, healthcare workers, and sports armour.

3. Functional textile market

The aerothermal concept has been adopted in functional textiles to control the heat containing airflow through fabrics. These functional textiles are manufactured and marketed by Adidas, and Peak Performance of the IC group. The Schoeller Technologies AG developed far infrared-based functional textiles' Energear' to collect the energy employed by the human body. Remarkable developments and innovations polymer and fibre science, coating and finishing technologies are the major driving forces for the growth of functional textiles. The initial demand for functional textiles was originated from various sports wear used in cycling, ski sports, swimming. The latest development in functional textiles turned the customer highly demanding for functional textiles. The higher cost of functional textiles stamped it textiles of premium class. Thus, the challenge of cost reduction in functional textiles manufacturing chain is the need for the present era to make it available for the common man and accelerate functional textiles' global market. Geographically, the functional textiles market is divided into five regions the Asia Pacific, Middle East and Africa, European Union, Latin America, United States of America. European Union secured the first position in manufacturing Functional Textiles mainly in Germany, France, Italy, and the United Kingdom. The USA also has a healthy manufacturing and market for functional textiles.

The Asia Pacific and Middle East Africa have the immense opportunity to grow the functional textile market. Japan, Malaysia, South Korea, Turkey, and Taiwan are predicted to be a promising market for functional textiles. The maximum number of functional textiles is manufactured by the application of various finishing agents on textiles. Thus, the international market for finishing agents is expected to grow to 4.52 billion US$ by 2025. The protective textile segment, including health care and protection, is an up-and-coming field of functional textile is anticipated to manoeuvre the market rise in the next years. Enhancing buying potential and expendable income in emerging countries like Brazil, China, India, Russia, Taiwan, and Indonesia is another driving force of the growth of the functional textile market. However, the strict regulatory norms to restrict harmful and toxic chemicals will remain the requisite provocation for functional textile manufactures.

Functional textile finishing agents are dominated by various repellant and release agents in recent years and expected to grow with CAGR of 4.8% till 2028. The application of flame retardant chemicals was 22% of global textile finishing chemicals in the year 2020.

The major functional finishing agent manufacturers are Archroma, BASF SE, Covestro, CHT Group, Evonik Industries, Huntsman Corporation, Sumitomo Chemicals, Dow Chemical Company, HT Fine Chemical, FCL, KAPP-CHEMIE, NICCA CHEMICAL CO., Ltd., OMNOVA Solutions, Tanatex Chemicals B.V., Wacker Chemie AG, Zydex Industries, Sarex and others.

4. Functionality in textiles

4.1 Functionality by micro and nano encapsulation

A technique in which an active substance is stored in tiny space covered or coated with a thin polymeric material to protect the core material along with controlled release, is called microencapsulation.

Particles obtained by this process are called microparticles, microcapsules and microspheres according to their morphologies and internal structure. For particles with a size range below one μm, the terms 'nanoparticles', 'nanocapsules' and 'nanospheres' are used, respectively. Furthermore, particles larger than 1000 μm are designed as microcapsules. The nomenclature used to define different parts of the encapsulated product includes terms for the shell, ie, 'wall', 'coating', 'membrane material'; and for the core material, ie, 'active agent', 'payload', 'internal phase', respectively. Different kinds of compounds such as dye, protein, fragrance, monomers, the catalyst can be encapsulated with various shell wall materials like natural polymer (gelatine, cellulose, chitosan, etc.), artificial polymers (cellulosic derivatives, etc.) and synthetic polymers (polyamide, polyester, etc.), with a loading content between 5% and 90% of the microparticles in weight (**Figure 2**).

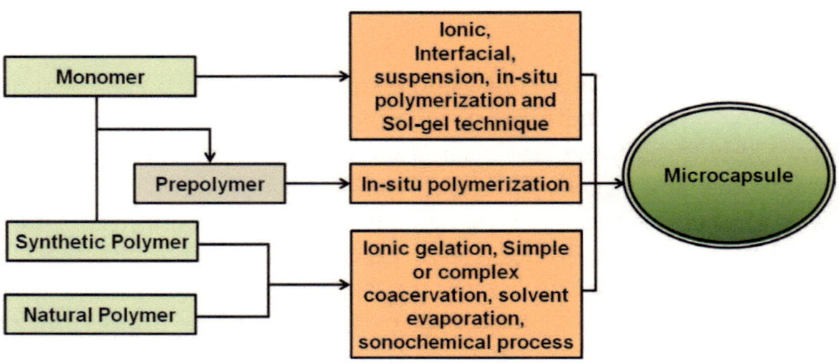

Figure 2.
Functionality by microencapsulation.

Sheath polymer	Core active material	Function	Reference
Poly (urea urethane) PUU	Sensitive dye	To get colour as a function of time	[4]
Poly (urea urethane) PUU	Cooling agent	To get cool feel during wearing	Salaün et al. [5]
Poly (urea urethane) PUU	Fragrances	To get controlled fragrance release	Teixeira et al. [6]
Poly urethane/ polyurea	Antimicrobial	To get bacterial protection	
Polyurea	Flame retardant	To get improved flame protection	[7], and Vroman et al. [8]
Polyurethane/chotopsan	Thermochromic	Colour change as a function of temperature	Fan et al. [9]
Cellulose derivative/PU	Phase change materials	To control environmental temperature flactuation	Salaün et al. [10]
Polyurethane	Cosmetic Ingradient	To get skin care functionality	Azizi and Chevalier [11]

Table 1.
Microencapsulation by interfacial polymerisation.

Micro and Nanoencapsulation is a technique in which a thin wall surrounds a tiny amount of active ingredient or droplets. The capsule wall's active material is called core material while the coating material is known as shell or membrane. Microcapsules have a diameter of few microns whereas nanocapsules have diameters of some hundreds of nanometres. Micro and nanoencapsulation is an up-and-coming technology to functionalise the textiles by potential core material for desires functionality. Micro and nanoencapsulation have permitted biocides, insecticides, essential oils, moisturisers, energisers, moisturisers, therapeutic oils, and vitamin E to be uploaded into fabrics. Some other applications are agrotextiles, cosmetics, industrial textiles, food additives, essences, herbicides, nutraceuticals, and sealants. The microencapsulation by interfacial polymerisation is given in **Table 1**.

4.2 Chemical grafting

The traditional binder application binds some functional compound on textile surfaces and creates a three-dimensional network and starts hindering the release of functional ingredient from the surface. The absence of strong chemical linkage between the capsule sheath and substrate surface exhibits inferior wash fastness and low air and moisture transmission. During pad-dry-cure, some of the capsules get burst due to the presence of applied pressure on substrate. Microcapsules are covalently connected onto textile surfaces by opting multifunctional crosslinking (coupling) reagents to enhance the fastness against wash and wear.

Microcapsules with ethylcellulose sheath are grafted on cotton fabric surface using 1,2,3,4-butane tetracarboxylic acid as crosslinking agents that react with a hydroxyl group cellulose to configure ester bonds [12].

Dimethyloldihydroxyethylene urea was opted as a coupling agent between chitosan and cellulose to form a covalent bond to enhance washing fastness [13].

Citric acid was another choice to crosslink chitosan microcapsules on cellulosic surfaces [9, 14].

Grafting strategies from the particle surface modification can be performed by introducing reactive chemical groups like α-Bromo-acrylic acid, adipic acid, 2,4,6-trichlorotriazine and dichloroquinoxaline, to react with the microcapsule sheath materials to offer further grafting possibilities onto natural, manmade fibres. Polyamide capsules and silica microspheres with 2,4,6-triochlorotriazine were functionalised to be deposited on cotton fibre surface [15].

Microcapsules were dispersed in water and glycidyl methacrylate monomer, and potassium persulfate was added to initiate an outer shell of poly(glycidyl methacrylate). The microcapsules were then applied by exhaustion in alkali medium to jersey cotton knitwear with a liquor ratio of 1:10 at 75 °C for 30 min. After rinsing, the sample was dried at 120 °C. The functional groups of the poly(glycidyl methacrylate) on the outer surface of the microcapsules are directly reacted with the functional groups of the fibres, which also conveyed durability of the PCM microcapsule-incorporated fabric even when subjected to physical processes involving frictional forces, or chemical processes such as domestic and industrial washing, or dry cleaning [16]. Gouveia used a sonication method to produce and simultaneously bind the microspheres onto textile materials [17].

Polyamide microcapsules are directly synthesised on cellulosic surfaces with 80% high encapsulation [18].

4.3 Functionality by layer by layer (LbL) deposition

Some studies were planned to modify the textile surfaces through layer by layer deposition to get nanocomposite textile fibre and protective clothing. Various functional molecules like enzymes, dyes and charged nanoparticles are deposited on textile surfaces in a controlled manner. Various finishing processes are based on Ag, TiO2, Zn nanoparticles to functionalised various textile surfaces [19]. Layer by layer deposition technique is a distinctive technique invented to develop ultrathin composite films on the surface of solid materials. A series of layer-by-layer deposition of polyanions and polycations on oppositely charged surfaces occurs in this method [20].

This process involves the charging of substrate sufficiently, followed by dipping in a conversely charged polyelectrolyte. The process begins by charging a substrate appropriately, followed by immersion in an oppositely charged polyelectrolyte solution and rinsing. Strong electrostatic bonds between charged surface and polyelectrolyte become the main instrumental in getting it to bind. The process begins with rinsing followed by monolayer coating of polyelectrolyte which gets bind by electrostatic bonds and process may repeat to be deposited 20 ultrathin layers [21–23].

Unlike pad-dry-cure, radiation, and thermal deposition methods, different finishing techniques are used to deposit various nanoparticles on textile substrates. The chemical coating on textile surfaces has some limitations due to higher thickness, which suppresses textiles' breathability. The LbL technology offers moderate chemical deposited surfaces to keep transmission, thickness and stability up to the desired extent. Various textile surfaces are modified by LbL technique.

Cotton, Kevlar, Nylon, Nomex, Silk and Wool fabrics were opted to functionalised for different applications by platinum atomic layer deposition. The platinum layer by layer deposition on textile surfaces was targeted to fabricate resistive heating devices with high stability and long life. The platinum deposition was found uniform layer by layer except for nylon fabric surface. [(1,2,5,6-η)-1,5-Hexadiene]-dimethyl-platinum(II) (HDMP; Tanaka Kikinzoku Kogyo K.K., Japan) and O2 gas were used as a Pt precursor and counter-reactant, respectively. The substrate temperature was maintained at 145 °C during the ALD process. Field emission scanning electron microscope integrated with energy dispersive spectroscopy (EDS) was used for quantitative and elemental deposition conformation on textile surfaces. The atomic-scale crystalline and chemical structures of treated fabric specimen were analysed opting transmission electron microscopy (TEM) and electron energy loss spectroscopy (EELS), respectively [24].

Lee et al. [25] developed a pressure sensor atomic layer deposition on the cotton surface by depositing [(1,2,5,6-η)-1,5-hexadiene]-dimethyl-platinum(II) (HDMP) and O2 as the Pt precursor and counter reactant. The research group developed a useful atomic layer deposition technique. The deposition process may repeatedly apply up to 10,000 times. This technique helps produce various E-textiles by combining and connecting the number of sensors in a textile item that may be proved as propitious applicants for a range of smart and wearable electronics.

Stawski et al. [26–28] deposited oppositely charged polyelectrolytes; poly(acrylic acid) and poly(allylamine hydrochloride) layer-by-layer on a polypropylene knitted fabric, which had areal density 80 g/m2, diameter per filament 14.9, 18 courses per centimetre, 14 wales per centimetre, 56 dtex multifilament yarn.

The polypropylene (PP) fabric was activated as per the protocol opted by Połowiński [29] and [30–32] by heat-treating the fabric samples in an aqueous solution (20 g/L) of ammonium persulfate for 30 minute at 80 °C, and saturated with nitrogen), rinsing thoroughly with distilled water, and grafting with concentrated acrylic acid (52 g/L) for 60 minutes at 80 °C, saturated with nitrogen. After completing the grafting process, the PP fabric samples were dipped in an aqueous polyelectrolyte solution (10–2 mol/L). This process may repeat many times as per the need of desired end applications. Before initiating a new layer, deposition fabric specimens were rinsed with distilled water. In this way number of polyelectrolyte layers was coated on PP fabric surface [28, 29]. The PP samples functionalised by LbL technique were found to significantly reduced the static charge half-disappearing time, from 46 to 5.7 minutes. The degradation temperature shifted from 330 to 420 °C. The capillary rise has increased from 50 to 400 mm in the case of surface-modified knitted fabric samples. SEM and wide-angle X-ray diffraction was used to confirm the layer by layer deposition on PP fibre surface. It was found that the layer-by-layer deposition of polyelectrolyte considerably modifies dyeability, electrostatic shielding potential, and hydrophilicity of PP fabric samples.

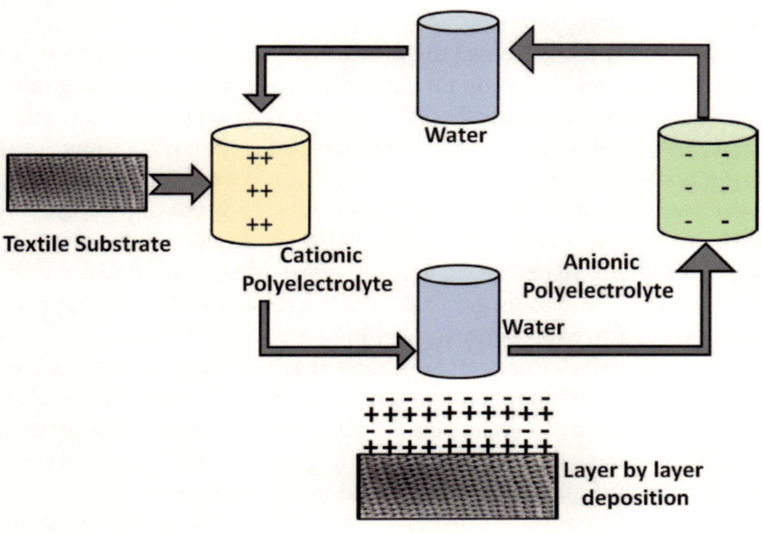

Figure 3.
Concept of layer by layer (LbL) Technology for Functional Textiles.

Highly durable hydrophobic polyester fabric surfaces were created by LbL nanoparticle coating on the fabric surface. Electrophoretic deposition (EPD) technique was used to modify the fabric surface with deposition of silica nanoparticles. The deposition of silica nanoparticles remains challenging on non-conductive surfaces like polyester due to undue cracks and poor adhesion. Thus electrostatic self-assembly layer-by-layer technique was opted to overcome these issues as shown in **Figure 3**. The polyester fabric modified by LbL silica NPs deposition offers a very high contact angle in static condition along with low contact angle hysteresis. The superhydrophobicity was remained intact ever after 500 h skin fiction [33]. This method provides fast and customisable deposition of superhydrophobic surface coatings. The coating thickness can be controlled by the electric field intensity and deposition time. Furthermore, the modified surface's morphology can be altered by changing the suspension stability during EPD [34]. Three significant routes achieve sustainable surface functionalisation; by alternating a charged substrate's immersion in aqueous solutions containing interacting charged particles, chemical vapour deposition, and spraying the interactive solutions on charges surfaces.

4.4 Functionalisation by fabric engineering

Fabric Comfort, good low-stress mechanical properties, pleasant aesthetic appeal, elasticity and recovery, favourable formability, desired crease-resistance are some attributes can be achieved by fabric engineering. These attributes are required for various applications like formal wear, party wear, ladies wear and sportswear.

The PBT (polybutylene terephthalate), yarns have good elastic potential with high recovery after heat treatment, have been adopted to manufacture highly elastic cotton-like fabric. The effect of PBT elastic yarn, weave and fabric structure was observed on physical, elastic, UPF, comfort) properties of the fabric. PBT yarns have found it appropriate to introduce elasticity in selected areas of the fabrics. The fabric had quick-drying, easier folding and storage and perfect fit to all dimensions [35].

3D printing is another way to introduce various functionalities in a different range of textiles. 3D printing has opted in case of defence, protective, sports, flexible electronics and safety clothing.

3D printing is used to customise the product for specific applications. Various types of additives can be loaded on fabric surfaces as a part of printing inks or pastes. A variety of hard and soft flexible materials can be printed directly on the textile surfaces. A combination of additives can also incorporate some functionality on a common textile substrate by 3D printing. The effect of family of twill weave and different weft densities on adhesion potential of printed objects on polyester/cotton fabric surfaces was studied. A range of 3D objects was printed with polylactic acid (PLA) filament on textile surfaces. T - Peel adhesion test was conducted by Instron dynamometer. It was found that the 1/3 broken twill fabric has the maximum impact on the adhesion perspective [36].

The worldwide health consciousness has enhanced natural dyes' use to avoid the threat of allergy, mutagenicity, and carcinogenicity. Cotton fabric was dyed with an extract of fourty different plants with zero mordant. Some plants were able to record great wash and lightfastness like pomegranate peel and turmeric for yellow, madder and quince for yellow, indigo for blue, myrobalan for green, white onion peel and catechu for brown colour. White onion peel or turmeric dyed cotton fabrics have registered significant improvement in ultraviolet protection functionality [37].

4.5 Functional textiles by enzyme immobilisation

Enzyme driven textile functionalisation has attracted the attention of textile manufacturers worldwide. Nonpolluting, non-toxic, and biodegradable nature of enzymes make it appropriate for the green processing of textiles. The enzyme production can be enhanced commercially anytime as per the need in industry. Enzymatic bleaching, scouring, bio-washing, and bio-polishing cotton fabric have become quite popular in the textile industry. Recently cotton fabrics are modified through transesterification by Proteinase subtilisin enzymes. Woollen fabrics are made shrink-proof by transesterification by the use of proteinase subtilisin enzymes. The Laccase enzyme is used to functionalised the wool in multiple order with antibacterial, antioxidant and water repellant for grafting alkyl galettes. The hydrophilicity and antistatic charge potential are introduced in polyester fabrics by treating it with cutinases and esterases enzymes. The nylon and acrylic fabrics are functionalised by treatment with amidase and nitrilase. The functionalisation of textile surfaces is made by 'enzyme immobilisation' on textile surfaces to introduce some special functions to textile surface. The immobilised enzymes work better than free enzymes on the textile surface to impart long term functionalities on textile substrates. As compared to free enzymes, immobilised enzymes are permanently attached to the textile, thereby adding unique functionalities to its surface.

5. Functional materials

5.1 Phase-change materials

The micro and nanoencapsulation of phase change materials can alter the core material from solid to liquid and liquid to stable by changing the entropy within a specific temperature span. This technique is used to suppress the effect of temperature variation on the targeted subject. The encapsulation of phase change materials is used to keep the temperature of clothing at a constant level. Microencapsulated phase change materials enhance the comfort delivery of blankets, duvets, mattresses, snowsuits, and vests [38].

5.2 Fragrance finishes

Fragrance finishes are applied straight on textiles, but aroma stability lasts maximum up to a couple of wash cycles. The micro and nanoencapsulation of fragrances are used to prolong the fabric's aroma functionality for a much more extended period. This technique is mainly adopted to encapsulate various essential oil flavours like lavender, rosemary, pine and others on for aromatherapy to treat headaches, insomnia, and prevent bad odour.

5.3 Fire retardants

Encapsulation of flame retardant materials does not allow sacrificing the softness and other low-stress mechanical properties seeded by direct application of flame retardant chemicals. Scientific selection of core and sheath materials for flame retardancy can offer synergistic effect; organophosphorus compound as core and nitrogenous compound as a sheath. Some intumescent flame retardant coatings can also be generated by micro and nanoencapsulation technique. This flame retardancy technique is widely used in the military sector to treat the tentage, upholstery and firefighting dresses.

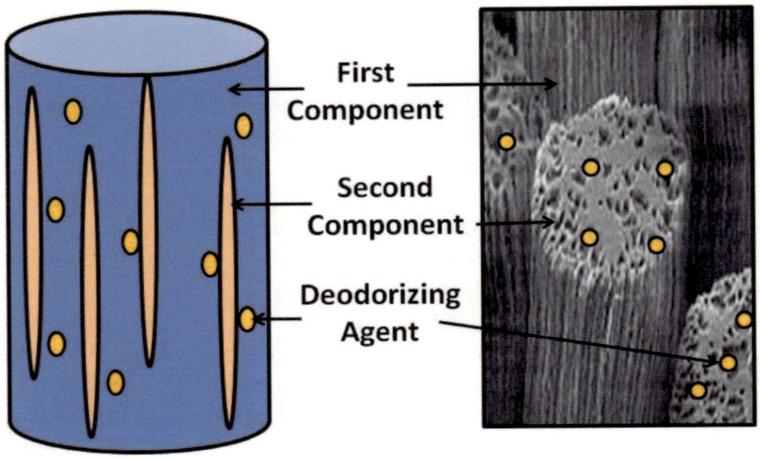

Figure 4.
Concept of inherent deodorant textile fibre.

5.4 Deodorant functional textiles

The deodorant fibres are manufactured by modifying the polymer molecular chain during polymerisation, by adding deodorant additives during fibre extrusion and by applying deodorant finish on fibre surface after spinning. The addition of deodorant additive as dope additive is simplest method to impart deodorancy in fibres as shown in **Figure 4**.

The polyester staple fibres were modified by photocatalyst and blended with cotton and bamboo fibre to produce several fabrics. The photocatalyst fibre contents were varied from 0 to 100% at a step of 20% increment in different samples. The deodorant potential of produced fabric samples was tested and examined. The conclusion is explained that as the photocatalyst fibre content reached 80% or 100%, the fabrics get better deodorant potential, but the photocatalyst content remains 40% to 60% in fabric samples, it shows low deodorant effect. It is established that at least 80% photocatalyst is essential to produce acceptable deodorant fabric [39, 40].

5.5 Polychromic microcapsules

The colours and dyes responsive against temperature are called as polychromic or thermochromatic dyes. The dyes and colours changes by a change in ultraviolet light are called as photochromatic dyes. The thermochromic or photochromic dyes are encapsulated inside the shell material used for product labelling, forensic purposes, and fashion applications. Many dyes and chemicals are available, which change colour by changing temperature and UV light exposure.

5.6 Antimicrobials

Microbes cause severe damage to various textile items. Some chemicals used to decay the microbial effects of fabrics are called as antimicrobials. These antimicrobials can be uploaded with textiles by microencapsulation technique in which antimicrobials remains in the core of the capsules. High-value textiles are treated by this method to prolong the life of these textiles.

5.7 Textiles for refreshing and relaxing

Refreshing and relaxing attributes are incorporated in textiles by uploading the *Aloe vera*, menthol, and essential oils with suitable emulsifiers in the capsule core. *Aloe vera* has found an appropriate finishing ingredient of textiles to achieve a refreshing and relaxing feel in textiles. These functional textiles are used to offer pleasant wear, wellness, and high energy levels.

Various plants and fruit-based aromas are also used by encapsulating and loading on fabrics for a functional point of view.

5.8 Textiles for cosmetics

A class of textiles responsible for imparting skincare, ageing combating, and wellness feeling is known as cosmetotextiles. Cosmetotextiles is one of the significant members of the functional textile family. The demand for cosmetotextiles increases every year due to the increase in self-wellbeing and purchasing power globally [41].

Cosmetotextiles are resultant of combining cosmetics and the textiles through different techniques, in which microencapsulation is the prime. Cosmetotextiles is a consumer textile product with a long lasting cosmetic ingredient released as a function of time.

5.9 Skincare functional textiles

Skincare is the most potent aspect of modern globally. The glowing skin is the desire of every person of the universe. The potential of skincare in textiles can be incorporated very easily by finishing the route. The functional textiles are capable of caring for human skin in various ways, and some are described here.

Skin soft 415 New:

This fabric finish is based on water-soluble phospholipid developed by Daiwa Chemical Inc., Japan. The finishing bath mainly consists of 2-methacryloyloxyethyl phosphorylcholine (MPC) with phosphatidylcholine polar groups to retain moisture on human skin for a long time [42]. A gentle softener is also used in Skinsoft 415 New to give a soft feeling to the wearer.

The skinsoft 415 finish has the potential to enhance the antibrowning and antistatic potential of textile surfaces [43].

Ohara Paragium Chemicals Kyoto, Japan, have floated a wide range of skincare and anti-ageing functional finishes to treat different fabrics. Some selected skincare finishes for textiles are;

- Parafine Skincare –1000: This finish was developed by Ohara Paragium Chem. Japan and that primly consists of silk-based amino acids. The amino acids are rich in moisture retaining properties that promote skin well-being by enhancing skin moisture.

- Parafine Skincare-3000: This finish offers cellulite reducing functionality by the presence of capsaicin, along with moisture-retaining and skincare effect by the presence of raspberry and squalane, respectively.

- Parafine Skincare-5000: The Parafine SC-5000 finish primly based on extracts of rice germ oil (ferulic acid and g -oryzanol) and vitamin E. The mixture of ferulic acid, g –oryzanol and vitamin E offers anti-oxidation attributes to impart skin anti-ageing. This finish accelerates the anti-oxidation, blood circulation, and bio-membrane stabilisation in human skin.

- EVOTM Care Vital: This skincare finish recipe mostly contains *Aloe Vera*, Jojoba oil and Vitamin E to impart anti-ageing functionality in the finished fabric. This skincare finish was developed by Dystar Auxiliaries GmbH, Frankfurt, Germany. Some natural bio-substances are added with silicone matrix, which is an essential component of most softeners to enhance the washfastness of treated fabric surfaces. This finish is applied as the final treatment in the standard exhaust and pad-dry-cure process sequence. DyStar has introduced a similar finish with the commercial name of Evo Care BeeWell with beewax, Evo Care AVP and Evo Care AVS. Primarily, these finishes are based on *Aloe Vera* and Jojoba oil. Evo Care finish can finish a wide range of textile materials to impart anti-ageing functionality in innerwear that come directly in skin contact.

5.10 Insect repellent

The purpose of insect repellant functionality on textiles is to protect the wearer and cloth both from insects. The insect repellant materials are used to finish the traditional textile surfaces either by natural materials like (essential oils) such as citronella, eucalyptus, lemon, and neem or synthetic materials such as picaridin (1-piperidine carboxylic acid 2-[2-hydroxyethyl]-1-methlypropylester) or permethrin and (*N*,*N*-diethyl-3-methylbenzamide) (DEET). Direct application of these ingredients possesses very poor wash fastness. Thus most of them are applied through microencapsulation route by melamine–formaldehyde, sodium alginate, and silica as sheath material by pad-dry-cure method [44–47]. The efficacy of DEET was tested under laboratory conditions to inhibit blood feeding and killing mosquitoes for six months. This formulation was found very useful for mosquitoes got resistance against pyrethroid.

Textiles loaded with microcapsules containing citronella as the active ingredient has found better insect repellency (more than 90% for three weeks) than fabric sprayed with citronella oil and ethanol solution directly on fabric surfaces [48]. Lemongrass oil extract was uploaded on polyester fabric in microcapsule form and found 92% insect repellency. The mosquito-repellency was 80% when the polyester fabric was treated through pad-dry-cure route microcapsules containing methanolic lemongrass leaves extract as an active ingredient in capsule core.

5.11 Wellness

A cosmetotextile is a textile containing a substance or preparation targeted to be released permanently on different epidermis parts. It claims one or several unique properties such as cleansing, fragrance, skin appearance change, protection and upkeep, or foul body odour correction.

Some multinational companies like Oracle (France) and Dim launch fabrics with moisturising agent containing microcapsules grafted on textile surfaces. Global companies; Cognis in 2001 and Invista in 2003 floated their products as cosmetotextiles solution. Some other companies like Lytess (France), supplier for L'Oréal since 2009 with Mixa (2010), Mennen Garnier (2011), Biotherm, and then Nivea (2014) are continuously involved in this business. These companies are exclusively dedicated to the development and commercialisation of cosmetotextiles and have become a European market leader as a textile brand in this area. The Cosmetotextiles market was estimated at 500 million Euros in 2013, in which the slimming garments contribute about 10% share. France is the first producer and consumer of cosmetotextiles, with 64% of the market in 2012. The development and manufacturing of new products will open new market opportunities, cosmetotextile manufacturers. Cosmetotextiles can be broadly divided into two major classes: (1) dermocosmetics (skin care) and (2) aromatherapy (release of essential oil and fragrances). Furthermore, a broad classification of cosmetotextiles is presented by Singh [49]. Grafting, padding, coating, spraying or screen printing are the major ways of applying microcapsules containing fragrance or cosmetic agents onto a textile [50].

5.12 Aromatherapy

Aromatherapy is considered an alternative route of medicine that uses volatile materials like essential oils by various peoples. Essential oil or fragrance is released from the microcapsule when any external stimuli actuate the fabric's microcapsule to promote the healing. The functional textiles for aromatherapy are found appropriate to affect the feelings, emotions and mood. Textile substrate works as a medium to deliver aroma conveniently at the desired moment. Various curtains, furnishings, handkerchiefs, and masks are treated with aroma finishes to incorporate functionalities. The microcapsules containing essential oils or fragrance are leaded on such textiles. Perfumed microcapsules are fixed on textile surfaces wither by the use of a binder or chemical grafting. Variety of essential oils like peppermint to get the exact thinking mood, lavender to get feeling of getting relax, similarly other oils for different purposes.

5.13 Photochromic textiles

Photochromic materials are uncoloured initially and do not absorbs light. These materials are activated only by high energy protons of ultraviolet rays present in the close surrounding of it. Organic substances like fulgides, spiropyrans, and spirooxanies are primly used as sensor in textiles.

For a textile application, organic compounds such as spiropyrans, spirooxanies and fulgides are mainly used to act as sensors [51]. The last category has found application in garments, toys and logos on T-shirts. Microencapsulation is used to improve the compounds' fatigue resistance as a result of deterioration after numerous repetitive cycles of irradiation and is affected by environmental factors [52, 53].

Photochromism is used in textiles to provide new functional smart fabrics such as garments capable of blocking UV radiation and sensing environmental changes, and also for aesthetics or functional effects such as camouflage, security printing, brand protection [51] and anticounterfeiting. Microencapsulated photochromic compounds can be applied by screen printing or grating onto a textile surface.

Di Credico et al. worked on a microencapsulation process to entrap a photochromic UV-sensitive dye dissolved in sunflower oil as a core material. After optimising the microcapsule shell's UV screening properties by tuning the core material's chemical composition, they demonstrated the use of such UV screening microcapsules in functional coatings for the nondestructive in situ visual detections of mechanical damage by colour change.

6. Benefits of functionalisation by microencapsulation

According to the desired functionality in textiles through microencapsulation route, microcapsules are planned and engineered. Nature of active ingredient core material, nature of sheath polymer, particle size, compatibility between core and sheath material are some prime parameters that drive microencapsulation. Microcapsules with porous, semiporous or impermeable shells are used for different applications.

6.1 Protection and shelf life enhancement

Most of the active substances available are volatile, chemically fragile, or chemically, physically or thermally unstable, and cannot be applied directly on the textile substrate without being covered inside a shell. The micro or nano encapsulation not only provide the safety to the active substance from environmental stimuli; acidity, alkalinity, heat, moisture, oxidation) To restrict interaction with other chemicals remains present in the system to enhance the functionality delivery period.

The capsule shell is used to block the evaporation of active ingredients.

The capsules can also prevent the dissipation of volatile compounds. Additionally, the microcapsules save human resources at manufacturing and users side from exposure to harmful substances. Microcapsules allow safe handling of active ingredients before processing and permit a soluble substance to be transformed in a temporarily insoluble form. This technique permits an unpleasant fragrance from active compounds to be masked before end application during manufacture.

6.2 Controlled release

The microencapsulation of active substances is one of the best routes to enhance the efficiency and minimise environmental damage by controlled release.

This technology prolongs an active ingredient's delivery until an external stimulus like heat, moisture; pressure is actuated at a specific rate, time or situation. The microcapsules are desired to escape the core ingredient to the wearer under a range of controlled situations, which mainly depend on the choice of shell materials, the microencapsulation process opted and final applications.

6.3 Compatibility

The compatibility of core and shell material assists in microencapsulation. A binder's efficiency in connecting microcapsules to the textile surface depends on compatibility between various interphases of each component's microcapsules and finishing and chemical nature.

7. Microencapsulation technologies

The micro and nano encapsulation involve three necessary steps: enclosure of active ingredient as core material, the formation of microparticles and hardening. Again, these processes are further divided into three main groups depending on microparticles formation mechanisms. These three mechanisms are mechanical, chemical and physicochemical. The selection of one mechanism depends on various factors in which processing cost and selection of organic solvent is a significant point of considerations (**Table 2**).

Some multifunctional ingredients impart a range of functionalities on textiles after uniformly coating on it. Zinc oxide can introduce collective functionalities like antimicrobial activity, electrical conductivity, flame retardancy, hydrophobicity, moisture management, photocatalytic self-cleaning, and UV protection. In the development of wearable electronics, the enhancement of the piezo-photocatalytic activity of ZnO NRs by controlling the structure grown on conducting textile

Basic Ingredient	Functionality	Reference
Pro-vitamin C soluble in sebum	Cosmeto-clothing: Pro-vitamin C converts into vitamin C in the presence of sebum and is applied on blouses, and men and women's shirts	[54]
Aloe Vera, and Chitosan with other PCMs	Aloe Vera, and Chitosan with other PCMs Leg wear and intimate clothing for both men, women and Yoga Lines: Delivering cosmetic and well-being benefits like freshness, moisturising and massage for leg wear and intimate apparel. Stretch and recovery function through the use of Lycra	[49]
Distilled oils of plants, fruits and leaves	Textile has the functionality to provide gentle care to tired feet and legs with the special effects of invigorating aromas	[55]
Ultra-thin cloth with extracts of Padina Pavonica	The cosmetically inspired fluid lingerie "Hydrabra" provides moisturising and firming effects	[56]
Seersucker Woven Fabrics with Therapeutic Properties	Seersucker woven fabrics provide anti-cellulite functional knitted. The fabrics were measured in the range of their structural, mechanical, comfort-related and functional properties. These fabrics offer sufficient air transmission and good thermal resistance with gentle micro-messaging functionality	[57]
Chitosan microencapsulation with essential oils and bio-surfactants on cotton fabric	Smaller size microcapsules are obtained in presence of bio-surfactants. The antibacterial activity of fabric increases with the increase the add-on of chitosan and essential oil concentrations. The presence of essential oil decreases the stiffness but has no effect on wrinkle recovery	[58]
Deodorising Textiles	Photocatalyst modified polyester staple fibre, cotton, bamboo fibre, and photocatalyst modified polyester blended woven fabrics were offered good deodorancy at 80% or 100% photocatalyst fibre content.	[39, 40]
Wearable and textile electronics	Wearable and textile electronics was developed by thermal atomic layer deposition (ALD) at 300 °C with highly reactive counter reactants, including plasma radicals and O_3. High functional cotton fabrics are developed.	[25]
Temperature responsive functional textiles	A series of stimuli sensitive polymers were applied on various fabric surfaces to make them thermal responsive	[59]
Radar absorbers, microwave	Polypyrrole coated fabrics and fibres becomes able to absorb various waves sensitive in radar range	[60]
Camouflage electrochromic functional textiles	Polymers have a higher ion exchange capacity, higher hydrophilic/intensely coloured in the charged state are used to coat the textile surfaces. Coated textile colour is dramatically altered by application of small quantities.	[61]
Flexible wearable pressure sensitive textiles	Piezoresistive properties are incorporated to detect the loacal pressure on the fabric. These functional textiles become useful for injury prevention, rehabilitation, sports and medical applications	[62, 63]

Table 2.
Details of some selected functional ingredients, functions in functional textiles.

substrates will be crucial [64]. Silver nanoparticles and titanium dioxide nanoparticles functionalised Cotton-cellulose-spandex fabrics with various weaving configurations like plain, twill and satin with ester crosslinking agent, silicone micro-emulsion. The treated fabric samples offered sufficient antibacterial, soft-handle, water/oil repellence, UV-protection and self-cleaning functionalities. The functionalised fabric samples retained these properties even after ten washing cycles [65]. Effective analytical techniques like scanning electron microscope with EDX confirmed the effective interaction between cellulosic surface and finishing nanoparticles.

The role of weave structure was also found crucial to enhance functionalities of treated fabric follows the descending sequence Satin (4) > Twill (2/2) > Plain (1/1) nevertheless of the used functional ingredients. The treated fabric samples showed bi-functional potentials like easy care-water and oil repellent, comfortable care-soft touch, or easy-care/antibacterial finish. The fabric finished with citric acid/ NaH2PO2/TiO2-Nanoparticles to get easy care/antibacterial/anti-UV/Self-cleaning effect was stable regardless of fabric weaves.

The in-line characterisation of flame retardant and polyvinyl acetate based stiffed polyester and cotton fabrics were scanned by a hyperspectral camera (1320–1900 nm) based on chemometric approaches using the partial least squares (PLS) algorithm. The finish was applied to enhance the areal fabric density from 10–50 $g.m^{-2}$. For both the fabrics, the root mean square prediction error (RMSEP) was estimated at 1.5–2gm − 2. These results were found a very close correlation with gravimetry results also. The near-infrared imaging technique was also opted to detect the finishing agents' traces after washing the treated fabric surface. It was proved that a very thin layer of areal density between 0.4 and 5.5 g m − 2 was found intact even after many wash cycles [66].

8. Functional textile assessment

Functionality is a broad term used to assess the specific needs of clothing customers. The assessment of functional textiles primly depends on the satisfaction of the customer. Customer demands water-proof breathable fabrics, flame-retardant deodorant textiles, antimicrobial-perfumed textiles, soft-skin glowing textiles and others. Creativity, reliability, and aesthetics are three significant points of consideration during functional textile assessment planning. Some features must be considered as a part of the assessment is:

- Low-Stress mechanical properties
- Breathability
- Thermal Transmission
- Air transmission
- Presence of active ingredients on the fabric surface
- Colour Index
- Scavenging potential of foul odours and toxic gases

8.1 Low-stress mechanical properties

The alacrity desire for comfortable fabrics has become steering of the increasing demand for functional textiles. A dramatic shift in apparel goods has registered from durability to functional aspect and increasing purchase power and customer awareness fueled it up. The rapid change in fashion trends and market demand has compelled the fabric manufacturer to follow the functional textiles design right from fibre manufacturing stage rather than relying upon experienced cloth manufacturing with conventional fibre design. The concept of high-quality apparel fabrics to achieve desired level appearance, handle and wearing comfort was finalised by Hand Evaluation Standardisation Committee (HESC).

Consumers'Consumers' purchasing decisions are usually based on feel fabrics for their tactile properties because, during daily wearing, low-stress mechanical action like bending, shear, compression, tensile, and hysteresis occurs on clothing. These common low-stress mechanical attributes significantly impact the feel, movement, sensory and tactile comfort. Other fabric and yarn properties like yarn counts; twist, coefficient of mass variation, neps, hairiness, thin and thick places, and strength and elongation also influence the clothing functionality.

Tactile properties of fabrics affect the functional aspect of apparel products and influence consumers' decision-making when purchasing textile clothing [67].

In the textile industry, tactile comfort is known as "handle" or in a broader sense "skin sensational wear comfort" or "sensorial comfort". Sensorial or tactile functionality, mostly identified by "hand or handle", is an inference of quantity of stress is generated in the fabric during use [68]. Tactile functional attributes are complex theories which incorporate dimensional alterations by small forces like bending, compression, shear, surface properties, and tensile. The feel of warmth and cool also influence the functionality of the fabrics.

Kawabata Evaluation system for fabrics (KES-FB) and Fabric Assurance by Simple Testing (FAST) systems are used to test the low-stress functional properties from a comfort point of view.

8.2 Breathability

The breathability of functional textiles is an essential parameter to be tested to assure the efficacy. The breathability of textiles is mostly referring to the moisture vapour transmission rate through the fabric. A series of instruments are available in the textile world, but moisture management tester of SDL Atlas is considered the prime instrument followed by some other concepts in which inverted cup method [69].

8.2.1 Sweating guarded hot plate tests

The sweating guarded hot plate's moisture vapour transmission resistance can be measured as per ISO 11092 testing standard [70]. This apparatus comprises the water supply unit and measuring unit in which the measuring unit is fixed with a metal block which consists of an appropriate heating element. The measuring unit is a permeable square metallic plate with area of 0.04 m2 and 3 mm. The specimen holder remains at the centre of heating plate which is surrounded by a guarded heating device. As shown in **Figure 5**, the guarded heating systems block any lateral heat escape from the samples' edges. The resist heater is fixed at the bottom of the heating plate to avert the descending heat loss from the specimen and guard section.

This positioning of various components operates heat or moisture transmission only upward along the specimen thickness direction. Distilled water is used to feed the surface of the porous plate through an appropriate dosing system. Water impermeable but water vapour permeable cellophane ultrathin membrane is fitted over the plate. The 300 X 300 mm2 sample is mounted over the membrane. The heating of square porous plate at the constant heating rate is started that mimics the

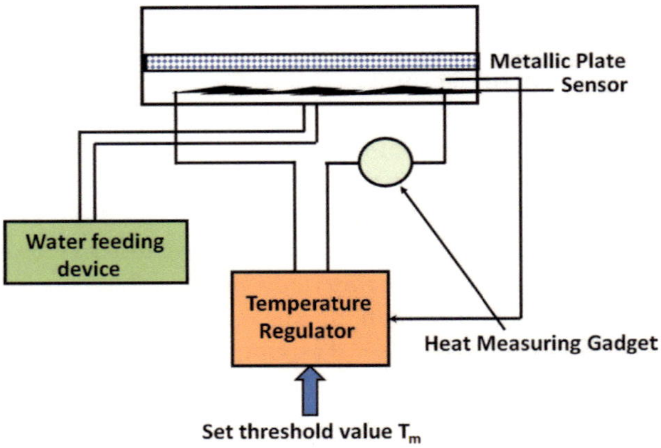

Figure 5.
Sweating guarded hot plate concept.

human body skin temperature, 35 °C, which is measured by a sandwiched sensor directly which is fixed underneath the plate surface. The entire system is closed in a chamber to control the micro-environment conditions very carefully. In order to simulate the actual condition, the air flow is kept at 1 m/s. The air temperature and relative humidity are maintained at 35 °C and 40% respectively.

8.2.2 Upright cup method

The water vapour transmission rates are measured as per ASTM E96, procedure B, and standard test methods by upright cup method. 100 ml distilled water is filled in a shallow cup and a specimen of size 74 mm is mounted on the cup by covering the gasket and fixing it in appropriate position. The cup and other accessories are housed inside an environmental chamber as shown in **Figure 6**. The temperature of circulating temperature is set to 23 °C with controlled relative humidity at 50%. The air flow is maintained with a velocity of 2.8 m/s. The cup assembly is

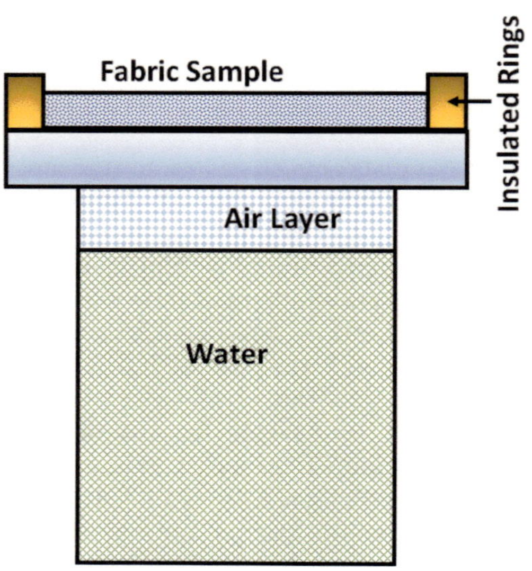

Figure 6.
Upright cup method.

weighed with accuracy of 0.001 g with the assistance of periodically top loading balance for 24 h. Finally the water vapour transmission rate is estimated by estimating the weight change g, in a time period of 24 h through a test area in m^2.

$$WVT = \mathrm{G\,X\,24\,/\,t\,X\,A} \quad (1)$$

WVT = water vapour transmission rate, $g/m^2/day$.
G = weight change, g.
t = time during which G occurred, h.
A = test area, m^2.

8.3 Antimicrobial assessment of functional textiles

Assessment of antimicrobial functional textiles can be completed by testing the following attributes as summarised in **Table 3**:

Colour.

The antimicrobial finish application on textile surfaces should not alter the colour, which becomes the cause of significant quality deterioration. It is preferred to add an antimicrobial agent in dye bath if the dye and antimicrobial agent's chemistry allows for this.

Chemical Effects.

The antimicrobial agent must have zero chemical effects on the functional textiles. The tensile strength, elongation, bending rigidity, fullness and surface smoothness must be maintained for a long time.

Efficacy.

The efficacy of bacteriostatic/fungistatic or bactericidal/fungicidal treatments must be appropriately checked. Variety of chemicals is available to destroy the microbes, but their logical selection is required to prolong the fabric's functionality in a controlled manner. Apart from that, the antimicrobial agent should be effective at a relatively small quantity to control the add-on and cost of the material under permissible limits.

Odour.

The antimicrobial finishes should not release an annoying odour to the finished product, especially in apparel class. Many antimicrobial agents are prone to transmit unpleasant odour, while some are entirely free from foul odour feature.

Fastness.

The fastness or stability of antimicrobial finish is calculated in terms of resistance to abrasion, heat, light, laundering, oxidising agents, and ultraviolet rays. The deficient number of antimicrobial agents possesses all the above features. The antimicrobial molecule must be stable as a compound in a manufacturing environment. It should be steady not only for the finished functional textiles' purposeful life but also for the long storage period.

Hand.

The antimicrobial treatment of functional fabrics should not deteriorate the functional fabric's low-stress mechanical behaviour, particularly in the apparel sector's manufacture. The fabric should not attain a rough hand after antimicrobial treatment.

Toxicity.

The antimicrobial functional textiles must be free from toxicity or of a short order of toxicity. The antimicrobial potential is an essential feature for kids clothing where any type of toxicity is not permissible.

Test	Test Description and Expected Inferences
AATCC 100–2004 Antibacterial Finishes	The microorganism growth is completed in liquid culture, followed by dilution in a sterile nutritive solution. Inoculation is essential for sample and glassware. Bacteria quantity is examined at "time zero" by elution in a neutralising broth, followed by dilution and plating. A standard sample run is essential to confirm the neutralisation/ elution method effectiveness. Suppression of microorganisms reference to initial concentrations and the control sample is estimated. Percent reduction of bacteria R = 100 (B – A)/B where: R is % reduction A is the number of bacteria recovered from the inoculated treated test sample, B is the number of bacteria recovered from the jar immediately after inoculation (at "0" contact time)
AATTC Test Method 147–2004 Parallel Streak Method	Control and treated both samples are placed in close contact with the inoculated agar surface with test bacteria. A logical zone of heckled growth below and along the sample sides represents antibacterial potential of the fabric sample. A usual bacterial strain must use for test. The mean width of a zone of inhibition along a streak on either side of the sample is calculated by: W = (T - D)/2 where: W is width of zone of inhibition (mm), T is total diameter (mm) of sample and D is diameter of the test specimen in mm.
AATTC: 30–2004 Antifungal Activity of Textiles	This test method includes four methods for antifungal assessment on textiles. • One method involves testing fabric properties after burial in soil that contains fungi. • Second includes the cellulosic fabrics exposed to Chaetomium globosum in an agar plate and examination of subsequent growth. • The third method exposes textiles to Aspergillus niger in an agar plate and visually determines any fungal growth. • The fourth method uses a humidity jar to expose textiles to mixture of fungi spores. Any growth on the textile is visually determined.
AATCC TM30 Test IV	A dry, 1 × 3 inches strip of nutrient saturated treated and untreated fabric, sprayed with a mixed-spore suspension of mildew is suspended and incubated with sterile water in the standard moist conditions. The percent fungal growth is recorded after the incubation period. This test allows more clear differentiation between treated and untreated samples.
SN 195920	Determination of antimicrobial activity on textile fabrics: Agar diffusion plate test
SN 195921	Determination of antimycotic activity on textile fabrics: Agar diffusion plate test
SN 195924	Determination of the antibacterial activity on textile fabrics: Germ Count Method
JIS L 1902	Testing for antibacterial activity and efficacy on textile surfaces
ISO 20743	Testing for antibacterial activity and efficacy on textile surfaces
BS EN ISO 20645	Determination of antibacterial activity: Agar diffusion Plate test
BS EN ISO 11721-1	Determination of resistance of cellulose-containing textiles to micro-organisms- Soil Burial Test – Assessment of rot-retardant finishing
ASTM D 4300	Antimicrobial Testing for ability of adhesive Films to support or resist the growth of Fungi
ASTM E2149	Determining the Antimicrobial Activity of Immobilised Antimicrobial Agents under Dynamic Contact Conditions
ASTM E2180	Determining the activity of Incorporated Antimicrobial Agents in Polymeric or Hydrophobic Materials

Table 3.
Various tests for antimicrobial assessment [71–74].

9. Enzyme immobilisation for functionality on textiles

Various enzymes have been immobilised on various textiles surfaces, cotton, polyester wool, and flax, to achieve additional functionalities. Enzymes are biological catalytic materials used to keep up biochemical reactions by expediting the catalytic potential. Enzymes are proteins, which remains available in cells of living entities that are proficient in reducing the stimulation energy needed by chemical reactions in organic medium and living creatures [75]. The demand for enzyme application has been triggered in all industrial segments, but the textile industry demands high performance, extremely stable enzyme at the uttermost pH and temperatures [76]. A physically attached enzyme on a water-insoluble surface, auxiliary material or carrier is called an immobilised enzyme [77]. These enzymes remain stable with the attached surface due to a proper linkage [78]. Various latest approaches appear regularly to immobilise various enzymes on surfaces to achieve improved efficiency, stability and applications. Many materials opted as carrier, substrate or support to immobilise the enzymes are inorganic, organic and organic synthetic carriers. Most inorganic carriers are aluminium oxide, activated carbon, bentonite, hydroxyapatite, kaolinite, nickel, titania, zirconia, silica gel, and glass. Inorganic carriers are less reactive with high stability and sound diffusion, and flowing potential. These carriers are very cost-effective also.

The organic carriers are mainly carbohydrates and proteins. The carbohydrate-based carriers are alginate, chitin chitosan, cellulose, dextran, and starch; however, the protein-based carriers are albumin, collagen, and keratin. These organic carriers offer little diffusion and flow potential. These organic carriers are effected easily by microbial contamination and pH.

Organic synthetic carriers are polyamide (PA), polypropylene (PP), polyvinyl, polyacrylate, polystyrene, copolymers of ethylene, polypeptides and polyaldehydes primly [79]. The organic synthetic carriers are appropriate for a wide range of enzymatic applications because they are not sensitive to microbial contamination [80].

9.1 Essential features for a substrate for enzyme immobilisation

The characteristics of immobilised enzymes are defined by the interaction between enzyme and substrate characteristics. Important characteristics to consider are the following.

Solubility.

The immobilised enzymatic system should be insoluble and rigid with the substrate surface to avoid biological contamination and enzymes' loss.

Functional groups.

The abundance, existence and activation of functionality are essential features of the matrix. These attributes are responsible for deciding the potential of the immobilised enzyme activity, stability under application situations. Generally, the immobilisation activity is performed via the nucleophilic reaction between the enzyme and substrate functional groups because the enzyme does not react with other organic reagents.

Dimensions and porosity/permeability

In general, the bigger the matrix surface area per mass unit, the greater the probability for the enzyme and substrate to get into contact. In terms of permeability, or porosity, the higher the porosity, the better the penetration of molecules between the enzyme and the substrate. Matrix pores bigger than 30 nm appear to support enzyme immobilisation by facilitating enzyme accessibility to the matrix's internal area [80].

Mechanical strength

This property takes importance depending on the reactor or the industrial vessel where the chemical reactions take place. When using immobilised enzymes in a stirring tank, the matrix is desired to be strong enough to prevent abrasion. Particle sizes lower than 50–100 μm may result in filters and sieve plugging.

Resistance to microbial contamination

The support material must be resistant and not affected by microbial attacks. It should be stable and inert to microbial contamination for an extended period.

Reuse

One of the benefits and thus a desirable feature of immobilised enzymes is their ability to be reused. This property makes them less expensive and compensative for any extra cost than soluble enzymes, which is especially important when using an expensive matrix or support materials in some specific applications. Proper orientation of immobilised enzymes and support from crosslinking agents improves reusability.

Hydrophobicity and/or hydrophilicity

The hydrophobicity and/or hydrophilicity of the carrier, supporting material, or matrix is vital because such characteristics affect the strength and affinity of enzyme-carrier interactions noncovalent interactions [78]. This characteristic can also affect surface assimilation, dissemination, and obtainability of the product and substrate.

Matrix size and shape

These properties are significant for operating times. In general, the size and shape are dependent on the applications. Commercially, the matrix size can be available in the range of 150–700 μm particle size. A spherical shape matrix with a particle size range of 150–300 μm is preferred for stirred tank batch productions.

9.2 Enzyme immobilisation for textile processing

Different enzymes are tried to immobilise textile surfaces to achieve various functionalities, as summarised in Table.

9.2.1 Cellulases

Cellulase enzymes recorded global identity in textile processing due to their potential to functionalised the cellulosic fibres in a regulated fashion, manufacturing improved quality fabrics without significant compromise in structural damage [81]. Cellulases, a group of enzymes, can cause cellulose hydrolysis via β-(1–4) linkages degradation of the biopolymer, consequently releasing reduced sugars [82–84]. The cellulase enzyme's multicomponent tendency finds its most application in removing fluffy and protruding fibres from cotton fabrics (biopolishing) and for a stonewashed look in fabrics by efficient abrading of indigo-dyed denim [85]. A soluble–insoluble reversible polymer, Eudragit L-100, was successfully opted for cellulase immobilisation on cotton fabric surfaces and found an alternative to be used in bio-polishing and/or bio scouring of cotton fabrics [86].

9.2.2 Pectinases

Based on their mode of action, polygalacturonases degrade the complicated pectins found in plant tissues into simpler molecules like galacturonic acids [87, 88]. Pectinase has found its way in textile processing in the 21st century; otherwise, it was a known enzyme for the food industry [89]. Many non-cellulosic impurities are found in the primary wall of cellulosic fibres and less in the secondary wall, restricting the penetration of dyes and other functional finishes in the fibre interior [87]. The bioscouring of cotton is applied to degrade the cuticle and primary wall constituents from the cotton fibre surface to improve the hydrophilicity [90].

Pectinase immobilised on the cotton fabric surface for bioscouring in a reverse micellar system with pectinase dose of 10% (2.8 IU/g of the fabric) on the weight of the fabric at60 °C for 120 min, pH 7 to produce a hydrophilic fabric [91].

9.2.3 Amylases

Amylases are the enzymes, which split the starch molecules and starch related compounds in either exo or endo positions by hydrolysing α- 1,4- and/or α-1,6-glucosidic linkages in either endo- or exo-locations [92]. The removal of starch from warp threads of the fabric in which unsized weft yarn also remains present is safely possible by the amylase enzyme's selective action [93]. The immobilised amylases

Enzyme	Support System	Immobilisation	Benefits	Ref.
Amylase	Alkylamine glass beads coated with zirconia	Adsorption followed by GLUTAL	Immobilised enzymes with better washing fastness till 100 launderings without any considerable loss of activity	Dhingra et al. [94]
Cellulase	Polyvinyl alcohol coated chitosan beads,	Epichlorohydrin-Adsorption	Acid cellulase became a neutral cellulase	Dinçer and Telefoncu [82, 83]
	Chemically modified pumices particles	$ZrOCl_2$ - Adsorption	Gives stonewashed finish on indigo-dyed denim fabrics by efficient abrading	Pazarlioglu et al. [81]
Catalase	Alumina pellets	Covalent-GLUTAL	Higher stabilities and surfactant inactivation	Costa et al. [99]
	Alumina pellets	Covalent-GLUTAL	93% protein bound and 87% activity retained	Paar et al. [100]
	α- and γ-Alumina balls, Novalox saddles and Raschigrings	Covalent-GLUTAL	Higher porosity and shape of the carriers are two main parameters to influence the enzyme immobilisation stability.	Fruhwirth et al. [101]
	Cotton fabric or Nylon 6	Adsorption and covalent-GLUTAL	Low cost and flexible construction	Opwis et al. [102]
	Poly(ethylene terephthalate) or polyamide 6.6	Covalent -Photo chemical	After 20 application cycles, the immobilised enzyme showed an integral activity around 3.5 higher than free catalase	Opwis et al. [103, 104]
	Poly(ethylene terephthalate)	Chemical and Covalent-Photochemical	Enzyme modification before the immobilisation; photochemical technique may be able to compete with conventional immobilisation procedures	Opwis et al. [105]
Peroxidase	Polyethylene	Covalent-GLUTAL	Reusabilty was studied for 15 cycles and the half-life was found to be 60 h	Shaffiqu et al. [106]
Laccase	poly amide 6,6	Cross linking-GLUTAL and spacer	Potential for application in the continuous decolorisation of textile effluents, where it can be applied into a membrane reactor	Silva et al. [98]
Glucose oxidase	Cotton	Covalent binding	Recycling of desizing liquors into bleaching liquors	Opwis et al. [107]
	polypropylene	plasma activated, -OH Bond	To produce enzymatically active films, activity prolong upto 30 days of storage	[108]

Table 4.
Enzyme immobilisation on textile surfaces, (GLUTAL: Gluteraldehyde).

enzyme has opted in the detachment of starch and detergents from cotton fabrics. All detergents' performance enhances in the presence of immobilised amylase on cotton fabric surfaces [94].

9.2.4 Proteases

Proteases enzymes are used to carry out the protein degradation through hydrolysis of the peptide bonds in the polypeptide molecular chains [56, 95]. The traditional chlorination process in woollen fabric to achieve shrink-proofness causes ecological issues due to chlorine release is successfully replaced by immobilisation of proteases enzyme. The proteases action on woollen fabric enhances the dyeability, whiteness index and hand behaviour [96]. Some researchers have reported excessive damage in strength and weight loss in woollen fabrics [97]. Immobilisation of proteases enzyme on the textile surface typically enhances the molecular size, constraining proteolytic attack to the cuticle.

The modified protease is immobilised on the cuticle layer region to hydrolyse just the cuticle layer, producing higher tensile strength and a lower felting of the wool fibres. Silva et al. [98] used a commercial protease (Esperase) covalently linked to Eudragit S-100 as summarised in **Table 4**. This novel approach is a promising alternative for wool shrink-resist finishing, replacing the conventional chlorine treatments. Under optimised conditions (Eudragit, 2.5% w/v, carbodiimide, 0.2% w/v, coupling time 1 h and blocking agent concentration, 0.05%), the conjugate activity yield was about 45%, and its operational stability at 60 °C was increased by 1.7 times. Recently different enteric polymers are coupled with Esperase using carbodiimide coupling. More recently, Smith et al. [96] demonstrated that different enteric polymers could also be successfully coupled with Esperase using carbodiimide coupling on woollen fabric.

9.2.5 Glucose oxidase

Glucose oxidase is a dimeric glycosylated flavoprotein enzyme that can accelerate the oxidation of glucose to gluconolactone, which in turn, spontaneously yields gluconic acid as H_2O_2 as a side- product [109]. Therefore, glucose oxidase has been considered a possible method for producing H_2O_2 for green-bleaching, targeted at enhancing the fibre performance before colouration by tear down the pigments initially present in the natural fibres that possess greyness. Enzymatically produced H_2O_2 also gives a comparative bleaching effect with chemical bleaching. The immobilised.

9.2.6 Catalases

Catalases enzymes are known to cleave H_2O_2 into water and oxygen. H_2O_2 is a powerful bleaching agent and oxidises reactive dyes if H_2O_2 does not remove properly from cotton fabric [110]. Catalases enzyme cannot withstand commercial bleaching conditions like temperature 60 °C and pH 9 and above [111]. Alkalothermophilic and thermophilic microorganisms generated catalases enzyme is used as a successful alternative to commercial chemical bleaching. The immobilisation of catalase enzyme on fabric surface counters this issue and offers enzyme for re-application, saving energy and water both. Catalase immobilisation has been practiced by various researchers [109, 112, 113], with different carriers like organic and inorganic materials such as porous glass, cellulose, alumina, silica gel and hydrogels. Some biopolymers like gelatin and chitosan; additionally, some synthetic polymers like polyacrylamide, were also used for bleaching treatments as summarised in **Table 4** [101]. In a remarkable work Kiehl et al. [114].

Opted catalase enzyme to immobilised on polyester, polyamide (Nylon 6, Nylon 66), cotton textile surfaces opting different strategies as mentioned in **Figure 7**. The catalase enzyme was loaded with 20–70 mg enzyme/g textile carrier to achieve reactivity upto 20% and excellent stability against enzyme desorption. The strategies like grafting, application of bifunctional coupling agents, monomeric and polymeric crosslinking agents were planned to achieve covalent fixation of the enzyme on textile carriers as shown in **Figure 7**.

9.3 Decolorisation of dyes and effluent treatment

9.3.1 Peroxidases

Peroxidases are oxidoreductases used to consume H_2O_2 to initiate the oxidation of a wide variety of organic and inorganic chemicals. Several studies covering general properties, biochemical and molecular characterisation, and industrial and environmental applications have been discussed and reviewed elsewhere [115–118].

The majority of the matrices currently used for the immobilisation of enzymes such silica, controlled pore glass, polyvinyl alcohol, polyacrylamide and chitosan beads were not suitable because of dye adsorption onto the matrices, probably inactivating the enzyme [106].

9.3.2 Laccases

Laccases enzymes are copper-containing oxidoreductases, which belong to the group of small blue oxidases. They are widely distributed in higher plants, fungi and bacteria [119, 120]. These enzymes are used to functionalised the various aromatic compounds (particularly phenols) and inorganic compounds, with concomitant reduction of oxygen to water. Laccases enzymes have found in various textile as well as other industries. The application laccase enzyme is expanding fast, decolourising textile effluents and bleaching textile substrates.

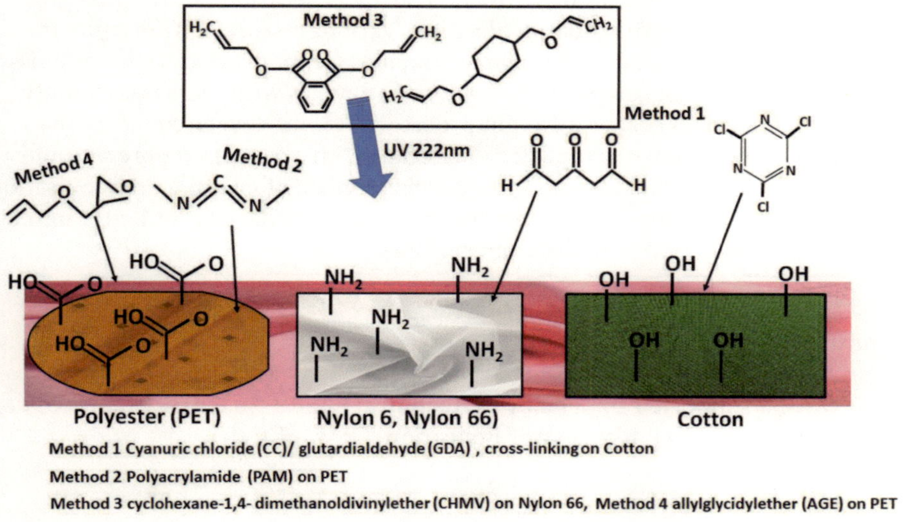

Figure 7.
Immobilisation of catalase enzyme on various textiles materials by different methods (Kiehl et al. [114]).

Most of the laccase enzymes are produced by white-rot fungi, which are efficient in decolourising dyeing effluents [121]. Research has shown that the subsequent coating of the alumina-immobilised laccase with polyelectrolyte layers considerably increased laccase stability.

In the initial stages of laccase action, decolorisation was primly due to the adsorption of the dyes molecules onto the support system, but the support enzymatic decolorisation was apparent after the saturation of support. Acid stable laccase enzyme works well in decolouration of low pH wool dyeing effluents with water recycling opportunity. Silva et al. [98] revealed the potential application of woven nylon 66 fabrics as a carrier for laccase immobilisation to be used in a membrane reactor.

10. Conclusions

Functional textiles are one of the most critical fields in the textile industry and textile materials science. They include breathable, heat and cold-resistant materials, ultra-strong fabrics (e.g., reinforcement for composites), new flame-retardant fabrics (e.g. intumescent materials), and optimisation of textile fabrics for acoustic properties. Functional textiles became more critical materials for various applications, and interest in them grew year by year.

Human skin offers the crucial first defence mechanism for the body to safeguard against external threats. Clothing fabrics and the human skin surface form a cushioning network that creates a thermal and sensorial state of comfort to keep a human being in the state of wellness.

The microencapsulation is the most versatile technique to impart various functionalities in textiles. Microencapsulation suppresses the compatibility between active ingredients and fibre surface by enhancing the functional durability, efficacy and sustainability. This technique's vast use can be witnessed in functional finish fabrics, medical and healthcare textiles, aromatherapy, cosmetic textiles, and many more functional textiles.

The moderate stability of these bio-catalysts primly restricts the immobilisation of enzymes on textile surfaces. The immobilisation of various enzymes on textile surfaces gives a sustainable solution of surface functionalisation for easy processing. Enzyme immobilisation twinning with other surface modifying techniques gives a synergistic effect in textile functionalisation. Various researchers are trying to enhance the temperature and pH range of enzymes for more effective immobilisation. The immobilisation allows the recovery of enzymes with increases stability to reduce the operation cost of different processes. Recent developments in the synthesis and fabrication of supporting materials with customised pore size and surface functionality have licenced more precise control of enzyme immobilisation. Perfectly oriented and highly rigid enzyme molecules are needed for better immobilisation and integration with different surfaces.

Acknowledgements

This work is supported financially by Technical Education Quality Improvement Program-III (TEQIP-III), Ministry of Education, Govt. of India.

DOI: http://dx.doi.org/10.5772/intechopen.96936

Author details

Mukesh Kumar Singh
Uttar Pradesh Textile Technology Institute, Kanpur Affiliated to Dr. APJAK Technical University, Lucknow, India

*Address all correspondence to: mukesh70ster@gmail.com

References

[1] Lee D, Rubner M F, and Cohen R E (2006), Nano Letters, 6(10), 2305-2312

[2] Michel M, Toniazzo V, Ruch D, and Ball V (2012), ISRN Mat. Sci. 1-7

[3] Gupta D (2011), Ind J of Fib and Text Res. 36 (4), 321-326

[4] Chang C P, Yamamoto T , Kimura M, Sato T, Ichikawa K, Dobashi T, (2003) J of Contr. Rel. 86 (2-3) , 207-211

[5] Salaün F, Bedek G , Devaux E , Dupont D L, (2011b) J of Memb. Sci., 370 (1-2), 23-33

[6] Teixeira M A , Rodríguez O, Rodrigues S, Martins I, Rodrigues A E (2012), AlChE J, 58 (6), 1939-1950

[7] Giraud S, Rochery B M, Vroman I , Tighzert L, Delobel R, PoutchF (2005) Poly. Degra. and Stab. 88 (1), 106-113

[8] Vroman I, Giraud S, Salaün F, Bourbigot S (2010) Poly. Degr. and Stab., 95 (9), 1716-1720

[9] Fan F, Zhang W, Wang C (2015) Cellulose, 22 (2), 1427-1438

[10] Salaün F, Devaux E, Bourbigot S, Rumeau P (2010), Carbohydrate Poly. 79 (4), 964-974

[11] Azizi N, Chevalier Y, Majdoub M (2014) Isosorbide-based microcapsules for cosmetotextiles Industrial Crops and Products, 52 (0) (2014), pp. 150-157

[12] Badulescu R, Vivod V, Jausovec D, Voncina B (2008) , Carbohydrate Polymers, 71 (1), 85-91

[13] Liu J, Liu C, Liu Y, Chen M, Hu Y, Yang Z (2018), Colloids and Surfaces B: Biointerfaces, 109 (0), 103-108

[14] Yang Z, Zeng Z, Xiao Z, Ji H. (2014) Flav. and Frag. J, 29 (2), 114-120

[15] Cuevas J M, Gonzalo B, Rodríguez C, Domínguez A, Galán D, Loscertales I G, (2014) , J of Exp. Nanosci., 10 (11), 868-879

[16] Naylor RGJI, Magalhaes VVRM, Pinto CBS (2006), Microcapsules with Functional Reactive Groups for Binding to Fibres and Process of Application and Fixation, WO 2006117702 A2

[17] Gouveia I C (2012) , Poly. for Adv. Tech., 23 (3), 350-356

[18] Salaün F, Vroman I, Elmajid I (2012) Chem. Eng. J, 213 (0) (2012), pp. 78-87

[19] Kathirvelu S, D'Souza L & Dhurai B, (2009) Materials Sci, 15,

[20] Uğur S S, Sarıışık M, Aktaş A H, Uçar M C & Erden E (2010) Nanoscale Res Lett, 5, 1204

[21] Bertrand P, Jonas A, Laschewsky A & Legras A,(2000), Macromol Rapid Commun, 21. 319

[22] Lvov Y, Price R, Gaber B & Ichinose I, (2002), Coll Surf A Physicochem Eng, 375, 198

[23] Ou R, Zhang J, Deng Y & Ragauskas A J,(2007) J Appl Polym Sci, 102, 1987

[24] Oh K, Park J S, Khan M R, Kim K, Lee Z, Shong B, and Lee H (2019) Chem. of Mat. 31, 8995–9002

[25] Lee J, Yoon J, Kim H G, Kang S, et al., (2016), NPG Asia Materials, 8, e331, 1-8

[26] Stawski D, Zielinska D, Simon F, Polowinski S, Puchalski M (2014) Industria Textila 65 (4), 2-9

[27] Stawski D, Połowiński S (2011). In: Vlákna a Textil, 18 (1), 16

[28] Sánchez L S, Rodríguez J F, Carmona M, Romero A, Sánchez

P (2011) J of Appl. Poly. Sci., 120 (1), 291-297

[29] Połowiński, S (2007) In: J of Appl. Poly. Sci., 103(3), 1-7

[30] Stawski D, Bellmann C (2009 In: Colloids and Surfaces A: Physicochemical and Engineering Aspects, 345 (1-3), 191

[31] Stawski D, Halacheva S, Bellmann C, Simon F, Połowiński S, Price G (2011 In: Journal of Adhesion Science and Technology, 25 (13), 1481

[32] Stawski D, Połowiński S, Herczyńska L, Sarna E, Rabiej S (2012), 123 (3), 1340

[33] Joung Y S and Buie C R (2015), ACS Appl. Mat. & Interfaces, A-I

[34] Joung Y S, Buie C R (2011), Langmuir, 27, 4156-4163

[35] Kostajnšek K, Dimitrovski K, Kadoglu H, Çelik P, Bayraktar G B, Üte T B, Duran D, Ertekin M, Demšar A and Bizjak M, (2013), Polymers, 13, 260, 1-18

[36] Čuk M, Bizjak M, Muck D, Kočevar T N (2020) 3D Printing and Functionalisation of Textiles, DOI: 10.24867/GRID-2020-p561-4

[37] Karabulut K and Atav R (2020) Fib. and Poly., 21(8), 1773-1782

[38] Nelson G (2002), Int. J of Pharma. 242, 55-62

[39] Zhang H, Ge C, Zhu C, Li Y, Tian W, Cheng D, and Pan Z, (2012a), Physics Procedia 25, 240-244

[40] Zhang H, Ge C, Zhu C, Li Y, Tian W, Cheng D, Pan Z, (2012b) Physics Procedia 25, 240-244

[41] Singh M K, Varun V K, and Behera B K (2011) Cosmetotextiles: State of Art, Fib. & Text. in East. Euro. 19(4), 27-33

[42] Kanjana S, Nalankilli G (2018), J of Text. Eng. & Fash. Tech. 4(4), 316-318

[43] Nakamura S, Nishioka K, Otsuki T, (2010) Interview available at http:// textileinfo.com/ en/chemicals/ daiwa/01_06.html.

[44] Anitha R, Ramachandran T, Rajendran R, Mahalakshmi M (2011), Elixir Bio Physics, 40, 5196-5200

[45] Boh B, Kardos D (2003) Microcapsule patents and products: innovation and trend analysis R. Arshady, B. Boh (Eds.), Microspheres, Microcapsules and Liposomes, Citus Books, London, 47-83

[46] Chan ASC, Valle J del, Lao K , Malapit C, Chua M, So RC, (2009), Philippine Journal of Science, 138 (1), 13-21

[47] İnceboz T, Erkan G, Türkoğlu G C , Sarıışık A M, Bakırcı S , Üner S, Üner A (2015) , Text. Res. J, 85 (19), 2071-2082

[48] Specos M M, Garcia JJ, Tornesello J, Marino P, Vecchia MD, Tesoriero MV, Hermida LG (2010) Trans. of the Roy. Soc. of Trop. Medi. and Hyg., 104 (10), 653-658

[49] Singh M K (2005) "Sun protective clothing" Jan-Feb, 91-97

[50] Theberge K, Goudreault I, Quirion F, Perron G (2010) Articles of Manufacture Releasing an Active Ingredient, US 20100226947 A1

[51] Billah S M, Christie R, Morgan M (2008), Coloration Tech. 124, 229-233

[52] Fan F, Wang C (2014), J of Appl. Poly. Sci., 131 (20), 211-216

[53] Zhou Y, Yan Y, Du Y, Chen J, Hou X J (2013) Sensors and Actuators B: Chem., 188 (0) 502-512

[54] Fuji Spinning Co. Ltd Japan in European Patent EP 1 251202

[55] Tiwari D, Upmanyu N, Malik J and shukla S (2017), Int. J of Pharma & Chem. Res.3(4), 814-827

[56] Singh J, Batra N, Sobti RC (2001) Serine alkaline protease from a newly isolated *Bacillus* sp. SSR1. Process Biochem 36: 781-785

[57] Matusiak M, Wilk E, Zieliński J (2018) Seersucker Woven Fabrics with Therapeutic Properties, Fib. & Text. in East. Euro. 26, 5(131), 54-58.

[58] Javid A, Raza Z A, Hussain T, and Rehman A (2014) J Microencapsul, 31(5): 461-468

[59] Save N S, Jassal M, Agrawal A K, (2005) J. Appl. Polym. Sci., 95(3), 672-680

[60] Kuhn H H, Child A D & Kimbrell WC (1995) Synth. Met., 71, 2139-2142.

[61] Invernale M A, Ding Y & Sotzing G A, (2011) Coloration Tech., 3, 167-172

[62] Mazzoldi A, De Rossi D, Lorussi F, Scillingo EP & Paradiso R, (2002) AUTEX Res. J., 2, 199-204

[63] De Rossi D, Della Santa A & Mazzoldi A, (1999), Mater. Sci. Eng., C7, 31-37

[64] Verbic A, Gorjanc M and Simoncic B (2019) Recent Advances, Coatings 9 (550), 1-25

[65] Salaün F, Bedek G, Devaux E D (2011a) Mat. Lett., 65 (2), 381-384

[66] Mirschela G, Daikosa O, Scherzera T, Steckert C (2018) Near-infrared chemical imaging used for in-line analysis of functional finishes on textiles, Talanta

[67] Behera B K, Ishtiaque S M, and Chand S, (1997) J of Text. Inst., 88 (3), 255-264

[68] Behera B K, and Singh M K (2014), J of Text. Inst.,105 (4), 365-376

[69] ASTM E96 (1995) Standard Test Methods for Water Vapour Transmission of Materials, in. "Annual Book of ASTM Stand- ards 4.06". American Society for Testing and Materials, West Conshohocken, PA, 1995

[70] ISO 11092 (1993), "Textiles Physiological Effects Measurement of Thermal and Water Vapour Resistance under Steady state Conditions (Sweating Guarded Hot Plate Test)". International Organization for Standardization, Geneva, Switzerland (1993)

[71] AATCC Test Method 100-2004 (2012), Amer. Ass. of Text. Chem. and Colo., Res. Tria. Park, NC, 148-149

[72] AATTC Test Method 30-2004 (2012), Amer. Ass. of Text. Chem. and Colo., Res. Tria. Park,, NC., 80-82

[73] Shinde S (2010), Man-made Text. of Ind., July, 247-248

[74] Sathianarayanan MP, Bhat NV, Kokate SS, Walunj VE. (2010), Ind. J of Fib. and Text. Res., 35, 50-58

[75] Hettiarachchy N S, Feliz D J, Edwards J S, Horax R (2018) Protiens in Food Processing, 569-590

[76] Conesa A, Punt PJ, van del Hondel CA, (2002) J Biotechnol 93:143-158

[77] Bayindirli A, (1995) Immobilisation of enzymes and potential applications in food industry.

[78] Datta S, Christera L R, Rajaram YRS, (2013) Biotech. 3, 1-9.

[79] Costa SA, Azevedo H S, Reis R L, (2005) Enzyme immobilisation in biodegradable polymers for biomedical applications. Biodegradable Systems in Tissue Engineering and Regenerative

Medicine. CRC Press/Taylor & Francis Group/LLC, Boca Raton, FL. 301-324

[80] Tischer W, Wedekind F, (1999) Topics in Current Chemistry. 200. Springer-Verlag, Heidelberg, Germany, 95-126.

[81] Pazarlioglu NK, Sariisik M, Telefoncu A. (2005) Process Biochem 40, 767-771

[82] Dinçer A, Telefoncu A. (2006a), J Mol Catal B Enzymatic 45, 10-14

[83] Dinçer A, Telefoncu A. (2006b) J Mol Catal B Enzymatic 45:10-14

[84] Sinegani AAS, Emtiazi G, Shariatmadari H, (2005) J Colloid Interface Sci. 290:39-44

[85] Gusakov AV, Sinitsyn AP, Markov AV, Sinitsyna OA, Ankudimova NV, Berlin AG. (2001) J Biotechnol 87:83-90

[86] Dourado F, Bastos M, Mota M, Gama FM, (2002) J Biotechnol 99:121-131

[87] Hoondal GS, Tiwari RP, Tewari R, Dahiya N, Beg QK, (2002) Appl. Microbiol Biotech. 59:409-418

[88] Kashyap DR, Vohra PK, Chopra S, Tewari R. (2001) Bioresour Technol 77:215-227

[89] Gummadi SN, Panda T. (2003) Process Biochem 38:987-996

[90] Wang Q, Fan X, Hua Z, Gao W, Chen J. (2007) Carbohydr Polym 67:572-575

[91] Joshi M, Badhe P, Adivareker R, (2013) J of Mole. Cata. B: Enzymatic, 98 (12), 106-113

[92] Hamilton LM, Kelly CT, Fogarty WM, (2000) Enzyme Microb Technol 26:561-567

[93] Gupta R, Gigras P, Mohaptra H, Goswami VK, Chaulan B, (2003) Process Biochem 38:1599-1616

[94] Dhingra S, Khanna M, Pundir CS, (2006) Ind. J Chem Technol 13, 119-121

[95] Gupta R, Beg QK, Lorenz P. (2002) Appl. Micro- biol Biotechnol 59:15-32

[96] Smith E, Schroeder M, Guebitz G, Shen J, (2010) Enzyme Microb Technol 47:105-111

[97] Queiroga A C, Pintado M M, Malcata F X, (2007) Enzyme Microb Technol 40:1491-1495

[98] Silva C, Silva CJ, Zille A, Guebitz GM, Cavaco-Paulo A, (2007) Enzyme Microb Technol 41:867-875

[99] Costa SA, Tzanov T, Paar A, Gudelj M, Gübitz GM, Cavaco-Paulo A. (2001) Enzyme Microb Technol 28:815-819

[100] Paar A, Costa S, Tzanov T, Gudelj M, Robra K-H, Cavaco-Paulo A, Gübitz GM. (2001) J Biotechnol 89:147-153

[101] Fruhwirth GO, Paar A, Gudelj M, Cavaco-Paulo A, Robra K-H, Gübitz GM, (2002) Appl Microb Biotechnol 60:313-319

[102] Opwis K, Knittel D, Schollmeyer E. (2004) Biotechnol J 2:347-352.

[103] Opwis K, Knittel D, Bahners T, Schollmeyer E, (2005a) Eng Life Sci 1:63-67.

[104] Opwis K, Knittel D, Bahners T, Schollmeyer E, (2005b) Eng Life Sci 1:63-67

[105] Opwis K, Knittel D, Schollmeyer E, (2007) Biotechnol J 2:347-352

[106] Shaffiqu T S, Roy J J, Nair R A, Abraham T E, (2002) Appl Biochem Biotech. 102:315-326

[107] Opwis K, Knittel D, Kele A, Schollmeyer E, (1999) Starch 51:348-353

[108] Vartiainen, J.; Rättö, M.; Paulussen, S (2005) Packag. Tech. Sci., 18, 243-251

[109] Betancor L, López-Gallego F, Hidalgo A, Alonso-Morales N, Dellamora-Ortiz G, Guisán JM, Fernández-Lafuente R. (2006). J Biotechnol 121:284-289

[110] Thompson VS, Schaller KD, Apel WA, (2003) Biotechnol Prog 19:1292-1299

[111] Oluoch KR, Welander U, Andersson MM, Mulaa FJ, Mattiasson B, Hatti-Kaul R. (2006) Biocatal Biotransform 24:215-222

[112] Opwis K, Knittel D, Schollmeyer E. (2004b) AATCC Rev 4:25-28.

[113] Wang Y, Caruso F, (2005) Chem Mater 17:953-961

[114] Kiehl K, Straube T, Opwis K, Gutmann J S, (2015) Strategies for permanent immobilization of enzymes on textile carriers, Eng. Life Sci., 15, 622-626

[115] Atack JM, Kelly DJ (2006) Adv Microb Physiol 52:73-106

[116] Moreira PR, Bouillenne F, Almeida-Vara E, Malcata FX, Frère JM, Duarte JC (2006a) Enzyme Microb Technol 38:28-33

[117] Matto M, Husain Q, (2006) J Chem Technol Biotechnol 81:1316-1323

[118] Moreira PR, Bouillenne F, Almeida-Vara E, Malcata FX, Frère JM, Duarte JC (2006b) Enzyme Microb Technol 38:28-33

[119] Alexandre G, Zhulin I B, (2000) Trends Biotechnol 18, 41-42

[120] Duran N, Rosa MA, d'Annibale A, Gianfreda L (2002), Appl Catal B Environ 28:83-99

[121] Kandelbauer A, Maute O, Kessler RW, Erlacher A, Gübitz GM, (2004) Biotechnol Bioeng 87:552-563

Chapter 4

Applications of Silk in Biomedical and Healthcare Textiles

Edison Omollo Oduor, Lucy Wanjiru Ciera and Edwin Kamalha

Abstract

Global trends are shifting towards environmental friendly materials and manufacturing methods. Therefore, natural fiber applications are gaining traction globally. Silk, a natural protein fiber is one of the textile fibers that have recently received more attention due to the new frontiers brought about by technological advancement that has expanded the use of silk fiber beyond the conventional textile industry. The simple and versatile nature of silk fibroin process-ability has made silk appealing in wide range of applications. Silk is biocompatible, biodegradable, easy to functionalize and has excellent mechanical properties, in addition to optical transparency. This review chapter explores the use of silk in biomedical applications and healthcare textiles. Future trends in silk applications are also highlighted.

Keywords: Silk, Silk fibroin, Bio-applications, Functional textiles

1. Introduction

Silk is a natural fibrous protein biopolymer, spun by arthropods like spiders, mites, fleas and silkworms [1]. The structure, composition and properties of silk differ depending on their specific function and source [2]. Silkworms are one of the silk spinning insects that has been researched in detail and finds wide applications in textiles [3]. Silk production, also known as sericulture, has a long history that is usually closely associated with China. Silk was discovered in 2640 B.C. by Hsi-Ling-Chi, who also found out that silk fiber loosened and unwound in hot water; and twisted to make thread that was used to weave a very strong cloth. Hsi-Ling-Chi later developed a means of raising silkworms and a method of reeling the fibers to make garments [4, 5]. As silk became a very precious commodity, sericulture spread within china and to other countries. Demand for silk products created a trade route that is famously known as Silk Road [6].

Silk filaments from silkworms are classified into two types; mulberry and non-mulberry (also called wild or vanya silk). Mulberry silk are generally produced by *Bombyx mori* which are insects belonging to the Bombycidae family. *Bombyx mori* feeds on mulberry plant leaves. Mulberry silk is further divided into bivoltine and multivoltine, depending on the number of silk cocoon crops harvested annually. Bivoltine is harvested twice a year while multivoltine is harvested throughout the year [7]. Non-mulberry silk on the other hand, is silk from Saturniidae family. Non-mulberry silk includes tasar silk, muga silk and eri silk. Tasar silk is secreted by *Antheraea* silkworms. They have hard and compact cocoons. Tasar silk can either be tropical tasar

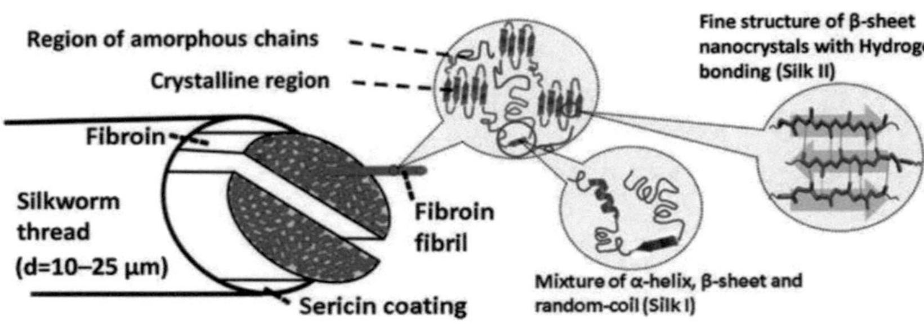

Figure 1.
Schematic presentation of the silk fibroin (SF) structure; d represents the diameter of a single silkworm thread [13]. Reproduced with permission.

or temperature tasar. Muga silk is produced by *Anteraea assamensis* silkworm; it has a unique natural golden color, with significant luster and durability. Eri silk is produced by *Philosamia synthia ricini* (also called *Samia cynthia)* silkworms, which usually feed on castor or papa plant leaves. Eri silk being in the wild silk category, has however been completely domesticated just like mulberry silkworms. Eri silk cocoons draw shorter silk fibers when compared to other silks which draw continuous filaments [8, 9].

Silk cocoons comprise over 95% proteins and about 5% impurities (mineral salts, waxes, ash). Raw silk consists of two proteins; sericin (gum) and fibroin (fibers). Sericin and fibroin are composed of amino acid chains. The types and composition of these amino acids are different for sericin and fibroin. Non-mulberry silk has lower sericin with higher levels of impurities compared to mulberry silk [10]. After degumming, sericin and other impurities are removed from the raw silk fibers. Therefore, degummed silk is composed of mainly fibroin protein [11, 12].

Several authors have reviewed, described and demonstrated the structure of silk fiber varieties, especially *Bombyx mori* in relation to several performance properties. These include: conformations of silk, heavy chains with possible chain folding and micelle assembly in water, primary structure, 12 repetitive and 11 amorphous regions, amino acid sequences of i, ii, and iii, hierarchal structure among others (**Figure 1**) [13].

Silk fibers are usually used for conventional textile applications after the removal of sericin and other impurities. Recently however, due to excellent mechanical and optical properties, as well as its biocompatibility, biodegradability and implant ability, silk has found increased applications in functional textiles [1, 14, 15]. This has been made possible by the simple and versatile nature of the silk fibroin process ability into various forms such as sponges, gels, strands, blocks, foam, films, and more recently, nanofibers [16–21]. Applications of silk in biomedical materials, drug delivery and in optics and sensing are therefore discussed in this chapter. The chapter underscores the forms and properties of silk making them suitable in these applications.

2. Common manufacturing processes for silk-based functional products

More recently, silk fibroin (SF) films with fineness ranging from hundreds nanometers to tens micrometres are obtained from regenerated solutions through liquid processing including: spin coating, inject printing, doctor blade, soft lithography, contact printing or nano- imprinting, among others; that support industrial scale production. Doping, blending and functionalization of SF has also been a route to achieve substrates for advanced technological use in organic electronic sensors based on field effect transistors, and with optically active dyes, particularly for biomedical applications [22–25]. Nano and micro-patterning, through spin coating and lasing, was used to obtain stilbene-doped silk film of significant mechanical performance and optical performance [24].

Innovative attempts during breeding and feeding of silk worms, through incorporation of dopants in the diet, have yielded modified SF substrates of functional value; e.g. silk threads with electrical conductivity through incorporation of silver nanoparticles in mulberry diet, fluorescence introduced in silk fiber through colorant compounds in mulberry feed, among others [26–29]. These approaches save on extra processes and time that would be required as after treatments, and enhance the durability of such functions. Optimization of silk-worm breeding is often required for control and reproducibility of functional substrates. For example, among others, the silkworm survival rates, temperature conditions and duration of the larval cycle are monitored.

Based on different varieties of *B. mori*, in 2019, a silk fibroin based technology was developed in order to optimize and support industrial bio-manufacturing [30]. The evaluation and standardization of extraction, purification, and characterization methods were reported; yielding biocompatible SF substrates with high purity and outstanding chemo-physical performance. The result was a validated bio-diagnostic microfluidic and photonic device (a lab-on-a-chip) (**Figure 2**).

Several conventional textile spinning and construction methods are used for production of functional silk yarns, fabrics; including a variety of finishing technologies through which active functional ingredients may also be introduced. Therefore, such might be applied during fiber spinning (e.g. for sutures) and after fabric construction through a variety of wet and dry finishing processes [31]. Innovative approaches include micropatterning, 3D printing and more nanotechnology based systems (**Figure 3**).

Electrospinning is a common method used in the production of nanofibers and microfibers from SF solutions. The ensuing fibers possess a high specific surface, favoring the use of such scaffolds in tissue regeneration [34, 35]. The mechanism of electrospinning (**Figure 4**) is based on a high electric voltage applied to the fiber polymer solution. The polymer solution is ejected when the electric force overcomes the surface tension of the polymer solution, forming a polymer jet. Electrospinning can be of needle or needleless. The needle electrospinning utilizes a high-voltage

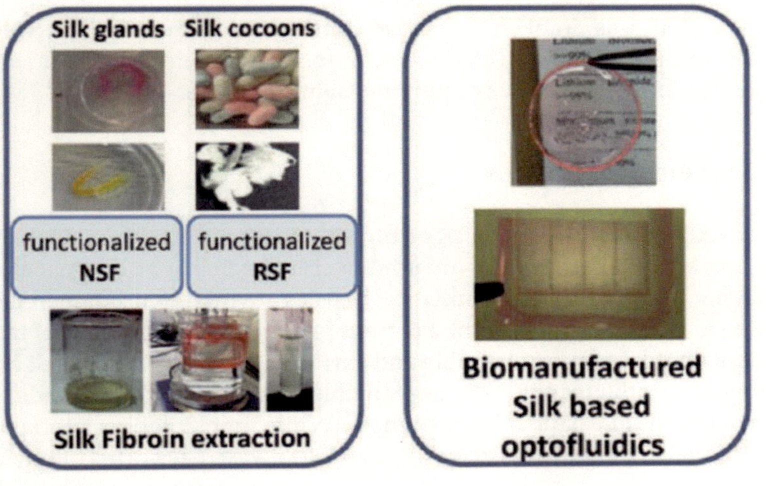

Figure 2.
Schematic picture of the biomanufacturing approach to obtain SF based technological substrates [30]. Reproduced with permission.

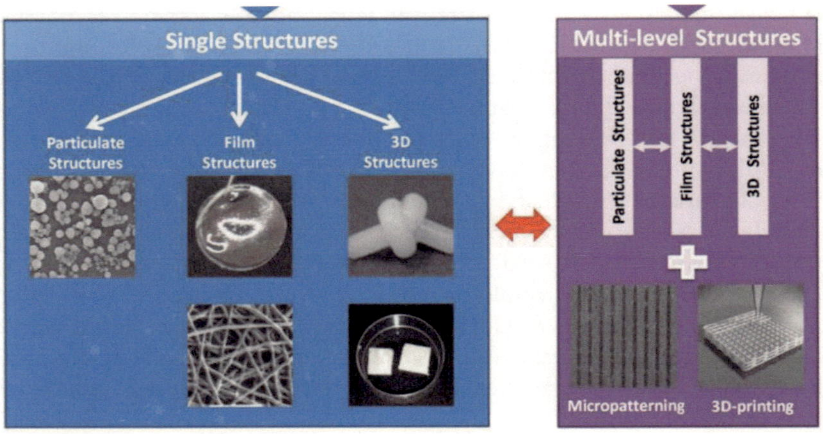

Figure 3.
Structural design of SF-based biomaterials from single structures to multi-level structure [32, 33]. Reproduced with permission.

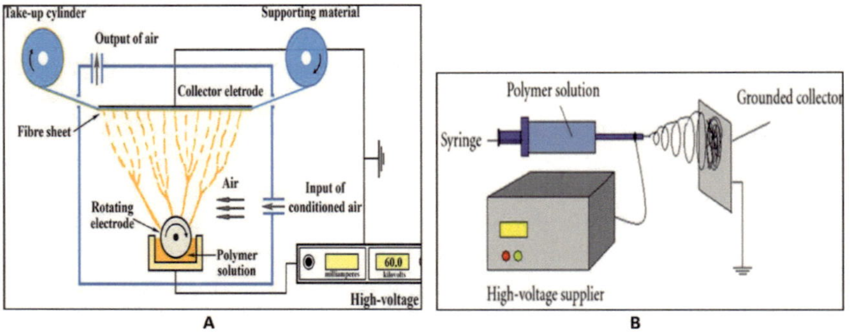

Figure 4.
Schematic images of needle-less (A) and Needle (B) electrospinning processes [40]. Reproduced with permission.

power supply, and a syringe needle connected to a power supply and pointed towards a collector. Needle electrospinning options pose a demerit of very low productivity, thus, unsuitable for practical commercial value. Needleless electrospinning setups have been innovated recently. In these systems, polymer jets form simultaneously from the surface of polymer solution by self-assembly [36–41].

3. Silk in biomedical textiles

Biomedical textiles are composed of fibrous units produced from natural or synthetic materials. These textiles are used in either external or internal environment of living organisms [42]. Biomedical textiles are further used, medically, to improve the medical condition of a patient [43, 44]. Some biomedical textiles include implantable, non-implantable and extracorporeal devices as well as hygiene and health care products [45]. Non-implantable materials/devices include wound dressings materials like bandages and gauzes. While, implantable materials/devices include artificial arteries, heart valves, sutures, and vascular grafts among others. Extracorporeal devices mainly include artificial body organs. Hygiene and health care products include sanitary towels, tissue paper, wipes, hospital gowns and uniforms, hospital bed covers, surgical covers, masks and caps. Other biomedical textiles include polymer sensors and wearable medical implants [46, 47].

Textile materials used in biomedical applications must be non-allergic, non-carcinogenic, non-toxic, biodegradable and biocompatible. Additionally, biomedical textile materials must be able to be stable to handling and use. For example, during sterilization and use, they should not change their physical or chemical properties (e.g. through oxidation or chemical reaction). Other important properties for these textiles include: high tensile strength, elasticity, high burst stream, low permeability and durability [42, 48, 49]. Different synthetic (e.g. polyester, nylon, acrylics, and polyethylene) and natural (e.g. silk and cotton) fibers are used in production of biomedical textiles. Silk fibers possess good toughness and ductility in terms of elongation at break, tensile modulus and tensile strength; suitable for biomedical applications[1, 50]. Additionally, regenerated silk solutions are gaining popularity in producing various biomaterials in form of gels, films, membranes and sponges [51].

3.1 Silk in wound dressing

Studies have shown that silk fabricated through non-weaving and electrospinning can be used in wound dressings, and as drug carriers [1, 46, 52, 53]. Xia *et al.* [54] reported that silk fibers functionalized with silver nanoparticles presented special antibacterial properties in a wound dressing material. A two-layered wound dressing developed from a wax-coated silk woven fabric, a sericin sponge and a bioactive layer of glutaraldehyde cross-linked silk fibroin gelatin was reported to reduce the size of the wound, collagen and epithelialization [55–57]. He *et al.* [58] asserts that fibroin hydrogel from *Bombyx mori* cocoons has good healing properties due to its biocompatibility nature, low biodegradability and immunogenic properties. On the other hand, Chouhan *et al.* [59] found that nanofibrous mats of silk, functionalized with Poly Vinyl Alcohol, (as a blend) mat supported diabetic wound healing. The mats were able to promote tissue re-modeling and also regulated extracellular matrix; thus the wound healing.

3.2 Silk garments for dermatology treatment

Atopic Dermatitis is a worldwide health concern, with a higher prevalence in developing countries, and occurring in among many age groups. Symptoms for Atopic Dermatitis include redness and itchiness of the skin. These symptoms can be severe leading to a chronically repeating flare characterized by serious eczema (distribution of skin lesions) [60]. Treatment and management of this condition requires skin stabilization, flare prevention, as well as the use of medication that can cure the symptoms [61]. Silk garments have been used as a textile-based therapy for Atopic Dermatitis owing to their hygienic properties including antibacterial properties. Additionally, silk filament fibers are strong and round in shape, and therefore fine and smooth. Wearers experience comfort to the skin as this structure prevents and scratching from friction and irritation to the skin [62]. Moreover, the fine and smooth fibers have no or very little abrasive effect on atopic skin. This enhances the recovery of the irritated skin unlike with rough fibers that irritate the skin. Due to a significant moisture regain, silk fibers are also able to maintain body humidity therefore reducing the sweat circulation and moisture loss that can make xerosis worse [63]. A study by Hung *et al.* [60] further the ability of silk garments to significantly decrease the severity level of dermatitis symptoms. The study emphasized the merits with the smoothness of silk which is friendly to the irritated skin. The fiber enhances collagen synthesis and also reduces inflammation which cures eczematous lesion [63, 64]. Moreover, hygienic properties of silk act as a skin barrier, protecting the skin from bacteria, viruses and other contamination that reduces the inflammation [60, 65]. Of importance is the sensory experience of patients with silk garments as highlighted by these studies; they contribute to the physical and emotional comfort of dermatitis patients which possibly aids the healing process. Therefore, silk garments can be used as a non-pharmacological therapy to impede the severity of Atopic Dermatitis and other related dermatology conditions.

3.3 Silk in hygiene and health care products

The good mechanical properties of silk, its softness and antibacterial properties partly account silk's application in producing hygiene and health care products. Some applications of silk in hygiene and health care include: materials used in hospital wards and operating theatre as well as materials used in care and safety of hospital staff and patients. Silk materials used in operating theatre are in form of patient drapes, and surgical gear (as gowns, caps, masks and cover cloths) [46]. Silk, functionalized with titanium dioxide nanoparticles was used to produce a photocatalytic silk mask paper. The mask was found to exhibit special protective functions— degrading volatile organic compounds achieved by combining the unique properties of silk fibers and nano-TiO_2 [66–69].

3.4 Silk-based tissue engineering

Tissue engineering applies principles of biological sciences and engineering to develop biological substitutes to replace, enhance and maintain damaged or defective tissues such as cartilage, bone, skin and even organs [70, 71]. The choice of the biomaterial and the methods used determines whether the resulting bio-substitute will be functional. Silk has good mechanical properties, has a slow degradation rate and a low inflammatory response which makes it fit for use in tissue engineering. However, Sericin can elicit immune response and must therefore be completely removed before being used [72]. The type of silk that is commonly used in tissue engineering is *Bombyx mori* silk. Other types of silk that are gaining popularity include silk fibers from; *P. ricini*, *A. assama*, A. *pernyi* and *A. mylitta* [71, 73]. Silk-based tissue engineering includes: Scaffolds in form of skin grafts/artificial skin, bone grafts, artificial pancreas, cardiac tissue, artificial liver, artificial Intervertebral Disc Intervertebral disc, among others [74].

3.4.1 Skin grafts/artificial skin

Skin, the largest body organ protects the body against infections from pathogens and microorganisms [75]. Due to certain illness, the skin may get damaged and may require some replacement in form of grafts. A good graft is supposed to cover and protect the intended place without causing any negative immune response. This promotes fast healing that reduces chances of scarring on the body [46].

In the recent past, different biomaterials like silk fibroin, cellulose alginate, collagen, polycaprolactone (PCL), polylactic acid (PLA), silicone, dextranelastin and polyethylene glycol(PEG) have been explored as possible cellular scaffolds for skin grafts and wound healing [73, 76, 77]. Among these biomaterials, silk has been used to mimic human skin as well as in wound healing. This is because silk has notable properties like low immune response, biodegradability, biocompatibility and is cost-effective [1].

Additionally, studies have proved that silk supports human keratinocytes and fibroblasts which are important in engineering artificial skin [46, 73, 78]. Studies by Chauhan et al. [59, 79] have reported successful use of electrospun silk fibroin from *A. assama* and *P. ricini* silk species in wound dressing. The studies also reported that a blend of electrospun silk fibron with polyvinyl alcohol promotes faster healing of wounds due to granulation during tissue formation. Other studies have demonstrated the use of electrospun silk fibroin from *A. assama* keratinocytes being successfully used in engineering artificial skin.

The therapeutic performance of SERI Surgical Scaffold has been studied; including open label clinical trials and case reports. A few studies have cited side effects such as poor scaffold integration (**Figure 5**), that have required surgical removal of the scaffold [80].

Comparing woven fabric, non-woven fabric and a film foam from silk fibroin in relation to cell culture responses by human oral keratinocytes, studies reported that water vapor-treated non-woven silk fibroin had better cell adhesion and dispersion of human fibroblasts and keratinocytes [46, 81]. This suggests that silk based biomaterials for tissue engineering requires a careful selection of fabrication techniques and material to blend with. Electrospinning is one of the preferred techniques for making non-woven nano-scale fiber mats for engineering artificial skin [51]. More results from electrospinning silk for tissue engineering include: electrospun silk fibroin scaffolds, 3D nonwoven scaffolds made from crosslinking silk fibroin with formic acid, and water vapor-treated silk fibroin nanofiber matrices among others. [46, 81, 82]. Reported blends that have been used successfully with silk for producing artificial skin include alginate, chitin, intermolecular cross-linked recombinant human-like collagen and biomimetic nanostructured collagen [46, 83–85].

3.4.2 Bone grafts

Today, various biomaterials are available for developing scaffold-based bone tissue. One of such material is silk fibroin which has good biological and physic-chemical properties— making it suitable for developing osteoinductive functional

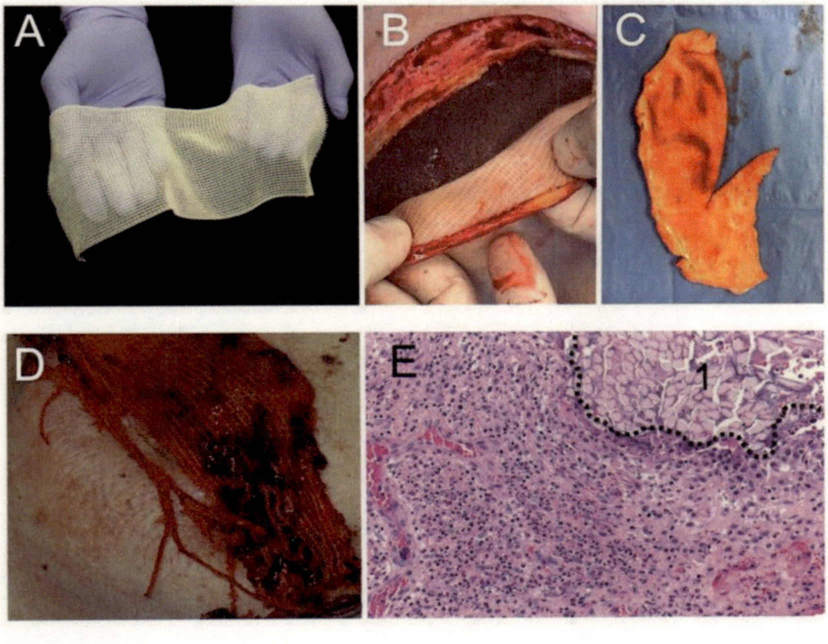

Figure 5.
Examples of SERI Surgical Scaffold implant loss in humans. A) Silk fibroin surgical mesh prior to implantation. B) Intraoperative view showing a free lying scaffold in the breast pocket. C) Retrieved scaffold surrounded with seroma. D) Interaoperative view of surgically removed scaffold with interpenetrated granulation tissue/scar plate (at >5 months), and E) histology of retrieved sample showing granulation tissue with neutrophiles and giant cells at the material (1) interface (dotted line). Reproduced with permission [80]. Copyright 2018, Elsevier.

bone grafts that resemble collagen [86, 87]. Silk fibroin from *A. mylitta* is reported to make porous scaffolds that mimic borne tissue [88]. Meinel et al. [89] induced osteogenic differentiation of human mesenchymal stromal cells in *B. Mori* silk fibroin to develop a bone graft. Other studies have explored blending *B. Mori* silk fibroin with hydroxyapatite to repair segmental bone defects [90, 91]. Findings by Reardon et al. [92] suggest that electrospun *B. mori* and *A. assama* silk fbroin blended with 70S bioactive glass repairs osteochondral tissue defects. Moreover, a study by Moses et al. [93] reports use of copper-doped bioactive glass silk composite matrices to repair large volume bone defects.

Fixation devices, including bone plates and bone screws have been manufactured from *B. mori* fibroin by casting in hexafluoroisopropanol, and formed into desired shaped (**Figure 6**) [94]. Silk screws tested in rats were well tolerated, showed early resorption and new bone formed around the threads of the screw. Such devices are easily malleable with hydration, allowing shaping for unique anatomical locations during surgery.

3.4.3 Artificial ligament and tendon

Tissue engineering for ligament and tendons requires biomaterials that are biodegradable, have good mechanical properties, good structural integrity,

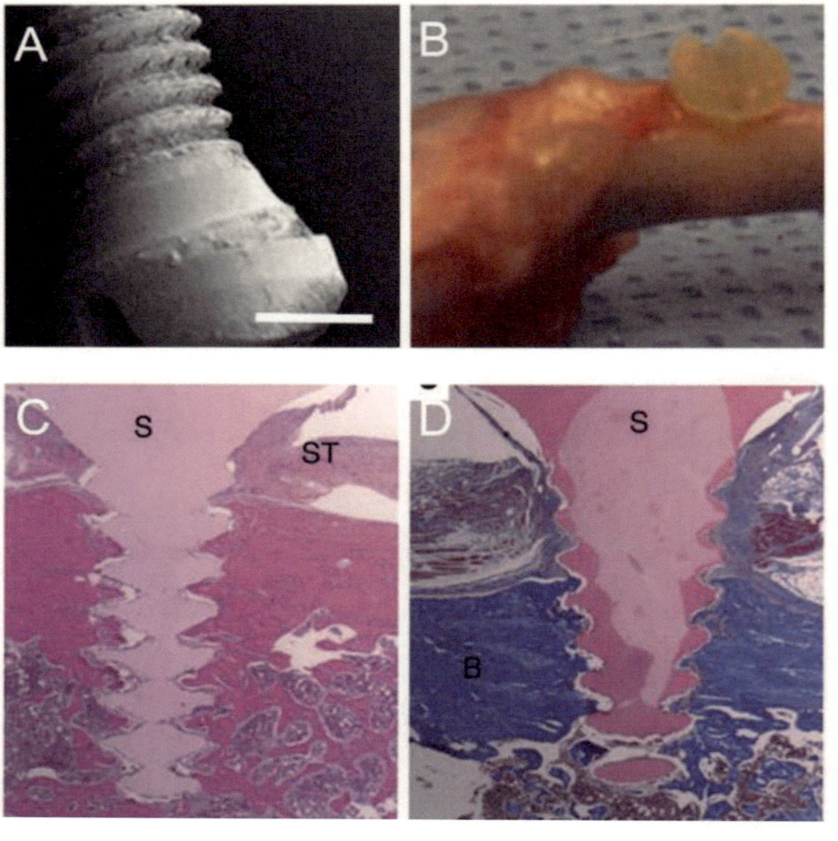

Figure 6.
Silk-based devices for fracture fixation. A) scanning electron microscopy image of a silk fibroin screw. Scale bar is 1 mm. B) Silk fibroin screw inserted into a rat femur at 4 weeks postsurgery. C, D) Cross-sec-tions of the silk fibroin screw inserted into a rat femur at 4 weeks post-surgery; sections stained with H&E and Masson's trichrome, respectively. Adapted with permission [94] Copyright 2014, Macmillan Publishers.

biocompatible and promote regeneration of new ligament and tendon tissues [46]. Silk is thus a suitable fiber that meets these requirements for performance and function [1, 93]. Weaving and braiding are reported as the most preferred techniques for making silk based ligament and tendon [95–97]. Other studies have reported crosslinking silk fibers with collagen matrix, coating poly(lacticco-glycolic acid) fibers with silk, blending silk fibers with fibroblast growth factor and transforming growth factor-β (TGF-β) in developing artificial ligament and tendon [46, 98–100].

3.4.4 Cardiac tissue

The most difficult part in cardiac tissue engineering is to perfectly mimic the original extracellular matrix. Patra et al. [101] and Stoppel et al. [102] reported to have successfully used scaffolds made *B. mori* and *A. mylitta* silk fibroins to treat myocardial infarction. Moreover, Mehrotra et al. [103] developed a 3-D cardiac construct made from stacking cell-laden silk films; the constructs proved to be good for cardiac tissue regeneration [73].

3.4.5 Liver modules

Different bio-artificial liver and cell therapies to treat liver diseases are available today. Cirillo et al. [104] developed a film from a blend of silk fibroin and collagen. A study by She et al. [105] examined a film made from silk fibroin, chitosan and heparin scaffolds that showed hepatocyte regeneration. Another study [106] reports scaffolds made from a blend of polylactic acid (PLA) and silk fibroin had a higher differentiation and proliferation as compared to scaffolds made from pure PLA alone. Likewise, a study by Janani et al. [107] reports that a functional liver can be fabricated from a blend of mulberry (*B. mori*) and non-mulberry (*A. assama*) silk fibroin.

3.4.6 Artificial pancreas

Different types of microspheres, hydrogels and nanoparticles have been developed to ensure a continuous release of insulin in diabetic patients [108]. In the recent past, islets have been encapsulated with biomaterials before they are transplanted to prevent immune response and to have a continuous insulin release [109]. A study by Davis et al. [110] reports encapsulating islets in silk hydrogel which improved the in vivo functions of the islets after transplanting. In another study, a bio-artificial pancreas was developed using silk alginate to encapsulate insulin secreting cells [73].

3.4.7 Artificial intervertebral disc

A perfect biomaterial for making artificial intervertebral disc must have high tensile strength, be biocompatible and be able to simulate the natural extracellular matrix [111]. A study by Park et al. [112] reports a biphasic hybrid scaffold that was developed from a blend of silk fibroin and hyaluronic acid to simulate the components of an intervertebral disc (nucleus pulposus (NP)) and an annulus fibrosus (AF)). In a related study, Du et al. [113] fabricated a 3-D biphasic silk fibroin scaffold to mimic the AF phase and phase separation technique for the NP phase. Moreover, Bhunia et al. [114] developed a bio-artificial AF construct with directional freezing technique involving concentric rings of lamellar silk scaffolds. The study further reports the proliferation of primary porcine AF cells using a mulberry and a non-mulberry silk combination which helped in cellular maturation, alignment and extracellular matrix deposition.

3.5 Silk in sutures

Sutures are an important material in surgical operations for primary wound closure. For this reason, various materials have been used in making sutures. These materials can be classified as either organic or synthetic according to their origin, or absorbable and non-absorbable according to their durability in the body [115]. Important properties for a good suture include: high tensile strength, elasticity, wound safety, knot safety and tissue reactivity [116, 117].

Silk, a natural non-absorbable suture material has been in use as a suture for several decades. However, other degradable synthetic sutures have dominated the market in the recent past. Nonetheless, silk suture is still preferred in cardiovascular, ocular and neural surgery because of its superior properties like good knot strength, ease of processing, minimum propensity to tear through tissue and biocompatibility [117].

Various modifications have been done on silk to improve its weak characteristics such as adding poly vinyl alcohol into silk fibroin to improve the tensile strength, elongation at break and the knot strength [118]. Bloch & Messores [119], reported coating silk filaments with fibroin and bounding them together to reduce the capillarity of silk sutures. Viju & Thilagavathi [120] coated silk-braided sutures with chitosan to improve the antimicrobial activity, tenacity and knot strength. Sudh et al. [121] developed a drug loaded antimicrobial silk suture for use in wound closure and wound healing meant to prevent surgical site infections. Choudhury et al. [122] developed a low-temperature O_2 plasma-treated (*Antheraea assama*) silk fibroin (AASF) yarn impregnated with amoxicillin trihydrate. This was aimed at producing a controlled antibiotic-releasing suture (AASF/O_2/AMOX) to prevent site bacterial infection and fasten wound healing. This shows the potential of silk in developing suture with special properties.

Type of Drug Delivery System/material	Associated active ingredient	Key results
Silk sponges	Erythromycin	Sustained drug release and prolonged antimicrobial activity against Staphilococcus Aureus
Silk films	Horseradish peroxidase (HRP)	Enhanced stability
	Glucose oxidase (GOx)	Increased enzymatic activity
	FITC-dextran	Controlled drug release
	Epirubicin	Controlled drug release
Silk lyogels	Hydrocortisone IgG	Enhanced efficacy Enhanced stability and sustained release
Insertable Silk discs	IgG and HIV inhibitor 5P12-RANTES	Enhanced stability and modified release profile
Silk nanoparticles	Curcumin	Modified release profile and enhanced cellular uptake
Silk microspheres	Horseradish peroxidase (HRP)	Modified the release profile
Silk coated PCL microspheres	Vancomycin	Modified the release profile
Silk coated liposomes	Ibuprofen	Enhanced adhesion to human corneal epithelial cells, tunable drug release
	Emodin	Selective targeting of keloid cells

Table 1.
Silk-based drug delivery systems [100].

3.6 Silk in drug delivery

Drug delivery through polymeric systems has gained popularity over the years [51]. These systems serve as reservoirs to active ingredient in drugs and improve the drug's physicochemical properties [123]. Polymeric drug delivery systems are also good in specific targeting, intracellular transport and some are biocompatible which help in improving efficiency of the treatment and the life quality of the patients [123, 124]. A good drug delivery system should be able to stabilize the active ingredient in drugs, be able to modulate the drug's release mechanism, be biocompatible and biodegradable, as well as minimize any side effects of tissue specific targeting of highly toxic drugs [125–127].

Silk fibroin is used in drug delivery systems owing to its properties such as good mechanical properties, mild aqueous processing conditions, biocompatibility, biodegradability and its ability to enhance the stability of active ingredient in drugs; as proteins and small molecules [46, 128]. That notwithstanding, silk fibroin solutions can be processed using various techniques to produce different forms of delivery systems like scaffolds, films, hydrogels, nanoparticles, microspheres, and microcapsules among others [129]. Additionally, silk fibroin has carboxyl and amino groups which allow bio-functionalization with different biomolecules for targeted drug delivery [130]. Silk based drug delivery systems include hydrogels, micro particles, lyophilized sponges, films, nano-fibers and nano-particles.

Formulation	Gene	Cell line
Recombinant silk–elastin-like polymer hydrogels (SELPs)	Ad[1]–CMV[2]–LacZ[3]	Head and neck cancer in mice
	pDNA[4] (pRL[5]-CMV-luc[6])	NA
	Ad–Luc–HSVtk[7]	Head and neck cancer in mice
3D porous scaffold	Adenovirus Ad-BMP7[8]	Human BMSCs
Bioengineered silk films	pDNA (GFP[9])	Human HEK cells
Spermine modified SF	pDNA and VEGF165–Ang-1[10]	In vivo-rat
SF-Coated PEI/DNA Complexes	pDNA (GFP)	HEK 293 and HCT 116 cells
SF layer-by-layer assembled microcapsules	pDNA-Cy5[11]	NIH/3 T3 fibroblasts
Bioengineered silk–polylysine–ppTG1 nanoparticles	pDNA	Human HEK and MDA-MB-435 cells
Magnetic-SF/polyethyleneimine core-shell nanoparticles	c-Myc[12] antisense ODNs[13y]	MDA-MB-231 cells

[1]*Adenovirus.*
[2]*Cytomegalovirus promoter gene.*
[3]*Beta galactosidase reporter gene.*
[4]*Plasmid DNA.*
[5]*Renilla luciferase.*
[6]*Luciferase reporter gene.*
[7]*Herpes simplex virus thymidine kinase gene.*
[8]*Bone morphogenic protein.*
[9]*Green fluorescent protein.*
[10]*Vascular endothelial growth factor and angiopoietin-1.*
[11]*Fluorescent probe.*
[12]*MYC Proto-Oncogene.*
[13]*Oligodeoxynucleotides [100].*

Table 2.
Silk –based gene delivery systems [100].

Different researches have reported successful use of silk fibroin in delivery systems for different drugs and genes [131–133]. **Tables 1** and **2** below presents some silk based drug and genes delivery systems.

4. Silk in protective clothing

Protective clothing are defined as textile structure designed to protect the human body from external threats such as fire, bullets, heat, cold, mechanical, biological, radiological, thermal and chemical hazards. Protective clothing are in different forms e.g. masks, gloves, vests, coats, aprons, hats, hoods or totally encapsulating chemical protective suits [134]. Some general characteristics of good protective clothing include: reliable barrier protection, durability, good fit, flexible, light weight, ease of care, maintenance and repair, ease of disposal and recycling.

Because of the interesting characteristics of silk, various research studies have examined its use in developing protective textiles. Some of these characteristics include hydrophobicity, antimicrobial and antiviral properties [135]. Recently, Parlin et al. [135] examined the potential of silk fabrics as a protective barrier for personal protective equipment and as a functional material for face coverings during the COVID-19 pandemic. Results of this study showed that the use of the commercially available 100% silk material can be used in producing protective coverings that can prolong the lifespan of N95 respirators. The study also found 100% silk fabrics suitable for developing face coverings for the general public to prevent COVID-19 [135]. Additionally, the study suggests that because silk has unique properties such as antimicrobial, antiviral, breathability, and slight hydrophobicity; prevention of penetration of droplets and antibacterial activity can imply potential use in developing respirator inserts [136, 137]. Moreover, other studies had showed that silk could be used as an antimicrobial barrier mask, with better filtration when multiple layers are used [135]. Besides, silk neither irritates the skin nor increases local humidity around the covered face, and prevents accidental stimulation of face touching; making it good for prolonged wear [138].

Another study by Zulan et al. [139] reports use of silk/graphene composite to make flame retardant protective clothing that can be used by fire fighters. Loh et al. [140] reports woven silk fabrics can be used for ballistic protection for aerospace, sports, military, marine and automotives. Mongkholrattanasit et al. [141] studied the ultraviolet (UV) protection properties of silk fabric dyed with eucalyptus leaf extract. Pad-dry and pad-batch techniques together with a metal mordant ($AlK(SO_4)_2$, $CuSO_4$, and $FeSO_4$) were used to apply a natural dye extracted from eucalyptus leaves on silk fabric. Results of his study showed that the UV protection factor of the silk fabric increased with an increase in the dye concentration and a darker shade gave the best UV-protective silk fabrics. Moreover, a study by Zhou et al. [142] also reports silk fabrics treated with red radish extracts provides good UV protection and that such fabrics can be used in making umbrellas, shade structures, awnings, and baby carrier covers among others.

5. Silk in optics and sensing

Synthetic biomaterials have been widely used in optics and sensing applications. For ophthalmic applications, which include lens replacement, retina reconstruction, vitreous replacement and ocular surface reconstruction, various materials such as poly-methylmethacrylate (PMMA), silicone, acrylics, poly-tetrafluoroethylene among others have been extensively used due to the biological inert nature of these materials [143, 144]. With technological advances, regenerative medicine strategies have shifted to relying on the ability of the biomaterial scaffolds in supporting human cell adhesion, growth and maintenance of the right cells that encourage tissue replacement as well as integration with adjacent tissues [145]. Since most synthetic biomaterials lacked the aforementioned abilities, emerging technologies attempted the modification of synthetic biomaterial surfaces [146]. However, a major limitation of surface modified synthetic materials was that such materials are not transparent, especially for applications in tissue grafts which need to be optically clear and they are not biodegradable [143]. Materials derived from nature have therefore become popular because they support cell attachment and proliferation. These materials include cross-linked collagen-chitosan hydrogels [147], keratin [148], cross-linked collagen gels [149], silk fibroin [150] etc.

Apart from ophthalmic applications, there has been an increased desire for real-time diagnostics, sensing and deep tissue light delivery, which has led to development of photonic medical devices from materials which are implantable and biocompatible. These devices can therefore be used within the body for therapeutics and long term health monitoring, where they are integrated into the living tissue in the human body [151]. Non-biodegradable inorganic materials such as silicon, gold and compound semiconductors have been traditionally used in photonic devices. However, their biocompatibility have been found to be dependent on the device size, the presence of coatings and mechanical properties [152]. Hydrogel materials from poly-vinylalcohol (PVA) and poly-ethylglycol (PEG), which are biocompatible have also been used in tissue engineering applications because of their ability to retain water and mimic the human body extracellular matrix [153]. However, they have not found extensive application in sensing because of their poor adhesion to substrates and poor mechanical properties [154]. Selection of the right material for implantable photonic devices requires consideration of biocompatibility properties as well as the structural stability, mechanical flexibility and optical clarity [151]; requirements that silk fibroin meet. Silk fibroin is thus gaining traction in optical interfaces and sensor applications in implantable biomedical fields owing to its good mechanical and optical transparency, coupled with its biodegradability and biocompatibility [14]. Silk in film form has a free standing structure with thickness ranging from 20 to 100 μm. The films are very transparent across the visible region of the spectrum and are mechanically robust with smooth surfaces. The films can also be patterned during fabrication to form traverse features that are tens of nanometers, making them attractive in optical device applications [155].

Substratum for corneal limbal epithelial cells has been developed from silk fibroin membranes, by casting dialyzed solutions of silk fibroin protein. The transparent silk membrane was found to support growth of human limbal epithelial (HLE) cell growth, which did not change even when the silk membranes were cast in the presence of fetal bovine serum (FBS) [156]. Such properties are favourable in the development of tissue engineered membranes for restoration of damaged ocular surfaces. Porous silk films have also been fabricated and shown to have potential in use as a carrier of cultivated epithelial sheets during regeneration of corneal epithelium [157].

Diffractive optical elements were fabricated by molding silk fibroin solution on poly-dimethylsiloxane (PDMS) moulds with ruled and holographic diffraction grating, producing nano-patterned silk optical elements of thickness ranging from 30 to 50 μm and a refractive index of $n = 1.55$. These nano-patterned silk gratings had a diffraction efficiency of 34%, at a wavelength of 633 nm in the first order, which compares to that of transmissive glass gratings. This led to successful formation of silk micro-lens arrays and silk lenses, which couple light into biological substrates [158]. Such silk gratings can also be functionalized, to maintain the biologically active optical elements. Therefore this could allow the use of these silk devices in delivering light to biological matrices and concentrate photons with doped substrates for biological function probes. In another study, silk diffraction gratings with desired patterns were fabricated through photo-induced polymerization of silk conjugates and a photo-initiator, producing good diffraction intensity [159].

Silk micro-prism arrays (MPAs) were prepared by micro-molding technique, resulting into a silk reflector film of 100 μm in thickness. The MPAs provided contrast and optical signal enhancement by retro-reflecting scattered photons through layers of tissue when used in vivo on BALB/c mice. The silk MPAs had no adverse biological effects, degraded slowly and were integrated into the native tissue. Functionalization of silk MPA with doxorubicin (a chemotherapeutic drug), was further reported to allow controlled delivery, storage and imaging of therapeutics, besides improving noninvasive tissue imaging [160].

Optical waveguides, which have the ability to transport and manipulate light in a controlled manner [161], have also been fabricated from silk. Silk fibroin ink, used in direct ink writing technique, has enabled creation of silk optical waveguides. These have been found to easily guide light of wavelength 633 nm. These waveguides were reported to exhibit comparable optical loss measurements to those of thin silk films, an indication that they can be applied in fabrication of functionalized, biocompatible and biodegradable biophotonic elements [155].

Silk fibroin hydrogels have also found use in surface plasmon resonances (SPR) sensors, fabricated by utilizing the principle of metal–insulator–metal (MIM) absorber. Inclusion of a thin insulator layer of 20 nm silk fibroin hydrogel between two 200 nm gold films enables the MIM structure produced to become highly sensitive to changes in thickness and refractive index of the insulating layer. Thus, the hydrogel properties of the silk spacer, which can accommodate water molecules by up to 60% in volume, increases sensitivity to analytes. Sensitivity is dependent on the refractive index and swelling ratio of the silk hydrogel. The silk polymer chains can also act as fluidic channels that facilitate flow of analytes in water, through a nano-sized layer, making silk plasmonic structures suitable for glucose sensor applications [162].

A wearable strain sensor was fabricated by carbonizing pristine silk georgette through high temperature treatment, followed by encapsulation in poly-dimethylsiloxane (PDMS), an elastic polymer. This has shown promising potential for applications in monitoring a wide range of motion based human activities [163]. Silk based wearable sensors utilizes the principle of transformation of silk fibroin through thermal treatment, into an electrically-conductive graphite nano-carbon [164]. Transparent and flexible silk nanofiber-derived carbon membranes have also been fabricated for multifunctional electronic skin with human physiological signal monitoring capabilities [165, 166]. Silk based self-powered pressure sensor films for use in wearable devices have also been fabricated through synthesis of silk and poly-vinylidene fluoride-co-trifluoroethylene [167]. In order to provide a strong interface between a biological surface and a sensor for epidermal electronics, calcium modified silk fibroin has been fabricated and shown to have strong adhesive properties with good stretchability, conductivity and reusability. Therefore calcium modified silk fibroin shows the potential to be applied as an adhesive for epidermal biomedical sensors [168].

Another promising application of silk is in the coating of otherwise non biocompatible optical fibers for bio-sensing inside the human body. Silica exposed core fiber are reported to have been coated with a thin layer of silk and thereafter, doped with fluorophore 5,6-carboxynapthofluorescein (CNF). The doped-silk layer was found to produce fluorescent signals that are coupled into the core of Silica exposed core fiber, allows for remote measurement of pH along the fiber length, when used in mice [169].

6. Conclusion

Silk fiber from different varieties has largely been used beyond the traditional textile scope. The widest and earlier use has been noted in biomedical use, especially as sutures, and protective wear due to the enviable properties highlighted for each function. The traditional classification of silk was tagged to luxury. Beyond this, research has been expounded on the functionality of silk. The various forms, including regeneration into nanofibers, nanofilms and nanomembranes provide surfaces for novel functionalization when processed with specific agents. Collagen has been reported the most as a functional material added to silk for, especially biomedical applications. Optics and sensing, present a unique and promising future for functional silk— especially in e-textiles and bio-sensing. However, it is also important to underscore that at different stages of regeneration, the silk structure seems to get altered; especially the loss of considerable strength resulting from altered crystallinity and re-orientation of β-sheets of silk fibroin. Owing to the low proportion of silk production on the market compared to cotton, and synthetic fibers, it is important to explore the annual global demand of silk in regard to future needs for silk in functional textiles. It is also important to explore statistics, on silk processed through novel methods like electrospinning, with respect to commercial viability. For instance, it is often required to strictly control biomaterial properties during processing, owing to the complexity of biomaterial molecules. Of important focus is the standardization of process/manufacturing parameters and equipment in the attempt to commercialize silk functional products. However, with increasing demand for more environmentally sustainable materials and products, more bio-based sectors and economies will emerge; hence, an increased uptake of natural biomaterials such as silk, in higher technology application needs.

Conflict of interest

The authors have no conflicts of interest to declare.

Author details

Edison Omollo Oduor[1*], Lucy Wanjiru Ciera[1] and Edwin Kamalha[2]

1 Technical University of Kenya, Nairobi, Kenya

2 Busitema University, Tororo, Uganda

*Address all correspondence to: edisonomollo@gmail.com

References

[1] Altman G, Diaz F, Jakuba C, Calabro T, Horan R, Chen J, et al. Silk-based biomaterials. Biomaterials. 2003;24(3):401-16.

[2] Craig C, Hsu M, Kaplan D, Pierce NE. A comparison of the composition of silk proteins produced by spiders and insects. International Journal of Biological Macromolecules. 1999;24(2-3):109-18.

[3] Babu M. Silk–production and future trends. Handbook of Natural Fibres: Elsevier; 2020. p. 121-45.

[4] Singh T. Sericulture. 2007.

[5] Capinera JL. Sericulture. Encyclopedia of Entomology: Springer; 2008. p. 3345-8.

[6] Merlange G. Silk in the orient. Journal of the Royal Central Asian Society. 1939;26(1):65-76.

[7] Lee Y-W. Silk Reeling and Testing Manual: Food and Agriculture Organization; 1999.

[8] Padaki NV, Das B, Basu A. 1 - Advances in understanding the properties of silk. In: Basu A, editor. Advances in Silk Science and Technology: Woodhead Publishing; 2015. p. 3-16.

[9] Hanumappa H. Sericulture for Rural development: Himalaya Publishing House; 1986.

[10] Padaki N, Das B, Basu A. Advances in understanding the properties of silk. Advances in silk science and technology: Elsevier; 2015. p. 3-16.

[11] Gupta V, Rajkhowa R, Kothari V. Physical characteristics and structure of Indian silk fibres. Indian Journal of Fibre and Textile research. 2000;25(1):14-9.

[12] Hearle JW, Morton WE. Physical properties of textile fibres: Elsevier; 2008.

[13] Volkov V, Ferreira A, Cavaco-Paulo A. On the Routines of Wild-Type Silk Fibroin Processing Toward Silk-Inspired Materials: A Review. Macromolecular Materials and Engineering. 2015;300(12):1199-216.

[14] Omenetto F, Kaplan D. A new route for silk. Nature Photonics. 2008;2(11):641-3.

[15] Yucel T, Lovett M, Kaplan D. Silk-based biomaterials for sustained drug delivery. Journal of Controlled Release. 2014;190:381-97.

[16] Jiang C, Wang X, Gunawidjaja R, Lin YH, Gupta M, Kaplan D, et al. Mechanical properties of robust ultrathin silk fibroin films. Advanced functional materials. 2007;17(13):2229-37.

[17] Wang X, Kim HJ, Xu P, Matsumoto A, Kaplan D. Biomaterial coatings by stepwise deposition of silk fibroin. Langmuir. 2005;21(24): 11335-41.

[18] Wang X, Kluge J, Leisk G, Kaplan D. Sonication-induced gelation of silk fibroin for cell encapsulation. Biomaterials. 2008;29(8):1054-64.

[19] Jin H-J, Fridrikh S, Rutledge G, Kaplan D. Electrospinning *Bombyx mori* silk with poly (ethylene oxide). Biomacromolecules. 2002;3(6):1233-9.

[20] Kim U-J, Park J, Li C, Jin H-J, Valluzzi R, Kaplan D. Structure and properties of silk hydrogels. Biomacromolecules. 2004;5(3):786-92.

[21] Jin HJ, Park J, Karageorgiou V, Kim UJ, Valluzzi R, Cebe P, et al. Water-stable silk films with reduced

β-sheet content. Advanced Functional Materials. 2005;15(8):1241-7.

[22] Bettinger C, Bao Z. Biomaterials-based organic electronic devices. Polymer international. 2010;59(5):563-7.

[23] Capelli R, Amsden J, Generali G, Toffanin S, Benfenati V, Muccini M, et al. Integration of silk protein in organic and light-emitting transistors. Organic electronics. 2011;12(7):1146-51.

[24] Toffanin S, Kim S, Cavallini S, Natali M, Benfenati V, Amsden J, et al. Low-threshold blue lasing from silk fibroin thin films. Applied Physics Letters. 2012;101(9):091110.

[25] Kim D-H, Kim Y-S, Amsden J, Panilaitis B, Kaplan D, Omenetto F, et al. Silicon electronics on silk as a path to bioresorbable, implantable devices. Applied physics letters. 2009;95(13):133701.

[26] Nambajjwe C, Musinguzi WB, Rwahwire S, Kasedde A, Namuga C, Nibikora I. Improving electricity from silk cocoons through feeding silkworms with silver nanoparticles. Materials Today: Proceedings. 2020;28:1221-6.

[27] Tansil N, Li Y, Teng CP, Zhang S, Win KY, Chen X, et al. Intrinsically colored and luminescent silk. Advanced Materials. 2011;23(12):1463-6.

[28] Tansil N, Koh LD, Han MY. Functional silk: colored and luminescent. Advanced Materials. 2012;24(11):1388-97.

[29] Sagnella A, Chieco C, Virgilio ND, Toffanin S, Posati T, Pistone A, et al. Bio-doping of regenerated silk fibroin solution and films: a green route for biomanufacturing. RSC Advances. 2014;4(64):33687-94.

[30] Benfenati V, Toffanin S, Chieco C, Sagnella A, Virgilio ND, Posati T, et al. Silk fibroin based technology for industrial biomanufacturing. Factories of the Future: Springer, Cham; 2019. p. 409-30.

[31] Koh L-D, Cheng Y, Teng C-P, Khin Y-W, Loh X-J, Tee S-Y, et al. Structures, mechanical properties and applications of silk fibroin materials. Progress in Polymer Science. 2015;46:86-110.

[32] Luo K, Yang Y, Shao Z. Physically crosslinked biocompatible silk-fibroin-based hydrogels with high mechanical performance. Advanced Functional Materials. 2016;26(6):872-80.

[33] Ghosh S, Parker S, Wang X, Kaplan D, Lewis J. Direct-write assembly of microperiodic silk fibroin scaffolds for tissue engineering applications. Advanced Functional Materials. 2008;18(13):1883-9.

[34] Brian LD. Processing of *Bombyx mori* silk for biomedical applications. In: Kundu SC, editor. Silk Biomaterials for Tissue Engineering and Regenerative Medicine: Elsevier; 2014. p. 78-99.

[35] Kamalha E, Zheng YS, Zeng YC, Mwasiagi JI. Effect of solvent concentration on morphology of electrospun *bombyx mori* silk. 2014.

[36] Niu H, Wang X, Lin T. Needleless electrospinning: developments and performances. Nanofibers-production, properties and functional applications. 2011:17-36.

[37] Lukáš D, Sarkar A, Martinová L, Vodseďálková K, Lubasová D, Chaloupek J, et al. Physical principles of electrospinning (Electrospinning as a nano-scale technology of the twenty-first century). Textile progress. 2009;41(2):59-140.

[38] Niu H, Lin T, Wang X. Needleless electrospinning. I. A comparison of cylinder and disk nozzles. Journal of applied polymer science. 2009;114(6):3524-30.

[39] Yarin A, Zussman E. Upward needleless electrospinning of multiple nanofibers. Polymer. 2004;45(9):2977-80.

[40] Sasithorn N, Martinová L, Horáková J, Mongkholrattanasit R. Fabrication of silk fibroin nanofibres by needleless electrospinning. In: Haider S, Haider A, editors. Electrospinning-Material, Techniques, and Biomedical Applications; Intech: London, UK. Croatia: InTech; 2016. p. 95-113.

[41] Wei L, Sun R, Liu C, Xiong J, Qin X. Mass production of nanofibers from needleless electrospinning by a novel annular spinneret. Materials & Design. 2019;179:107885.

[42] Holland C, Numata K, Rnjak-Kovacina J, Seib P. The biomedical use of silk: past, present, future. Advanced healthcare materials. 2019;8(1):1800465.

[43] Qin Y. 2–An overview of medical textile products. Medical Textile Materials Jiaxing: Woodhead Publishing. 2016:13-22.

[44] Kennedy JF, Bunko K, Santhini E, Vadodaria K, Rajasekar S. The use of 'smart'textiles for wound care. Advanced Textiles for Wound Care: Elsevier; 2019. p. 289-311.

[45] Daniele M, Boyd D, Adams A, Ligler F. Microfluidic strategies for design and assembly of microfibers and nanofibers with tissue engineering and regenerative medicine applications. Advanced healthcare materials. 2015;4(1):11-28.

[46] Li G, Li Y, Chen G, He J, Han Y, Wang X, et al. Silk-based biomaterials in biomedical textiles and fiber-based implants. Advanced healthcare materials. 2015;4(8):1134-51.

[47] Rajendran S, Anand SC. Woven textiles for medical applications. Woven Textiles: Elsevier; 2020. p. 441-70.

[48] Chen X, Guan Y, Wang L, Sanbhal NA, Zhao F, Zou Q, et al. Antimicrobial textiles for sutures, implants, and scaffolds. Antimicrobial Textiles: Elsevier; 2016. p. 263-85.

[49] Syed M, Khan M, Sefat F, Khurshid Z, Zafar M, Khan A. Bioactive glass and glass fiber composite: biomedical/dental applications. Biomedical, Therapeutic and Clinical Applications of Bioactive Glasses: Elsevier; 2019. p. 467-95.

[50] Kluge J, Li A, Kahn B, Michaud D, Omenetto F, Kaplan D. Silk-based blood stabilization for diagnostics. Proceedings of the National Academy of Sciences. 2016;113(21):5892-7.

[51] Kamalha E, Zheng YS, Zeng YC, Mutua F, editors. FTIR and WAXD study of regenerated silk fibroin. Advanced Materials Research; 2013: Trans Tech Publ.

[52] Wharram S, Zhang X, Kaplan D, McCarthy S. Electrospun silk material systems for wound healing. Macromolecular Bioscience. 2010;10(3):246-57.

[53] Farokhi M, Mottaghitalab F, Fatahi Y, Khademhosseini A, Kaplan D. Overview of silk fibroin use in wound dressings. Trends in biotechnology. 2018;36(9):907-22.

[54] Xia Y, Gao G, Li Y. Preparation and properties of nanometer titanium dioxide/silk fibroin blend membrane. Journal of Biomedical Materials Research Part B: Applied Biomaterials: An Official Journal of The Society for Biomaterials, The Japanese Society for Biomaterials, and The Australian Society for Biomaterials and the Korean Society for Biomaterials. 2009;90(2):653-8.

[55] Wang M, Yu J, Kaplan D, Rutledge G. Production of submicron diameter silk fibers under benign

processing conditions by two-fluid electrospinning. Macromolecules. 2006;39(3):1102-7.

[56] Kamalathevan P, Ooi P, Loo Y. Silk-based biomaterials in cutaneous wound healing: a systematic review. Advances in skin & wound care. 2018;31(12):565-73.

[57] Chouhan D, Mandal B. Silk biomaterials in wound healing and skin regeneration therapeutics: From bench to bedside. Acta Biomaterialia. 2020;103:24-51.

[58] He S, Shi D, Han Z, Dong Z, Xie Y, Zhang F, et al. Heparinized silk fibroin hydrogels loading FGF1 promote the wound healing in rats with full-thickness skin excision. Biomedical engineering online. 2019;18(1):1-12.

[59] Chouhan D, Janani G, Chakraborty B, Nandi S, Mandal B. Functionalized PVA–silk blended nanofibrous mats promote diabetic wound healing via regulation of extracellular matrix and tissue remodelling. Journal of tissue engineering and regenerative medicine. 2018;12(3):e1559-e70.

[60] Hung M-H, Sartika D, Chang S-J, Chen S-J, Wang C-C, Hung Y-J, et al. Influence of silk clothing therapy in patients with atopic dermatitis. Dermatology Reports. 2019;11(2).

[61] Thomas K, Bradshaw L, Sach T, Batchelor J, Lawton S, Harrison E, et al. Silk garments plus standard care compared with standard care for treating eczema in children: A randomised, controlled, observer-blind, pragmatic trial (CLOTHES Trial). PLoS medicine. 2017;14(4):e1002280.

[62] Criton S, Gangadharan G. Nonpharmacological management of atopic dermatitis. Indian Journal of Paediatric Dermatology. 2017;18(3):166.

[63] Hermanns J-F, Goffin V, Arrese J, Rodriguez C, Piérard G. Beneficial effects of softened fabrics on atopic skin. Dermatology. 2001;202(2):167-70.

[64] Agner T. Staphylococcal-mediated worsening of atopic dermatitis: many players involved. The British journal of dermatology. 2010;163(6):1147.

[65] Macias E, Pereira F, Rietkerk W, Safai B. Superantigens in dermatology. Journal of the American Academy of Dermatology. 2011;64(3):455-72.

[66] Sha L-Z, Zhao H-F, Xiao G-N. Photocatalytic degradation of formaldehyde by silk mask paper loading nanometer titanium dioxide. Fibers and Polymers. 2013;14(6):976-81.

[67] Sha L, Zhao H. Preparation and properties of Nano-TiO 2 photo-catalytic silk respirator paper. Fibers and Polymers. 2012;13(9):1159-64.

[68] Lawrence B, Pan Z, Liu A, Kaplan D, Mark Rosenblatt. Human corneal limbal epithelial cell response to varying silk film geometric topography in vitro. Acta biomaterialia. 2012;8(10):3732-43.

[69] Lawrence B, Wharram S, Kluge J, Leisk G, Omenetto F, Rosenblatt M, et al. Effect of hydration on silk film material properties. Macromolecular bioscience. 2010;10(4):393-403.

[70] Levenberg S, Khademhosseini A, Langer R. Embryonic stem cells in tissue engineering. Essentials of Stem Cell Biology: Elsevier; 2009. p. 571-81.

[71] Caddeo S, Boffito M, Sartori S. Tissue engineering approaches in the design of healthy and pathological in vitro tissue models. Frontiers in bioengineering and biotechnology. 2017;5:40.

[72] Song G. Improving comfort in clothing: Elsevier; 2011.

[73] Bandyopadhyay A, Chowdhury SK, Dey S, Moses JC, Mandal B. Silk: a promising biomaterial opening new vistas towards affordable healthcare solutions. Journal of the Indian Institute of Science. 2019:1-43.

[74] Behrens MR, Ruder WC. Biopolymers in Regenerative Medicine: Overview, Current Advances, and Future Trends. Biopolymers for Biomedical and Biotechnological Applications. 2021:357-80.

[75] Groeber F, Holeiter M, Hampel M, Hinderer S, Schenke-Layland K. Skin tissue engineering—in vivo and in vitro applications. Advanced drug delivery reviews. 2011;63(4-5):352-66.

[76] Vasconcelos A, Cavaco-Paulo A. Wound dressings for a proteolytic-rich environment. Applied microbiology and biotechnology. 2011;90(2):445-60.

[77] Abrigo M, McArthur S, Kingshott P. Electrospun nanofibers as dressings for chronic wound care: advances, challenges, and future prospects. Macromolecular bioscience. 2014;14(6):772-92.

[78] Zhang X, Reagan M, Kaplan D. Electrospun silk biomaterial scaffolds for regenerative medicine. Advanced drug delivery reviews. 2009;61(12):988-1006.

[79] Chouhan D, Chakraborty B, Nandi S, Mandal B. Role of non-mulberry silk fibroin in deposition and regulation of extracellular matrix towards accelerated wound healing. Acta biomaterialia. 2017;48:157-74.

[80] Arjen van Turnhout, Franke C, Vriens-Nieuwenhuis E, Sluis Wvd. The use of SERI™ Surgical Scaffolds in direct-to-implant reconstruction after skin-sparing mastectomy: A retrospective study on surgical outcomes and a systematic review of current literature. Journal of Plastic, Reconstructive & Aesthetic Surgery. 2018;71(5):644-50.

[81] Min B-M, Lee G, Kim SH, Nam YS, Lee TS, Park WH. Electrospinning of silk fibroin nanofibers and its effect on the adhesion and spreading of normal human keratinocytes and fibroblasts in vitro. Biomaterials. 2004;25(7-8): 1289-97.

[82] Min BM, Jeong L, Lee KY, Park WH. Regenerated silk fibroin nanofibers: Water vapor-induced structural changes and their effects on the behavior of normal human cells. Macromolecular Bioscience. 2006;6(4):285-92.

[83] Yoo CR, Yeo I-S, Park KE, Park JH, Lee SJ, Park WH, et al. Effect of chitin/silk fibroin nanofibrous bicomponent structures on interaction with human epidermal keratinocytes. International journal of biological macromolecules. 2008;42(4):324-34.

[84] Roh D-H, Kang S-Y, Kim J-Y, Kwon Y-B, Kweon HY, Lee K-G, et al. Wound healing effect of silk fibroin/alginate-blended sponge in full thickness skin defect of rat. Journal of Materials Science: Materials in Medicine. 2006;17(6):547-52.

[85] Wang G, Hu X, Lin W, Dong C, Wu H. Electrospun PLGA–silk fibroin–collagen nanofibrous scaffolds for nerve tissue engineering. In Vitro Cellular & Developmental Biology-Animal. 2011;47(3):234-40.

[86] Wittmer C, Claudepierre T, Reber M, Wiedemann P, Garlick J, Kaplan D, et al. Multifunctionalized electrospun silk fibers promote axon regeneration in the central nervous system. Advanced functional materials. 2011;21(22):4232-42.

[87] Salgado A, Coutinho O, Reis R. Bone tissue engineering: state of the art and future trends. Macromolecular bioscience. 2004;4(8):743-65.

[88] Mandal B, Kundu S. Osteogenic and adipogenic differentiation of rat bone marrow cells on non-mulberry and mulberry silk gland fibroin 3D scaffolds. Biomaterials. 2009;30(28):5019-30.

[89] Meinel L, Hofmann S, Betz O, Fajardo R, Merkle H, Langer R, et al. Osteogenesis by human mesenchymal stem cells cultured on silk biomaterials: comparison of adenovirus mediated gene transfer and protein delivery of BMP-2. Biomaterials. 2006;27(28):4993-5002.

[90] Leukers B, Gülkan H, Irsen S, Milz S, Tille C, Schieker M, et al. Hydroxyapatite scaffolds for bone tissue engineering made by 3D printing. Journal of Materials Science: Materials in Medicine. 2005;16(12):1121-4.

[91] Liu H, Fan H, Toh S, Goh J. A comparison of rabbit mesenchymal stem cells and anterior cruciate ligament fibroblasts responses on combined silk scaffolds. Biomaterials. 2008;29(10):1443-53.

[92] Reardon PJ, Konwarh R, Knowles J, Mandal B. Mimicking hierarchical complexity of the osteochondral interface using electrospun silk-bioactive glass composites. ACS applied materials & interfaces. 2017;9(9): 8000-13.

[93] Wang Y, Kim H-J, Vunjak-Novakovic G, Kaplan D. Stem cell-based tissue engineering with silk biomaterials. Biomaterials. 2006;27(36):6064-82.

[94] Perrone G, Leisk G, Lo T, Moreau JE, Haas DS, Papenburg BJ, et al. The use of silk-based devices for fracture fixation. Nature communications. 2014;5(1):1-9.

[95] Altman G, Horan R, Lu H, Moreau J, Martin I, Richmond J, et al. Silk matrix for tissue engineered anterior cruciate ligaments. Biomaterials. 2002;23(20):4131-41.

[96] Hairfield-Stein M, England C, Paek H, Gilbraith K, Dennis R, Boland E, et al. Development of self-assembled, tissue-engineered ligament from bone marrow stromal cells. Tissue engineering. 2007;13(4):703-10.

[97] Liu L, Liu J, Wang M, Min S, Cai Y, Zhu L, et al. Preparation and characterization of nano-hydroxyapatite/silk fibroin porous scaffolds. Journal of Biomaterials Science, Polymer Edition. 2008;19(3):325-38.

[98] Chen J, Altman G, Karageorgiou V, Horan R, Collette A, Volloch V, et al. Human bone marrow stromal cell and ligament fibroblast responses on RGD-modified silk fibers. Journal of Biomedical Materials Research Part A: An Official Journal of The Society for Biomaterials, The Japanese Society for Biomaterials, and The Australian Society for Biomaterials and the Korean Society for Biomaterials. 2003;67(2):559-70.

[99] Moreau J, Chen J, Horan R, Kaplan D, Altman G. Sequential growth factor application in bone marrow stromal cell ligament engineering. Tissue engineering. 2005;11(11-12):1887-97.

[100] Panas-Perez E, Gatt C, Dunn M. Development of a silk and collagen fiber scaffold for anterior cruciate ligament reconstruction. Journal of Materials Science: Materials in Medicine. 2013;24(1):257-65.

[101] Stoppel W, Hu D, Domian I, Kaplan D, III LB. Anisotropic silk biomaterials containing cardiac extracellular matrix for cardiac tissue engineering. Biomedical materials. 2015;10(3):034105.

[102] Patra C, Talukdar S, Novoyatleva T, Velagala S, Mühlfeld C, Kundu B, et al. Silk protein fibroin from Antheraea mylitta for cardiac tissue engineering. Biomaterials. 2012;33(9):2673-80.

[103] Mehrotra S, Nandi SK, Mandal B. Stacked silk-cell monolayers as a biomimetic three dimensional construct for cardiac tissue reconstruction. Journal of Materials Chemistry B. 2017;5(31):6325-38.

[104] Cirillo B, Morra M, Catapano G. Adhesion and function of rat liver cells adherent to silk fibroin/collagen blend films. The International journal of artificial organs. 2004;27(1):60-8.

[105] She Z, Liu W, Feng Q. Silk fibroin/chitosan/heparin scaffold: preparation, antithrombogenicity and culture with hepatocytes. Polymer International. 2010;59(1):55-61.

[106] She Z, Jin C, Huang Z, Zhang B, Feng Q, Xu Y. Silk fibroin/chitosan scaffold: preparation, characterization, and culture with HepG2 cell. Journal of Materials Science: Materials in Medicine. 2008;19(12):3545-53.

[107] Janani G, Nandi S, Mandal B. Functional hepatocyte clusters on bioactive blend silk matrices towards generating bioartificial liver constructs. Acta biomaterialia. 2018;67:167-82.

[108] Qiu Y, Park K. Environment-sensitive hydrogels for drug delivery. Advanced drug delivery reviews. 2001;53(3):321-39.

[109] Vaithilingam V, Tuch B. Islet transplantation and encapsulation: an update on recent developments. The review of diabetic studies: RDS. 2011;8(1):51.

[110] Davis N, Beenken-Rothkopf L, Mirsoian A, Kojic N, Kaplan D, Barron A, et al. Enhanced function of pancreatic islets co-encapsulated with ECM proteins and mesenchymal stromal cells in a silk hydrogel. Biomaterials. 2012;33(28):6691-7.

[111] Stergar J, Gradisnik L, Velnar T, Maver U. Intervertebral disc tissue engineering: A brief review. Bosnian journal of basic medical sciences. 2019;19(2):130.

[112] Park S-H, Gil ES, Cho H, Mandal B, Tien L, Min B-H, et al. Intervertebral disk tissue engineering using biphasic silk composite scaffolds. Tissue Engineering Part A. 2012;18(5-6):447-58.

[113] Du L, Zhu M, Yang Q, Zhang J, Ma X, Kong D, et al. A novel integrated biphasic silk fibroin scaffold for intervertebral disc tissue engineering. Materials Letters. 2014;117:237-40.

[114] Bhunia B, Kaplan D, Mandal B. Silk-based multilayered angle-ply annulus fibrosus construct to recapitulate form and function of the intervertebral disc. Proceedings of the National Academy of Sciences. 2018;115(3):477-82.

[115] Silverstein L, Kurtzman G, Shatz P. Suturing for optimal soft-tissue management. Journal of Oral Implantology. 2009;35(2):82-90.

[116] Lilly G, Armstrong J, Salem J, Cutcher J. Reaction of oral tissues to suture materials: Part II. Oral Surgery, Oral Medicine, Oral Pathology. 1968;26(4):592-9.

[117] Chu C. Types and properties of surgical sutures. Biotextiles as medical implants: Elsevier; 2013. p. 231-73.

[118] Lee KH, Baek DH, Ki CS, Park YH. Preparation and characterization of wet spun silk fibroin/poly (vinyl alcohol) blend filaments. International journal of biological macromolecules. 2007;41(2):168-72.

[119] Bloch A, Messores A, inventors; Ethicon Inc., assignee. Silk suture. United States of America1969.

[120] Viju S, Thilagavathi G. Effect of chitosan coating on the characteristics

of silk-braided sutures. Journal of Industrial Textiles. 2013;42(3):256-68.

[121] Sudha D, Dhurai B, Ponthangam T. Development of herbal drug loaded antimicrobial silk suture. Indian Journal of Fibre & Textile Research (IJFTR). 2017;42(3):286-90.

[122] Choudhury AJ, Gogoi D, Chutia J, Kandimalla R, Kalita S, Kotoky J, et al. Controlled antibiotic-releasing Antheraea assama silk fibroin suture for infection prevention and fast wound healing. Surgery. 2016;159(2):539-47.

[123] Tomeh MA, Hadianamrei R, Zhao X. Silk fibroin as a functional biomaterial for drug and gene delivery. Pharmaceutics. 2019;11(10):494.

[124] Luo Z, Li J, Qu J, Sheng W, Yang J, Li M. Cationized *Bombyx mori* silk fibroin as a delivery carrier of the VEGF165–Ang-1 coexpression plasmid for dermal tissue regeneration. Journal of Materials Chemistry B. 2019;7(1):80-94.

[125] Torchilin V. Multifunctional, stimuli-sensitive nanoparticulate systems for drug delivery. Nature reviews Drug discovery. 2014;13(11):813-27.

[126] Yin Z, Kuang D, Wang S, Zheng Z, Yadavalli V, Lu S. Swellable silk fibroin microneedles for transdermal drug delivery. International journal of biological macromolecules. 2018;106:48-56.

[127] Rezaei F, Damoogh S, Reis R, Kundu S, Mottaghitalab F, Farokhi M. Dual drug delivery system based on pH-sensitive silk fibroin/alginate nanoparticles entrapped in PNIPAM hydrogel for treating severe infected burn wound. Biofabrication. 2020;13(1):015005.

[128] Choi M, Choi D, Hong J. Multilayered controlled drug release silk fibroin nanofilm by manipulating secondary structure. Biomacromolecules. 2018;19(7):3096-103.

[129] Seib P. Reverse-engineered silk hydrogels for cell and drug delivery. Therapeutic delivery. 2018;9(6):469-87.

[130] Vepari C, Kaplan D. Silk as a biomaterial. Progress in polymer science. 2007;32(8-9):991-1007.

[131] Moses JC, Nandi SK, Mandal B. Multifunctional cell instructive silk-bioactive glass composite reinforced scaffolds toward osteoinductive, proangiogenic, and resorbable bone grafts. Advanced healthcare materials. 2018;7(10):1701418.

[132] Lv Q, Hu K, Feng Q, Cui F, Cao C. Preparation and characterization of PLA/fibroin composite and culture of HepG2 (human hepatocellular liver carcinoma cell line) cells. Composites Science and Technology. 2007;67(14):3023-30.

[133] Li A, Kluge J, Guziewicz N, Omenetto F, Kaplan D. Silk-based stabilization of biomacromolecules. Journal of Controlled Release. 2015;219:416-30.

[134] Gorji M, Bagherzadeh R, Fashandi H. Electrospun nanofibers in protective clothing. Electrospun nanofibers: Elsevier; 2017. p. 571-98.

[135] Parlin A, Stratton S, Culley T, Guerra P. A laboratory-based study examining the properties of silk fabric to evaluate its potential as a protective barrier for personal protective equipment and as a functional material for face coverings during the COVID-19 pandemic. PloS one. 2020;15(9):e0239531.

[136] Dong Z, Song Q, Zhang Y, Chen S, Zhang X, Zhao P, et al. Structure,

evolution, and expression of antimicrobial silk proteins, seroins in Lepidoptera. Insect biochemistry and molecular biology. 2016;75:24-31.

[137] Borkow G, Zhou S, Page T, Gabbay J. A novel anti-influenza copper oxide containing respiratory face mask. PLoS One. 2010;5(6): e11295.

[138] Parlin A, Stratton S, Culley T, Guerra PA. Silk fabric as a protective barrier for personal protective equipment and as a functional material for face coverings during the COVID-19 pandemic. medRxiv. 2020.

[139] Zulan L, Zhi L, Lan C, Sihao C, Dayang W, Fangyin D. Reduced graphene oxide coated silk fabrics with conductive property for wearable electronic textiles application. Advanced Electronic Materials. 2019;5(4):1800648.

[140] Loh K, Tan W, Oh R, editors. Developing Woven Enhanced Silk Fabric for Ballistic Protection. Solid State Phenomena; 2012: Trans Tech Publ.

[141] Mongkholrattanasit R, Kryštůfek J, Wiener J, Viková M. UV protection properties of silk fabric dyed with eucalyptus leaf extract. The Journal of The Textile Institute. 2011;102(3):272-9.

[142] Zhou Y, Yang Z-Y, Tang R-C. Facile and green preparation of bioactive and UV protective silk materials using the extract from red radish (*Raphanus sativus* L.) through adsorption technique. Arabian Journal of Chemistry. 2020;13(1):3276-85.

[143] Lace R, Murray-Dunning C, Williams R. Biomaterials for ocular reconstruction. Journal of Materials Science. 2015;50(4):1523-34.

[144] Williams R, Wong D. Ophthalmic biomaterials. Biomedical Materials: Springer; 2009. p. 327-47.

[145] Mason S, Rosalind Stewart, Kearns V, Williams R, Sheridan C. Ocular epithelial transplantation: current uses and future potential. Regenerative medicine. 2011;6(6):767-82.

[146] Nguyen P, Yiu S. Ocular surface reconstruction: recent innovations, surgical candidate selection and postoperative management. Expert Review of Ophthalmology. 2008;3(5):567-84.

[147] Rafat M, Li F, Fagerholm P, Lagali N, Watsky M, Munger R, et al. PEG-stabilized carbodiimide crosslinked collagen–chitosan hydrogels for corneal tissue engineering. Biomaterials. 2008;29(29):3960-72.

[148] Reichl S, Borrelli M, Geerling G. Keratin films for ocular surface reconstruction. Biomaterials. 2011;32(13):3375-86.

[149] Dravida S, Gaddipati S, Griffith M, Merrett K, Madhira SL, Sangwan V, et al. A biomimetic scaffold for culturing limbal stem cells: a promising alternative for clinical transplantation. Journal of tissue engineering and regenerative medicine. 2008;2(5):263-71.

[150] Harkin D, George K, Madden P, Ivan Schwab, Hutmacher D, Chirila T. Silk fibroin in ocular tissue reconstruction. Biomaterials. 2011;32(10):2445-58.

[151] Humar M, Kwok S, Choi M, Yetisen A, Cho S, Yun S-H. Toward biomaterial-based implantable photonic devices. Nanophotonics. 2017;6(2):414-34.

[152] Bar-Ilan O, Albrecht R, Fako V, Furgeson D. Toxicity assessments of multisized gold and silver nanoparticles in zebrafish embryos. Small. 2009;5(16):1897-910.

[153] Hoffman A. Hydrogels for biomedical applications. Advanced drug delivery reviews. 2012;64:18-23.

[154] Morais J, Papadimitrakopoulos F, Burgess D. Biomaterials/tissue interactions: possible solutions to overcome foreign body response. The AAPS journal. 2010;12(2):188-96.

[155] Parker S, Peter D, Jason A, Jason B, Jennifer L, David K, et al. Biocompatible silk printed optical waveguides. Advanced Materials. 2009;21(23):2411-5.

[156] Chirila T, Barnard Z, Harkin D, Schwab I, Hirst *L. Bombyx* mori silk fibroin membranes as potential substrata for epithelial constructs used in the management of ocular surface disorders. Tissue Engineering Part A. 2008;14(7):1203-11.

[157] Higa K, Takeshima N, Moro F, Kawakita T, Kawashima M, Demura M, et al. Porous silk fibroin film as a transparent carrier for cultivated corneal epithelial sheets. Journal of Biomaterials Science, Polymer Edition. 2011;22(17):2261-76.

[158] Lawrence BD, Cronin-Golomb M, Georgakoudi I, Kaplan DL, Omenetto FG. Bioactive silk protein biomaterial systems for optical devices. Biomacromolecules. 2008;9(4):1214-20.

[159] Pal R, Kurland N, Wang C, Kundu S, Yadavalli V. Biopatterning of silk proteins for soft micro-optics. ACS Applied Materials & Interfaces. 2015;7(16):8809-16.

[160] Tao H, Jana K, Sean S, Eleanor P, Angelo S, Bruce P, et al. Implantable, multifunctional, bioresorbable optics. Proceedings of the National Academy of Sciences. 2012;109(48):19584-9.

[161] Hofmann S, Henri H, Annette K, Ralph M, Gordana V-N, David K, et al. Control of in vitro tissue-engineered bone-like structures using human mesenchymal stem cells and porous silk scaffolds. Biomaterials. 2007;28(6):1152-62.

[162] Lee M, Jeon H, Kim S. A highly tunable and fully biocompatible silk nanoplasmonic optical sensor. Nano letters. 2015;15(5):3358-63.

[163] Wang C, Xia K, Jian M, Wang H, Zhang M, Zhang Y. Carbonized silk georgette as an ultrasensitive wearable strain sensor for full-range human activity monitoring. Journal of Materials Chemistry C. 2017;5(30):7604-11.

[164] Cho SY, Yun YS, Lee S, Jang D, Park K-Y, Kim JK, et al. Carbonization of a stable β-sheet-rich silk protein into a pseudographitic pyroprotein. Nature communications. 2015; 6(1):1-7.

[165] Wang C, Xia K, Zhang M, Jian M, Zhang Y. An all-silk-derived dual-mode E-skin for simultaneous temperature–pressure detection. ACS applied materials & interfaces. 2017;9(45):39484-92.

[166] Wang Q, Jian M, Wang C, Zhang Y. Carbonized silk nanofiber membrane for transparent and sensitive electronic skin. Advanced Functional Materials. 2017;27(9):1605657.

[167] Jung M, Lee K-J, Kang J-W, Jeon S, editors. Silk-Based Self Powered Pressure Sensor for Applications in Wearable Device. 2020 International Conference on Electronics, Information, and Communication (ICEIC); 2020: IEEE.

[168] Seo JW, Kim H, Kim K, Choi S, Lee H. Calcium-Modified Silk as a Biocompatible and Strong Adhesive for Epidermal Electronics. Advanced Functional Materials. 2018;28(36):1800802.

[169] Khalid A, Peng L, Arman A, Warren-Smith S, Schartner E, Sylvia G, et al. Silk: A bio-derived coating for optical fiber sensing applications. Sensors and Actuators B: Chemical. 2020;311:127864.

Chapter 5

Application Technologies for Functional Finishing of Textile Materials

Zeynep Omerogullari Basyigit

Abstract

Nowadays, the primary energy resources and existing water reserves in the world are gradually decreasing. Because of global warming and high consumption of energy and water, researches have focused on new technologies and methods which aim of optimum use of resources while applying functionalites to the material. When the energy and water consumption of industries is examined, it could be obviously determined that the textile industry is seen to be at a substantial level. For this reason, in this chapter broad information of application systems including conventional and low-liquor application techniques with updated versions which show notable improvements in textile industry lately, have been detailed in a way of properties, parameters and running mechanisms on textile materials.

Keywords: low-liquor application, energy, water, finishing, application, textile

1. Introduction

Today, consumption of fossil fuels is dramatically increasing along with improvements of life such as industrialization processes with the increase of the world population. It has long been recognized that this excessive fossil fuel consumption not only leads to an increase in the rate of diminishing fossil fuel reserves, but it also has a significant adverse impact on the environment, resulting in increased health risks and the threat of global climate change [1]. The potentially most important environmental problem relating to energy and water resources is global climate change (global warming or the greenhouse effect). The increasing concentration of greenhouse gases such as carbon dioxide, methane, chlorofluorocarbons, halons, nitrous oxide, ozone, and peroxyacetylnitrate in the atmosphere is acting to trap heat radiated from Earth's surface and is raising the surface temperature of Earth [2]. This climate change has an increasingly negative impact on water resources; causing a serious decrease in available water reserves in the world [3]. In order to solve this problem, researchers have been focused on new methods for optimum use of resources with new technologies which save energy and reduce water consumption.

When the energy consumption in textile enterprises are compared to other branches of industry, it can be seen that the textile industry is seen to be at a substantial level. It could also be note that finishing departments are the most energy consuming ones among the other parts of textile manufactories. 45–75% of the energy consumed in finishing departments is listed as wet processes, 15–40% for drying and fixing processes, 8–18% for other processes and ventilation. Electricity consumption is not much in finishing departments; however, heat energy and water consumption is at very high level [4, 5]. For this reason, energy and water-saving technologies play an important role in the machines and application methods used in finishing processes.

Although textile finishing processes can be applied in different material forms (fiber, tops, yarn, fabric etc.), the most common is fabric finishing. General expectations for all these finishing are having homogenous effect, non-damaged fibers, non-broken fabric, repeatable and economical process, low environmental impact and reduced energy and water consumption. Different application techniques are used in the studies related to give desired properties to textile materials. Most of these application techniques are wet processes. These wet processes include exhaustion, impregnation, vacuum application, maximum application techniques as well as spraying, coating, transfer and foam application methods which are among the low-liquor application techniques [6]. In addition to these methods, microencapsulation, plasma application, sol–gel technology and lamination techniques, which have become increasingly important in recent years, are also included in finishing applications.

Nowadays, the methods and techniques used in the textile industry are desired to be environmentally-friendly and to save water and energy besides the other requirements such as functionality, durability, repeatability and being cost-effective. With the increase of diversification of today's industrial requirements, one functionality on the fabric may be insufficient to meet these requirements, therefore, although it varies according to area of utilization, being multifunctionality becomes more of an issue. In some cases, the fact that the fabric has more than one functionality on the whole surface entirely, regardless of front or back sides of the material, causes an increase in cost unnecessarily and also prevents showing sufficient efficiency in the area of use. For instance; for a sportswear outfit, the interior structure of the fabric is desired to be hydrophilic to absorb the sweat and water occurred during movement of the body, while the outer structure of the material is expected to water–oil-soil repellent. If the water repellency functionality is applied to the entire fabric, water repellant chemicals will act functional on the outer side of the fabric while it will serve as blocking barriers by preventing the absorption of sweat and water in the interior side of the material. This will not only bring on difficulties in use but also cause increased unnecessary cost during the finishing process of the material.

Since the conventional padding application methods, which are still in use widely today, do not allow the transfer of different chemicals to different sides, both sides of the fabric are treated with the same chemical substance and due to unnecessary material transfer, both the expected requirements cannot be fully met and cause an increase in costs. For this reason, it has emerged that some functionality needs to be applied to a single surface of the fabric.

Providing multifunctionality in a single-bath could have disadvantages in many respects that all the requested functionalities are mixed with each other in a single recipe and in the same bath. The first of these disadvantages is that all basic and auxiliary chemicals used in the same bath, belonging to different functionalities, may not be compatible with each other. Since the chemical structures of materials belonging to different functionalities are different, their mechanism of action is also different, and therefore problems may be encountered in providing a homogeneous mixture. The second of these disadvantages is that since all the chemicals are mixed with each other, the functionalities will be given in a mixed order regardless of the back or front face of the fabric, unfavorable functionality may be occurred on the undesired side or the requested efficiency is not achieved as expected or no functionality is obtained at a sufficient level. Therefore, due to all these problems and requirements, achieving multifunctionality by transferring more than one functionality to the same and/or different sides of the fabric in an effective and permanent way in accordance with the field of use has become a necessity both in industrial and academic terms.

There is a way of combining fabric layers with different functionalities by lamination methods or so, for obtaining multifunctionality, but because of each separate layer has own functionality, the endproduct takes up more space in terms of volume and increases the cost of the material. For this reason, multifunctionality in single-layer fabrics should be carried out by considering the requirements of providing mobility to the user, being useful and flexible, and saving in material costs.

Due to all the above-mentioned requirements in textile industries; advantages and disadvantages of existing conventional methods besides new generation finishing processes which focuses on reducing water and energy consumption mostly, are defined in this chapter.

2. Wet finishing processes

Most of the textile finishing application techniques are wet processes. These wet processes can be listed as follows: exhausting process, impregnation, coating, transfer, spraying and foam application. It should be noted there is an optimum level of liquor application which is just high enough to ensure adequate distribution of the chemicals within the fabric. This critical add-on value (CAV) depends on fiber type, fabric construction and fabric pretreatment [6–11]. The wet processes have been in use for a long time however; they have been updated in many ways nowadays; such as using new technologies in impregnation machines with lower wet pick-up ratios, providing homogenous application in new version of chemical foam application, removing the blockage occurred in nozzles of spraying machines. Minimum application methods, which focuses on reducing water and energy consumption, have an increased importance recently in finishing processes with a wide range of functionalities provided such as water–oil-soil repellency, flame retardancy, antibacterial efficiency and so many other functionalities due to their significant advantages over conventional methods. The application technologies including conventional ones and updated low-add-on finishing applications have been detailed in this review.

2.1 Exhausting process

The essence of the application with the exhaustion method is that the product to be treated is in wet-process for a long time at the long liquor ratio. The liquor ratio in the studies according to the exhaustion method is in the range of 1:2–1:100. This method is also called full bath method or discontinue method. The fabric to be treated is placed in a bath and allowed to absorb the finishing agent from the bath until a chemical-balance is reached between the finishing agent on the fiber and the one in the bath. In order to provide sufficient and economical results in this method, it is essential that the finishing material used has an affinity towards the textile material. In other words, the finishing material dissolved or homogeneously dispersed in the bath should have a high desire to be withdrawn from the bath by the fibers.

Dyeing process can be carried out with textile fibers, yarns, fabrics or garments. However, there are reasons for dyeing different fiber forms. Fiber dyeing is used as a styling technique; natural fibers or staple synthetics are dyed in bundles or baskets. Dope or solution dyeing is the process where color is mixed into the polymer solution prior to fiber extrusion. Certain synthetics fibers such as polyethylene can only be colored using this technique. Yarn dyeing which is also a styling technique, is used to produce stripes, plaids, and some complex designs with 100% fiber content products. Beam dyeing is a technique where multiple yarns are wound side by side onto a single perforated beam. This can be a few hundred or even a few thousand yarns wound onto a single beam. Fabric or piece dyeing is the most cost efficient and highest productivity technique. Fabric dyeing machines include jet machines, dye becks, fabric beam, and jig dyeing machines. All of these machines as well as most of the yarn dyeing machines are batch or exhaust machines [12]. Some of the machines used in exhaustion method are shown in **Figures 1** and **2** [10, 15].

Advantages:

- Processing time, temperature and composition of the liquor (pH of the liquor, amount of electrolyte, auxiliary substances) can be adjusted as desired. Thus, the application speed (proper application of the dyestuff from the beginning) and the application amount (color fixation in dyeing) can be adjusted.

- The process can be easily intervened and additional toning can be made. Therefore, it is easier to attach the result than the impregnation method.

- It provides ease of operation as washing, bleaching, dyeing and finishing processes are carried out in the same machine.

Disadvantages:

- The most important disadvantage is that the liquor ratio is long, so the consumption of water, finishing chemicals and energy (required for both heating and moving the liquor) is so high.

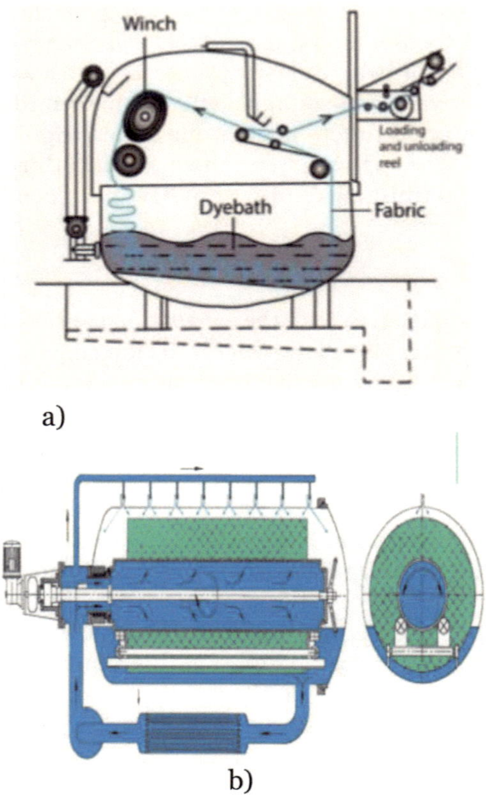

Figure 1.
Exhausting process machines in textile finishing (a) winch dyeing (b) beam dyeing [13, 14].

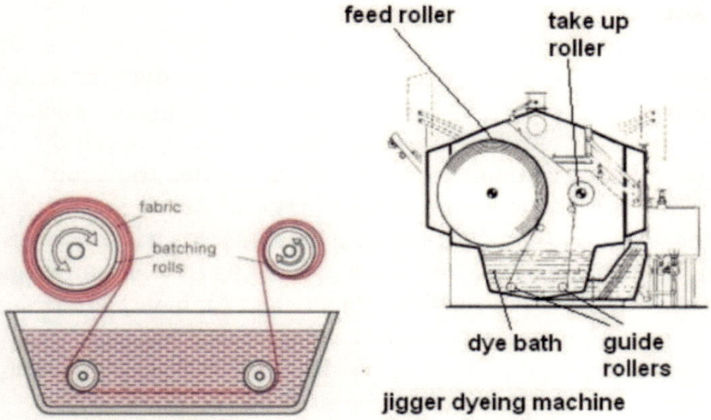

c)

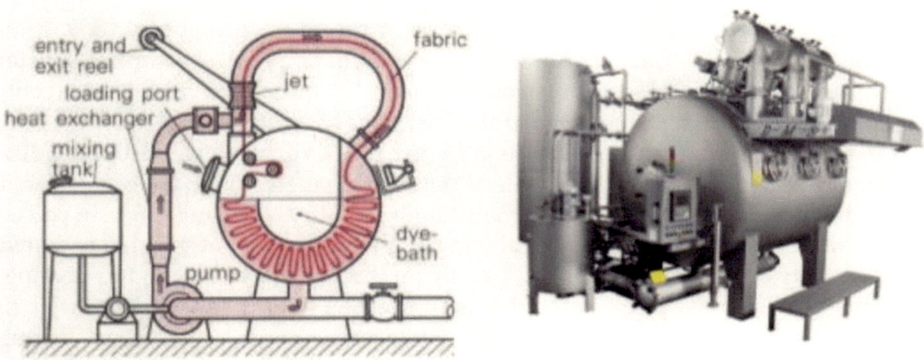

d)

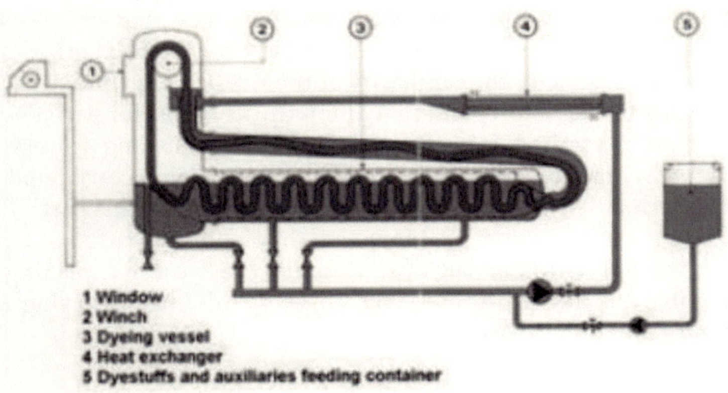

e)

Figure 2.
Exhausting process machines in textile finishing (c) jigger dyeing (d) jet dyeing (e) over-flow dyeing machines [13, 14].

- These are the discontinue methods that have long processing time (usually longer than 30 minutes).
- It requires a hand work and long time for operating the machine before application [5, 6, 8–10, 15].

2.2 Impregnation method

The application process, which is done by passing the textile product through the liquor in a bath within a short time (t < 30 seconds) and squeezing, is called impregnation (padding) method. After the fabric has been padded through the liquor and prior to being squeezed through the rollers of the padder, the liquor is distributed as follows: within the fibers; in the capillary regions-between the fibers; in the spaces between the yams; on the fabric surface [11, 15]. Two characteristics stand out in this method: short liquor ratio and short processing time. In impregnation method, it is important to not to use finishing chemicals that have affinities to the textile materials, or it should be at a very low level. As this is a continuous method, the concentration of the liquor absorbed by the product at the beginning should be same in the bath and at the end of the process. However, if finishing materials with high substantivity are used, the process results in with a tailing effect (when dyeing is done by padding method, gradual color change occurs along the fabric length because of the decrease in the concentration within time) [6, 9, 15]. Since most of the materials used in textile pretreatment do not show much substantivity towards fibers, unlike those used in dyeing, the most used application technique in these processes is impregnation. The types of impregnation machines work due to padding mechanism are shown in **Figure 3**. This method has also some disadvantages varied according to the types of the system, such as high concentration of finishing chemicals and long-time processing in pad-batch, tailing effect in pad-roll dyeing which is also labor intensive process, high energy consumption in pad-steam and necessity of an extra machine in pad-jig method which results in over costing investment. It should be also noted that high wet-pick-up ratios associated with the padding system are disadvantageous, not only because of the high thermal-drying and water consumption costs, but also because, during the evaporation of the liquor in thermal drying, the molecules of the applied substances tend to migrate from the inside to the outside of the fabric, leaving behind an uneven chemical distribution which leads to a decreased fabric durability of the functionality [5, 7, 15].

This method is divided into two as dry to wet impregnation and wet to wet impregnation. Since it is easier for a dry textile product to be absorbed in a bath containing finishing material (because the capillarity of the fibers is high due to its absorption ability), the dry-to-wet impregnation method works in shorter periods than wet-to-wet impregnation method. However; if a second wet finishing process is to be carried out after a wet finishing process, when the drying step is considered to be a very expensive intermediate process, the advantage of removing this part and saving energy indirectly will be achieved. Moreover, the risk of migration of the finishing chemicals which occur in drying process before the dyestuff/finishing agents are fixed on the fibers, could be prevented by removing this interim drying process in wet to wet impregnation [5, 15].

2.3 Transfer method

In this minimum (low wet-pick-up) application method made in special foulard, the fabric itself is not dipped into the bath. The liquor containing the finishing agent is taken by a rolling roller and transferred to the back surface of the fabric. That's why we could see this type of finishing systems under the name of "Lick/Kiss Roll Applicators" [19]. Transferring is an application method that can be applied with high viscosity finishing liquors. Excess liquor on the transfer roller or fabric is scraped off with the help of doctor blades. The **Figure 4** shows 4 different transfer systems, which differ in terms of the number of rollers, the location of the paddle

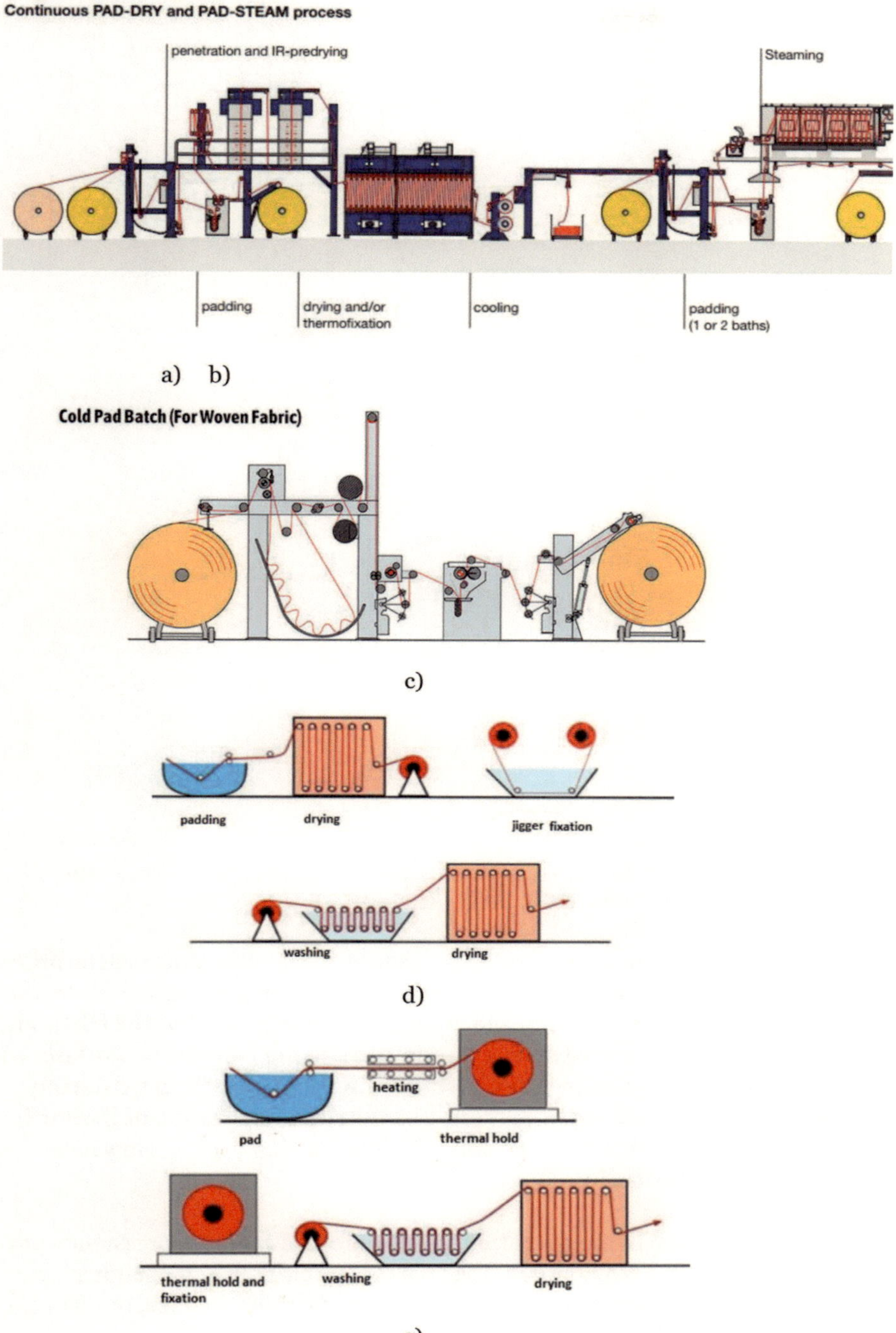

Figure 3.
Impregnation machines (a) pad-dry-cure (b) pad-steam (c) pad-batch (d) pad-jig (e) pad-roll [16–18].

and measurement techniques [6, 7, 9, 11, 15, 19]. The schematic representation of Triatex MA machine which uses on-line monitoring to control the wet pick-up values, has been shown in **Figure 4d**. The kiss roll rotational speed is automatically adjusted (with the help of β-rays measured by detectors) [15] relative to the fabric speed to maintain the desired wet pickup.

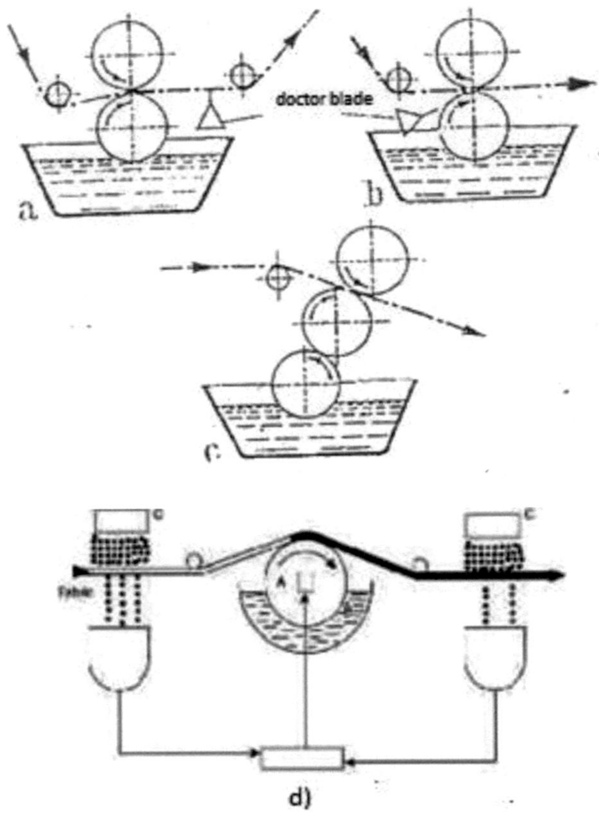

Figure 4.
Different types of transfer (with lick/kiss roll applicators) finishing methods (a) and (b) two-roller with doctor blade type (c) three-roller transfer type (d) Triatex MA [15, 19].

The amount of finishing agent transferred to the textile product is determined by following parameters; the structure of the transfer roller, condition of doctor blade and other rollers, viscosity of the liquid, the passing speed of the fabric and the rotation speed of the transfer roller. There are two big benefits of working with less liquor; firstly, energy savings as less water will need to be removed during subsequent drying and secondly, reduction of the risk of migration of dyestuffs or finishing substances that have not yet been fixed during drying. However; this method has also some disadvantages.

- The main problem in the operations of this method is the inconsistency of the formation of a smooth liquor on the transfer roller. This homogenous formation is not only dependent on the structure of the roller surface, but it's also closely related to the composition of the liquor.

- The second problem is the disability to ensure that the same amount of liquor continuously applied to the textile material [11, 15, 19].

2.4 Spraying method

Spraying method in finishing process has been known and applied on textiles for a long time however; it has not been improved much for many years because of some difficulties in conventional (nozzle) spraying machines mentioned below:

- It is difficult to apply the same amount of liquor all over the fabric continuously,
- Nozzles are frequently clogged, especially when working with viscous chemicals.
- A part of liquor sprayed in very fine particles is placed on other parts of the machine instead of the fabric, causes excessive pollution and unnecessary chemical loss.

Especially after the oil crisis in 1974, the spraying method has become updated with the increase in the importance of the application processes which has low wet-pick-up values. To overcome the difficulties in conventional spraying methods, indirect spray applications have been demonstrated by the Farmer Norton and Weko applicators. In these systems, the spray is generated by pumping the finishing solution by a proportioning pump from a well onto the center of a rapidly rotating set of spinning discs (Farmer Norton) or rotors (Weko) [7, 15, 19, 20]. In addition to have the main advantages of being in the minimum application system, spraying technology has some other advantages such as; being suitable for wet-to-wet applications, no tailing effect even if the chemicals have the affinity, ability to be applied on one or two sides of the fabric upon request [15, 20]. Weko Fluid application system has been shown in **Figure 5**.

2.5 Coating method

Finishing liquors with high viscosity can be applied directly to one side of the fabric. As a result of such application, a large amount of finishing material can be transferred to the surface of the material and this process is also called "coating", since the finishing agent mostly covers the surface of the material. The coating method is frequently used especially for producing artificial leather and waterproof finishing process. Multi-layered functional materials can be produced

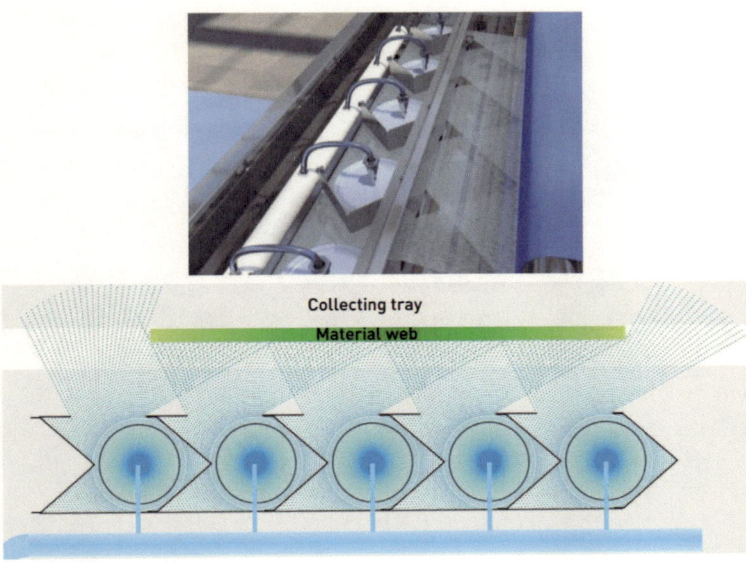

Figure 5.
WEKO-fluid application system [20].

with different coating methods including direct (directly coating on to the fabric) and indirect (using for exp. silicone paper during coating) system [15, 21–23]. In **Figure 6** illustrations of direct and indirect coating systems have been shown [23]. The basic mechanism of a direct coating method is spreading polymer on the textile material, in the form of thick liquid (viscous) or paste using a special knife called doctor blade [21]. The smoothness and the thickness of the applied layer are adjusted with the help of doctor blades.

In "blade on air" coating system, the fabric passing under the blade does not lean anywhere. Therefore, it is not possible to make thick and very smooth coatings with this type of coating. It is preferred for light coatings of loose woven fabrics. However; the coating material that passes to the other side of the fabric due to the loose texture, it contaminates the rubber band or roller under the blade and causes uneven coating. In "blade over the roller" system, the fabric that passes under the doctor blade is based on a rotating rubber or steel cylinder. The thickness of the layer applied in the coatings can be adjusted precisely. But on the other hand, since the cylinder cannot stretch; dust or fly can be trapped under the blade, causing soiling and forming drag lines [21].

There are some other coating processes used in the film and paper industries which require expensive equipment and which must be carried out on a large scale

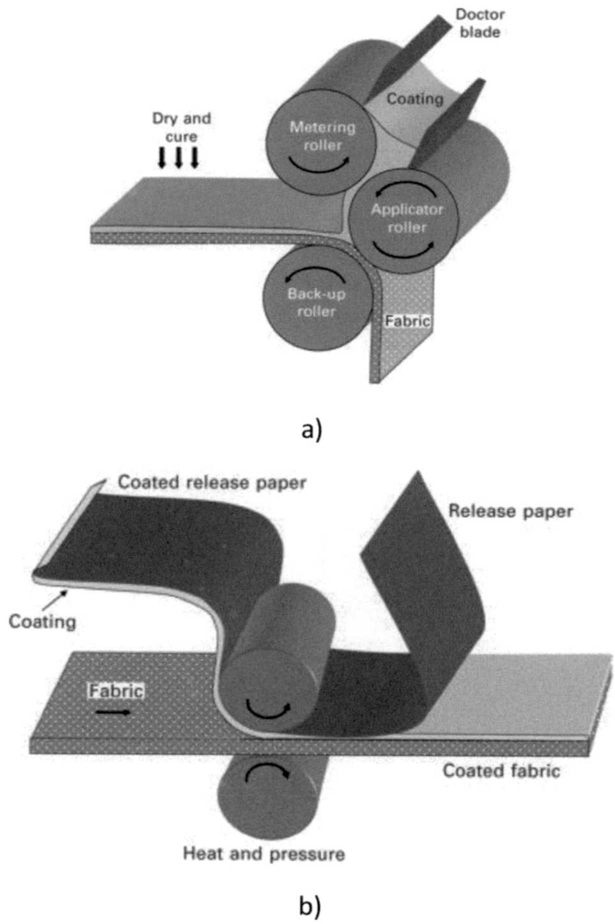

Figure 6.
(a) Direct coating system (three-roll coatings: Nip feed coating) (b) indirect coating system (transfer coating) [23].

to be commercially viable. Among these are powder coating, vacuum deposition, electrostatic deposition and sputtering techniques. It is possible, for example, to produce highly reflective surfaces by these methods, but a smooth continuous surface is required and fabrics may not be suitable [21, 22].

2.6 Foam application method

The most important and interesting development in the application of finishing agents to textile products in the mid-1970s is undoubtedly the application methods with foam. Machines that enable working with foams instead of aqueous solutions have been put on the market because of increased energy costs in the textile industries. However, the application studies with foam, which developed very rapidly in the beginning, have entered a pause and had no significant improvement until 2010 [24].

Foam is a microheterogeneous colloidal mixture, short or long-lasting a metastable system in which the surface area is increased nearly 1000 times by inflating a liquid with a suitable gas (air), and therefore contains less liquid. Foam is obtained by dispersing the air in water as fine particles with the help of surfactants. If a surfactant is dissolved in aqueous solution and air bubbles are present in this solution, then a surfactant film covers the air bubbles. Air bubbles move towards the upper surface of the liquid which covered with a tenside film. Thereby, a second tenside film is formed around the upward air bubbles (**Figure 7**) In this way, the air bubble in which the liquid is located between the two tenside films surrounding, called a "foam cell" [7, 15, 24, 25].

There are many types of foam application such as; open foam method (Horizontal pad foam, Knife-roll-over foam, Autofoam systems), offset open foam methods (Küsters Janus contact roller system and Monforts vacuum drum system), closed foam methods (FFT Foam Finishing Technology-Gaston County Dyeing Machine, CFS Chemical Foam System-Gaston System, Stork rotary screen foam applicator and Stork CFT Coating and Finishing Technology) have been shown in **Figures 8–10** [24, 29, 30]. As seen in the figures, there have been much improvement in the profiles of foam applicators in time in order to prevent the problems such as non-uniform and uneven applications during processes. The case in point can be the improvement of CFS parabolic profile which has been developed to solve "dead foam" issue occurred in FFT. With the help of parabolic chamber, equal distance paths are covered from the point of foam inlet to the fabric surface so that the problem of wet and unfoamed parts occurred partly in FFT foamed fabrics have been solved [24, 29].

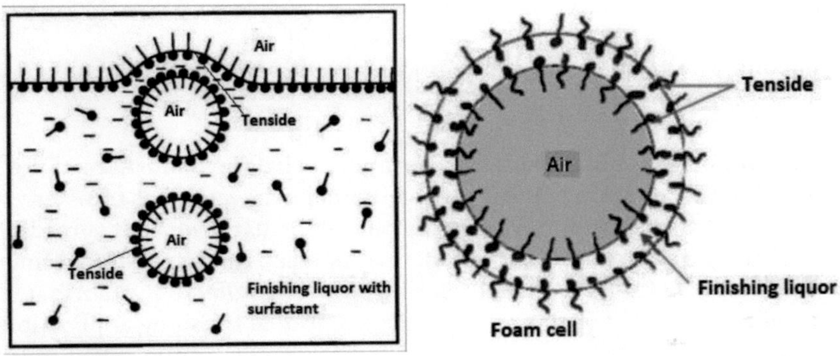

Figure 7.
Formation of a foam cell.

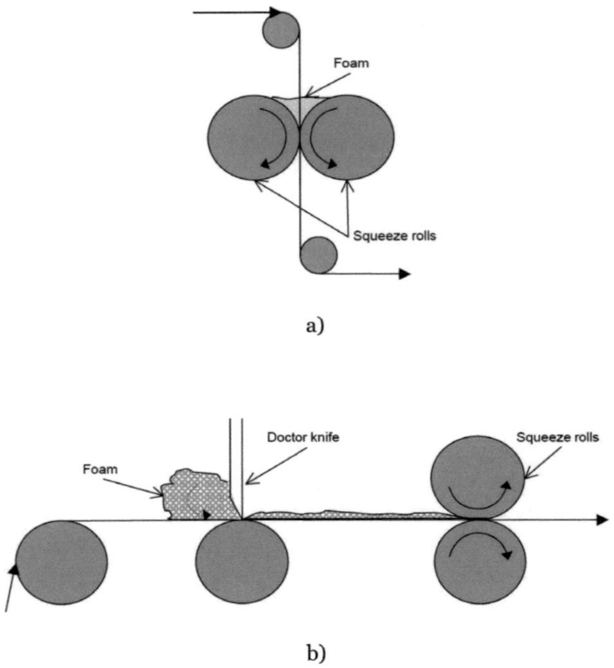

Figure 8.
Types of foam application (a) horizontal pad foam (b) knife-roll-over foam [24, 26–28].

For an effective foam application process, the followings should be taken into consideration:

- The foam should not be collapsed during the time between it is taken from the foam generator and transferred to the fabric.

- On the other hand, when the foam reaches out the textile product, it should collapse into the fabric as quickly as possible. Foam stability constitutes an important place in foam applications. Very stable foams play role in decreasing the penetration efficiency into the fabric whereas unstable foams cause uneven applications because of collapsing before the application. Therefore, foam cell should be in semi-stable form.

- The foam and the tenside used in the application must have good compatibility with other chemical substances in the bath.

- The foam to be used in a finishing process should always has the same form and the same concentration.

- Another important point in foam application is that the foams should not have much water. If the foam that does not contain much water, it is transferred directly onto the fabric surface moving perpendicularly without spreading around the fabric surface. On the other hand, when the conventional finishing liquors first penetrated to the fabric, they spread parallel to the fabric surface by capillarity effect. For this reason, in conventional finishing applications such as in padding methods non-functional caked chemical residuals remained at the fibers intersections cause uneven application and nonhomogeneous penetration [15, 27].

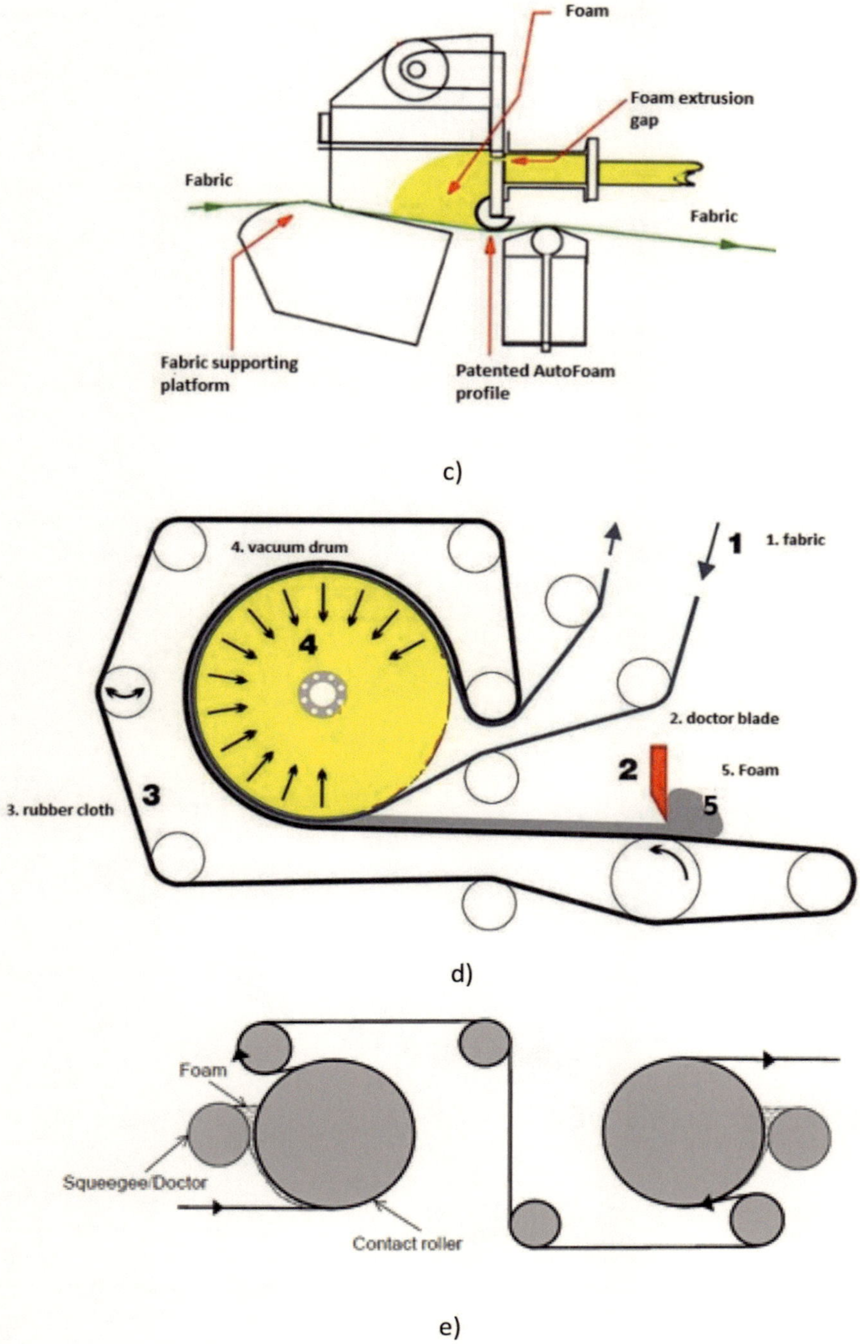

Figure 9.
Types of foam application (c) autofoam systems (d) Monforts vacuum drum system (e) Küsters Janus contact roller system [24, 26–28].

2.6.1 Advantages of foam application

Foam finishing is a versatile application system which could be very effective for bleaching, dyeing and varied finishing processes. It can be also used to apply different applications to the face and back side of the fabric in a single pass with

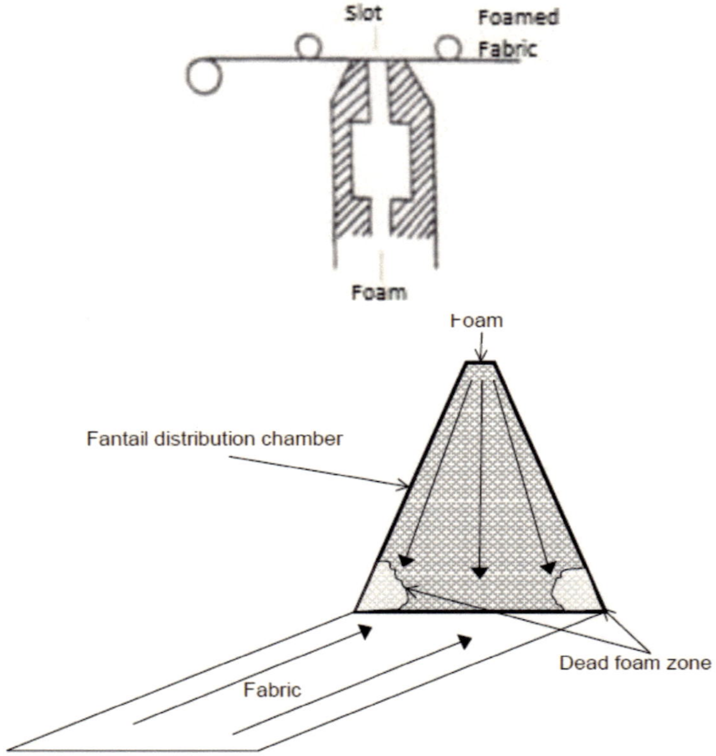

f)

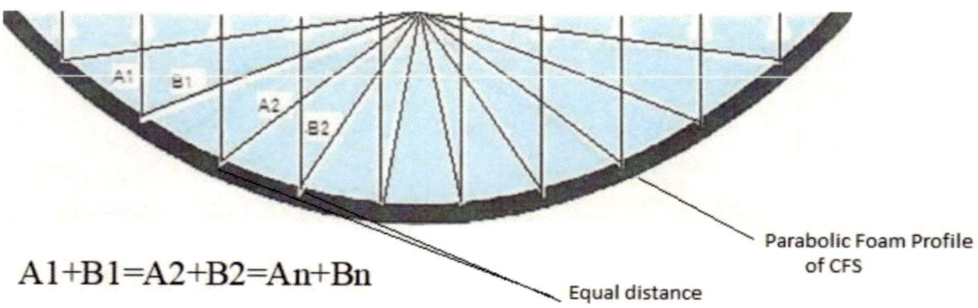

g)

Figure 10.
Types of foam application (f) foam finishing technology (FFT) (g) chemical foaming system (CFS) [24, 26–28].

dual-applicators. Multifunctional fabrics with increased durability against repeated washing and drying processes could be produced. As reported in the literature [24] that many combinations of functionalities such as face side flame retardant and water repellent whereas antibacterial back side of textile materials could be provided via foam finishing application. The most important advantage of this system is providing significant decrease in water consumption (up to 80%) due its lower wet-pick-up system. For cotton-rich fabrics, the wet pickup in foam finishing is typically between 15% and 35%, depending on the process, compared with 60%

to 100% for a conventional pad application [27, 29, 30]. For synthetic fabrics, wet pick up ratio of foam application is in the range of 3–10% whereas it is 35–60% in conventional padding. Moreover, the energy consumed for heating and venting the air is significantly decreased due to lower wet-pick-up values. This reduction in time needed for pre-drying step (could be eliminated), also minimizes the migration of the chemicals. Since the amount of liquor taken is small in foam finishing, excessive swelling of the fibers is not possible. Since the fibers do not swell and the capillary channels do not expand too much, the transferred chemical substance is not carried to the surface with water during drying and remains where it is transferred. This is particularly effective in preventing the reduction of the abrasion resistance in the wrinkle recovery finishing. Since migration, that is, the accumulation of the chemical substance on the surface more than necessary during drying, is effective in the decrease in friction resistance. Foam application also potentially results in less waste water pollution than with traditional application methods. On the contrary of an aqueous pad system, the small liquor volumes required for foam application result in less waste during a changeover [15, 24, 27, 29, 31–34].

Updated versions of foam application systems has offered better solutions to the basic problems encountered with the other low-add-on systems. The main problem of the low-add-on systems is the difficulty of distributing a relatively small quantity of liquor uniformly over a large surface of fabric, especially on hydrophilic fibers. In the case of low-add-on expression systems, a basic limitation is the inability to achieve wet-pick-up levels below the critical application value of the component fibers [7]. When compared to conventional methods, foam finishing provides homogeneous and effective chemical applications via controlled, uniform and repeatable foam formations.

2.6.2 Chemical foaming system (CFS)

If CFS foam machine is examined, it could be clearly seen that uniform and homogenous applications are achieved via performing correct systematic on the distance between the foam generator and applicator.

As shown in the **Figure 11**, the foam diameter increases as the foam formed in the CFS foam generator is transferred from high pressure to the low pressure on the way to the foam applicator. The pressure gradually decreases on the fabric surface and the foam penetrates the fabric. The distance between the foam generator and its applicator is critical for uniform foam formation.

DG: Initially, the radius of the foam produced in the foam generator.

DA: The radius of the foam being transferred from the foam generator to the foam applicator.

DB: The radius of the foam at the foam applicator just before it penetrates to the fabric.

WG: Initially, the area covered by the liquid contained in the foam produced in the foam generator.

WA: The area covered by the liquid in the foam being transferred from the foam generator to the foam applicator.

WB: The area covered by the liquid in the foam just before it penetrates into the fabric, at the foam applicator.

Before the transfer process starts, the state between these parameters is $WG > WA > WB$ and $DG < DA < DB$, while these equations are reversed as the pressure decreases gradually, and $WG < WA < WB$ and $DG > DA > DB$ becomes. Therefore, on the foam applicator, penetration to the fabric takes place evenly with maximum radius and minimum liquid area of the foam [27].

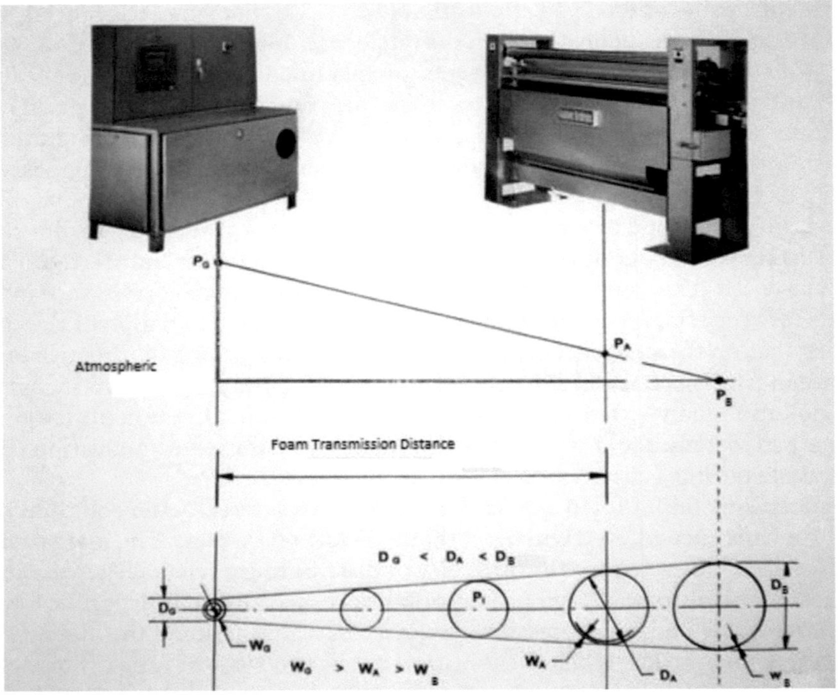

Figure 11.
Chemical foaming system (CFS) [27].

3. Conclusion

Considering the finishing methods used in the textile industry in general, conventional impregnation method is commonly used in the industry due to the low affinity of the chemicals used in the finishing processes. However, it has disadvantages such as high water and energy consumption, inability to apply different functionalities to face and back sides of the material and significant waste load. Extraction method is not included among the preferred methods because the chemicals used have affinity to the fabric, the operation times are long and the amount of water consumed is very high due to the long liquor ratio. Apart from these methods, transfer and coating methods are also available. Transfer methods with various types such as roller transfer or with doctor blades takes place in the minimum application methods. Likewise coating methods is in the low-liquor finishing applications with the types of blade coating, calendar coating, transfer and printing technique. However, both transfer and coating methods are not suitable for low viscosity chemicals. In coating and transferring methods, the effectivity of the application is directly related to the viscosity of the finishing agents, the construction and surface structure of textile material, production method of the material (woven, non-woven or knitted), weight of the textile, speed of the finishing method etc. So, it could be noted that they are not very flexible application techniques in a view of finishing agents and textile materials. Even if the direct spraying method had some problems in the past such as clogging of nozzles or excessive pollution on the machine, with the use of indirect systems such as discs or rotors, it has been much more improved. When the history of the foam application is examined, it could be clearly noticed that significant improvements have been carried out by time in order to make functional or multifunctional (via using dual-applicators) textiles via uniform applications (with developed applicator profiles) which provides reduction in water and energy consumption. Apart from these, there are various lamination techniques that can be used to obtain multifunctionality however; since the lamination process is based on the principle of combining fabric and polymer layers to form a composite material, any factor that prevents adhesion between the structures, low heat resistance of the materials or no resistivity for water and moisture can cause problems during applications.

Apart from these methods, there is also environmentally friendly techniques such as plasma technology, which is used in the textile industry and academic pilot studies, and gives functionality to the fabric with partly ionized gas or monomers without using water [35–37]. With this technology many researches have been carried out to provide functional textiles such as antibacterial cotton fabric supported by silver nanoparticules [38], water repellent and antimicrobial cotton/polyester blend [39], anti-bacterial and anti-static polyester fabric [40]. However, in this technology, the vacuum plasma method, which is quite effective, is a discontinuous method and could not be industrialized because it allows a very low size in terms of fabric length and width. The atmospheric plasma method, which is suitable for industrial use as the operating dimensions, is not as effective as vacuum plasma on porous textile surfaces. Microencapsulation technique, is also one of the effective methods used in textile applications. Microencapsulation technology involves the process by which small particles, mostly bioactive, are encapsulated in a wall consisting of a heterogeneous or homologous polymer matrix, which forms a complex known as a microcapsule [41, 42]. Microencapsulation helps to improve functional textile products such as fabrics with durable fragrances, UV-ray absorbing shirts, thermo-regulation vehicle seats, thermo-changeable dyed apparels, vitamin loaded fabrics as cosmototextiles or military uniforms with microcapsulated insecticides [42]. However, in order to transfer performed microcapsules onto the textile material, the capsulation method is continued with a conventional finishing method frequently (mostly padding) so that two-step applications are carried out with no significant reduction in water and energy consumption. Sol–gel technology, which is a method that can obtain macromolecules by taking advantage of the growth and development of polymers in a solvent, can also be an effective alternative in terms of giving functionality to textile materials. There have been lots of studies on sol–gel functionality such as self-cleaning superhydrophobic films [43], flame retardant and hydrophobic coatings on cotton fabrics [44], hydrophilic, antistatic and antimicrobial cotton and polyester fabrics [45]. However, it should be noted that the requirement of using solvents brings environmental threats and applying some special polymers increase input costs [46]. Nano-technology seems to be a significant alternative for achieving functional and multifunctional textile materials [35] however; in some cases, there have been still some issues of industrialization of nanoparticles because of producing them only in laboratory scale experiments. Studies about using liposomes in dyeing processes [47–49] photocatalytic reactions for bleaching process [50, 51] and layer-by-layer self-assembly technique for producing multifunctional multilayers [52, 53] have also been taken place in the literature but there has been no scientific clue reported in the literature for industrialization of these methods in textile manufactories.

Author details

Zeynep Omerogullari Basyigit
Department of Textile, Apparel, Shoes and Leather, Textile Technology Programme, Inegol High Vocational School of Higher Education, Uludag University, Bursa, Turkey

*Address all correspondence to: zeynepbasyigit@uludag.edu.tr

References

[1] Farhad, S., Saffar-Avval, M., & Younessi-Sinaki, M. (2008). Efficient design of feedwater heaters network in steam power plants using pinch technology and exergy analysis. International journal of energy research, 32(1), 1-11.

[2] Panwar, N. L., Kaushik, S. C., & Kothari, S. (2011). Role of renewable energy sources in environmental protection: A review. Renewable and sustainable energy reviews, 15(3), 1513-1524.

[3] Lu, S., Bai, X., Li, W., & Wang, N. (2019). Impacts of climate change on water resources and grain production. Technological Forecasting and Social Change, 143, 76-84.

[4] Namboodri, C. G., & Duke, M. W. 1979. Foam finishing of cotton-containing textiles. Textile Research Journal, 49(3): 156-162.

[5] Tarakçıoğlu I. 1984. Tekstil terbiye işletmelerinde enerji tüketimi ve tasarrufu. Uludağ Üniversitesi Basım evi, Bursa, 102 s.

[6] Çoban S.1999. Genel tekstil terbiyesi ve bitim işlemleri, E.Ü. Tekstil Konfeksiyon Araştırma Merkezi, İzmir, 314 s.

[7] Elbadawi, A. M., & Pearson, J. S. 2003. Foam technology in textile finishing. Textile Progress, 33(4):1-31.

[8] Schindler, W. D., & Hauser, P. J. (2004). Chemical finishing of textiles. Elsevier.

[9] Choudhury, A. K. R. (2017). Principles of textile finishing. Woodhead Publishing.

[10] Cay, A., Tarakçıoğlu, I., & Hepbasli, A. (2009). Assessment of finishing processes by exhaustion principle for textile fabrics: An exergetic approach. Applied Thermal Engineering, 29(11-12), 2554-2561.

[11] Van der Walt, G. H. J., & Van Rensburg, N. J. J. (1984). A review of low add-on and foam application techniques.

[12] 12.https://www.cottonworks.com/wpcontent/uploads/2018/01/Dyeing_Booklet.pdf

[13] https://www.slideshare.net/sheshir/jigger-dyeing-machine-33040185

[14] https://www.slideshare.net/asifalilakho3/exhaust-dyeing-process

[15] Tarakçıoğlu I. 1998. Tekstil Terbiyesinde Aplikasyon Yöntem ve Makinaları, E.Ü. Basımevi, Bornova, İzmir, 97s.

[16] https://textileinsight.blogspot.com/2014/08/pad-steam-dyeing-machineprocess.html

[17] https://www.shreeshaktikrupa.com/cold-pad-batch/

[18] https://tekstilbilgi.net/pad-jig-yontemi-ile-boyama.html

[19] Madhu, A., & Pal, S. (2012). Low Wet Pick-up Techniques In Textile Finishing. Man-Made Textiles in India, 40(1).

[20] http://parandchemie.ir/pdfs/DR.PETRY-SALES-CONFERENCE 2016/WEKO_denim_PRESENTATION_en.pdf

[21] Fung, W. (2002). Coated and laminated textiles (Vol. 23). Woodhead Publishing.

[22] Kovačević, S., Ujević, D., & Brnada, S. (2010). Coated textile materials. Woven Fabric Engineering, 241.

[23] Shim, E. (2013). Bonding requirements in coating and laminating

of textiles. In Joining Textiles (pp. 309-351). Woodhead Publishing.

[24] Başyiğit Ömeroğulları, Z. (2016). "Improving functional properties of cellulosic based textile materials via foam application" PhD thesis, Uludag University, Institute of Science and Technology, Bursa, TR.

[25] Bafford, R. A., & Namboodri, C. G. 1984. Foam Processing of Textiles. Sixth Annual Industrial Energy Technology Conference Volume I, Houston, TX.

[26] http://iqbaltextile3022.blogspot.com/2009/06/

[27] 27.https://www.gastonsystems.com/system/system.html#:~:text=CFS%20is%20a%20highly%20controlled,nonwovens%2C%20paper%2C%20and%20others.&text=This%2C%20combined%20with%20CFS%20technology,control%20of%20chemical%20penetration%20depth.

[28] http://www.autofoam.co.uk/en/app_std.html

[29] https://www.cottoninc.com/wp-content/uploads/2017/12/TRI-3008-Processing-100-Cotton-Woven-Fabrics-for-Filling-Stretch.pdf

[30] Gregorian, R. S., Namboodri, C. G., Young, R. E., & Baitinger, W. F. 1983. Foam application of phosphonium salt flame retardants. Textile Research Journal, 53(3):148-152.

[31] Song, M. S., Hou, J. B., Lu, Y. H., Lin, J., & Cheng, D. H. (2013). Performance of foam and application in foam finishing of textile. In Advanced materials research (Vol. 821, pp. 661-664). Trans Tech Publications Ltd.

[32] Baker, K. L., Bryant, G. M., Camp, J. G., & Kelsey, W. B. 1982. Foam finishing technology. Textile Research Journal, 52(6):395-403.

[33] Brown, R. L., & Bryant, G. M. 1984. Influence of Resin Add-on and Co-applied Water in Foam Finishing of Cotton Fabric. Textile Research Journal, 54(12): 807-812.

[34] Camp, J. G. 1990. Recent Developments in Foam Application Systems for Nonwoven Substrates. Journal of Industrial Textiles, 19(4): 252-260.

[35] Omerogullari Basyigit Z. 2017. "Functional Finishing For Textiles" Researches On Science And Art In 21st Century Turkey. Book Chapter 257, pp: 2297-2308, Gece Publishing, TR.

[36] Omerogullari Basyigit Z. 2018. "Plasma Applications in Textile" Researches On Science And Art In 21st Century Turkey. Book Chapter 114, pp: 1468-1480. Gece Publishing, TR.

[37] Gulrajani, M. L., & Deepti, G. (2011). Emerging techniques for functional finishing of textiles.

[38] Deng, X., Nikiforov, A., Vujosevic, D., Vuksanovic, V., Mugoša, B., Cvelbar, U. & Leys, C. (2015). Antibacterial activity of nano-silver non-woven fabric prepared by atmospheric pressure plasma deposition. Materials Letters, 149, 95-99.

[39] Davis, R., El-Shafei, A., & Hauser, P. (2011). Use of atmospheric pressure plasma to confer durable water repellent functionality and antimicrobial functionality on cotton/polyester blend. Surface and Coatings Technology, 205(20), 4791-4797.

[40] Zhang, Z., Zhao, Z., Zheng, Z., Liu, S., Mao, S., Li, X., & Li, G. (2019). Functionalization of polyethylene terephthalate fabrics using nitrogen plasma and silk fibroin/chitosan microspheres. Applied Surface Science, 495, 143481.

[41] Muhoza, B., Xia, S., Wang, X., Zhang, X., Li, Y., & Zhang, S. (2020).

Microencapsulation of essential oils by complex coacervation method: preparation, thermal stability, release properties and applications. Critical Reviews in Food Science and Nutrition, 1-20.

[42] Šiler-Marinković, S., Bezbradica, D., & Škundrić, P. (2006). Micro-encapsulation in the textile industry. Chemical Industry and Chemical Engineering Quarterly, 12(1), 58-62.

[43] Zhang, X. F., Zhao, J. P., & Hu, J. M. (2017). Abrasion-Resistant, Hot Water-Repellent and Self-Cleaning Superhydrophobic Surfaces Fabricated by Electrophoresis of Nanoparticles in Electrodeposited Sol–Gel Films. Advanced Materials Interfaces, 4(13), 1700177.

[44] Zhang, D., Williams, B. L., Shrestha, S. B., Nasir, Z., Becher, E. M., Lofink, B. J., & Sun, L. (2017). Flame retardant and hydrophobic coatings on cotton fabrics via sol-gel and self-assembly techniques. Journal of colloid and interface science, 505, 892-899.

[45] Chen, G., Haase, H., & Mahltig, B. (2019). Chitosan-modified silica sol applications for the treatment of textile fabrics: a view on hydrophilic, antistatic and antimicrobial properties. Journal of Sol-Gel Science and Technology, 91(3), 461-470.

[46] Tobiszewski, M., Namieśnik, J., & Pena-Pereira, F. (2017). Environmental risk-based ranking of solvents using the combination of a multimedia model and multi-criteria decision analysis. Green Chemistry, 19(4), 1034-1042.

[47] Barani, H., & Montazer, M. (2008). A review on applications of liposomes in textile processing. Journal of liposome research, 18(3), 249-262.,

[48] Montazer, M., Taghavi, F. A., Toliyat, T., & Moghadam, M. B. (2007), Optimization of dyeing of wool with madder and liposomes by central composite design. Journal of applied polymer science, 106(3), 1614-1621.

[49] Campardelli, R., Trucillo, P., Iorio, M., & Reverchon, E. (2020). Leather dyeing using a new liposome-based process assisted by dense gas technology. Dyes and Pigments, 173, 107985.

[50] GEDİK, G. (2020). Bleaching of Cotton/Lyocell Fabrics with Heterogeneous Photocatalysis with Titanium (IV) Oxide Under UV Light and Investigation of the Effect of Oxygen Radicals on Bleaching Process. 2020 (Volume: 27), 118.

[51] Barros, M. A., Conceição, D. S., Silva, C. G., Sampaio, M. J., & Faria, J. L. (2020). Sustainable Bleaching Process of Raw Cotton by TiO2 Light-Activated Nanoparticles. U. Porto Journal of Engineering, 6(2), 11-21.)

[52] Safi, K., Kant, K., Bramhecha, I., Mathur, P., & Sheikh, J. (2020). Multifunctional modification of cotton using layer-by-layer finishing with chitosan, sodium lignin sulphonate and boric acid. International Journal of Biological Macromolecules.

[53] Lu, L., Hu, C., Zhu, Y., Zhang, H., Li, R., & Xing, Y. (2018). Multi-functional finishing of cotton fabrics by water-based layer-by-layer assembly of metal–organic framework. Cellulose, 25(7), 4223-4238.

Chapter 6

Surface Design Technique through Tradition Technique

Harozila Ramli and Tajul Shuhaizam Said

Abstract

This aim of this study is to examine the application of tritik technique in creating exquisite batik pattern design. Essentially, tritik is a technique in batik pattern making that is almost similar to the "tie-and-dip" (ikat-dan- celup) technique; however, the subtle difference between the two techniques lies in the aspect of fabric treading, with the former being able to produce elegant and appealing patterns. This study used a qualitative approach using an observation method in which the researchers observed the creation of such art through studio practice. Essentially, the examination of the practice of such a technique was carried out based on direct observation and unstructured interviews, and the collection of textile products. The research findings showed the experimentation of the tritik technique in the textile pattern designs was highly effective, as evidenced by the exquisite aesthetical effects on the surface of the fabrics, such as the formation of elegant lines consisting of dots and of dashed lines and 3-dimensional texture. In addition, the research findings revealed that the quality of fabrics, the type of colors, and sewing polarity heavily influenced such exquisite tritik pattern design. Collectively, all the above elements were intertwined that helped create appealing, beautiful tritik pattern design infused with high aesthetical values.

Keywords: tritik technique, Batik, Textile pattern designs

1. Introduction

Textile art in the Malay world, especially in Malaysia have been detected since the start of the historical development of the culture of the archipelago. Since the Sultanate has recorded about how different types of fabric and textiles is taken as an omen to the status of goods in individual position in society and benchmark the progress of Malay civilization.

According to Raffles in The history of Java [1] has described how different types of clothing and fabrics are unique with the technique of patterning the surface of the fabric such as tie and *tritik*, as well as illustrations of batik patterns produced. Skeat [2] also describe how the Malays use natural coloring during dyeing silk and cotton which are obtained from Singapore.

In a note, Winsteadt [3] Malay Industries Part I, Art and Craft, he also describes the surface design techniques on fabric produced by the Malay community at the time. Techniques that have been applied are such as coloring, weaving, embroidering, embroidering and knitting techniques for the production of textile patterns and ornaments. Obviously tradition surface design of decorative fabric or Malays textile surface have long practiced and it has been developed and passed down from generation to generation until now in the development of traditional cultural arts in Malaysia.

Essentially, batik making is a method of creating beautiful textile materials or cloth involving the use of candles and coloring materials based on natural or synthetic colors. In creating batik, candles are the main medium used to create the required pattern and, at the same time, serve as the medium to separate the colors. To help create artistic batik, several techniques can be used, including tradition technique, the use of *canting*, metal block, wood block, screen, and, lately, dedicated computer software, to create digital batik products.

Tritik technique is one of the traditional decorative techniques that have long been practiced by textile craftsmen in Malaysia. The adaptation of this technique has become one of the uniqueness of batik design in Malaysia apart from the technique of dyeing, canting, and the use of batik blocks and also screen printing as well as the use of brush techniques on fabric.

Tritik is indeed not a new discovery in textile history. This technique has existed for a long time when society began to explore fabrics and colors in dyes for fabric coloring. Instead, there was previously in India called Bandhani and Japan called Shibori, in Malaysia and Indonesia called Tritik. In fact, there is a much older tie-dye motif found in Peru in 500. The designs found include circles and small lines with bright colors, such as red, yellow, blue, and green.

But in Japan and China have developed tie-dye techniques since the sixth century using silk cloth. Silk fabric is evaluated as a suitable material for a more perfect color absorption process. These skills are also likely to have evolved in the Malay Archipelago as a result of trading activities involving the exchange of goods in the past. Skills staining on fabrics, ornaments and decorations technique is adapted according to the nature of Malay culture and become a work of art in the textile design community in the archipelago.

Tritik or *sasirangan* batik is one of the high fashions that help project the uniqueness and beauty in terms of its creation, such as the type of polarity or motifs created with the method of sewing and pull. In the early history of textile, this technique was used by the Banjarmasin society in which the early design of tritik batik only used simple motifs deemed moderately sufficient to meet the fashion needs of the people dwelling in the district of Banjarmasin. However, in tandem with the advancement in fashion designs taking place in the world, tritik batik has undergone a series of innovative transformations through which the patterns and motifs created by such technique have been reshaped and redesigned with diverse geometrical and organic patterns that helps project their artistic beauty.

Moreover, the application and combination of colors also play an important role in establishing the required motif and pattern on the surface of the batik design. Surely, the knowledge and skills in pattern design of fabric surfaces are a critical element in designing exquisite motifs on such surfaces [4].

Consistent with the current trend in fashion designs, the new, contemporary tritik batik, with its exquisite aesthetical effects visibly appearing on the surface of the fabrics, helps make its wearers look elegant and attractive. Despite the uniqueness in such pattern design, tritik technique has gradually being neglected in today's batik pattern design, which is partly attributed to the complicated process involved in making such design.

To help sustain the use of batik in Malaysia as a national attire, the Malaysian government had made it compulsory for the public servant to wear batik shirts or *baju kurung* (women Dress). Apparently, the rapid development of fashions has been instrumental in influencing the design of fashions throughout the world.

Despite such development, however, some of the traditional designs, such as batik blocks, batik drawing, and batik printing, have managed to survive the test of

DOI: http://dx.doi.org/10.5772/intechopen.97069

time, with many fashion fans keeping their loyalty with such designs. As such, the use of tritik technique can be re-energized to create batik that has a new appealing design with high aesthetical and artistic values and exquisite pattern design that projects unique beauty. Admittedly, due to the rapid development of the fashion world taking place at an unprecedented rate, the tritik technique has started to decline in its use in the making of batik textile. Unmitigated, such a decline will see such a unique technique becomes obsolete – a thing of the past – in batik-textile making. Obviously, more efforts have to be put in place to address this pressing predicament by encouraging practitioners to adopt the tritik technique in designing intricate batik patterns. Another problem that contributes to the declining use of such a technique lies in the lack of proper learning or training in pattern design of batik textile, especially with respect to the structure of patterns that needs to be discerning learned. For example, the knowledge regarding the closely aligned stitches to create intricate patterns with amazing characteristics, such as sharp teeth, base, *dovo, regulon,* and *gadan*, and the application of red, green, and yellow have to be mastered by practitioners.

Seen from the socio-cultural viewpoint, such a problem is the manifestation of the lack of knowledge among the members of the society, in particular, Art students, with respect to the societal impact of the tritik technique, effecting a decline in the awareness or appreciation of such a culturally enriched method of producing traditional batik. Clearly, to help overcome such a problem, the tritik technique needs to be used in the pattern design process to produce elegant and immaculate patterns, which are on par with those created by other techniques, such as tie-*and-dip* (ikat celup) technique.

Based on a practical studio experimenting with the tritik technique in the making pattern - design process of batik motifs on the surface of a fabric. In addition, the effect of this technique on the surface of the fabric, also has been examined which began from the creation of the Napthol color through the mixing of Diazo salt and Remazol coloring dye to the complete tritik process performed on the fabric. Through practiced studio process, focusing on the process prior to sewing was carried out, the inherent constraints encountered during the process of sewing a particular polar of a pattern and the effects of untying knots on the fabric, also able to identify the outcome of the pattern design of the tritik technique.

Definitely, the selection of suitable fabrics in creating tritik batik is of paramount importance. Obviously, the use of quality fabrics will improve the rate of absorption, enabling the coloring materials to penetrate deeply into the fabrics to produces stunningly attractive, intricate, and appealing effects of the tritik technique. In this regard, the use of suitable fabrics has a profound impact on the effectiveness of the tritik technique that helps the Naphthol color to seep deep into every fabric of the batik materials. To date, several types of fabrics have been widely used with this technique, such as cotton fabric, rayon fabric, and silk fabric, which are clothes made from natural sources. Essentially, such fabrics contain natural fibers with good "working characteristics", with which the tritik technique can produce amazing effects.

In Malaysia, the majority of people prefer to wear clothes made of cotton. Such a preference is not surprising as cotton can easily absorb sweats produced by the human body in countries in the tropical region of the world, such as Malaysia. In essence, this type of cloths is made from cotton fibers that are used to make short, soft, and fluffy fibers In general, these cotton fibers are used as the primary material in making shirts, robes, bedspreads, and others. Given their delicate characteristics, cotton fabrics are suitable for batik practitioners who manually use their hands with some degree of force in making batik materials (**Figure 1**).

Figure 1.
Cotton fabrics are suitable for batik practitioners.

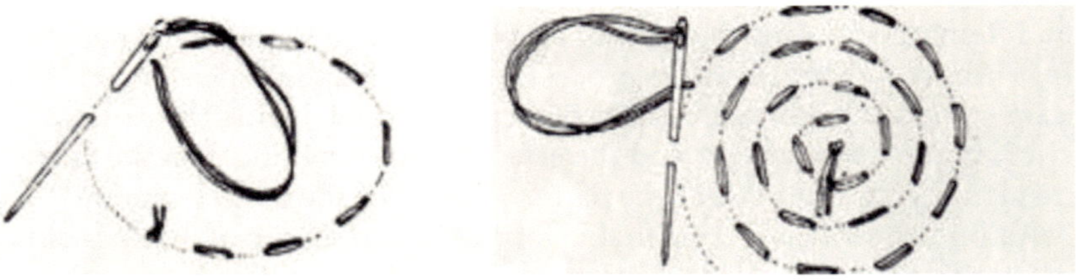

Figure 2.
Illustration of sewing or stitching process in a spiral.

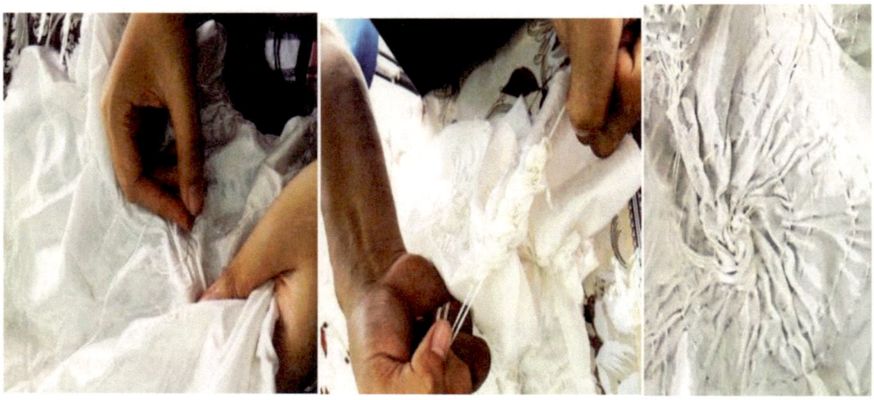

Figure 3.
Sewing process in a spiral moves according to a prescribed spiral pattern.

2. The method of stitching

In general, this type of stitching has a number of diverse sewing techniques, but to create a pattern on the fabric will entail the needle to move in a spiral. Effectively, such a spiral motion of the needle, in which it moves according to a prescribed pattern based generally on a distance of I cm, can help achieve the desired effects.

Furthermore, the threads need to be tightened when the sewing or stitching process has completed. Subsequently, colorings will be swiped over the entire surface of the fabrics that have been completely sewn (**Figures 2** and **3**).

3. The tritik cotton-fabric patterns

The followings are some of the patterns of the cotton cloth created by the effects of the stitching technique used. Clearly, such forms and shapes of the patterns were the results of a sewing or stitching processing a particular direction or polarity, effecting the desired effects that helped create such amazing pattern designs (**Figure 4**).

4. The rayon fabrics

Principally, Rayon is a fabric that can be weaved or merged, depending on its diverse use. In fact, the effectiveness of stitching Rayon is relatively higher than those of other fabric materials, such as *taf* cloth of cotton cloth. In the batik-making industry, the Rayon fabric is categorized as a semi-soft fabric that most batik

Figure 4.
Spiral patterns of the cotton cloth created by the effects of the stitching technique.

Figure 5.
The rayon fabrics.

practitioners find easy to manually work on. As such, the use of this fabric should be emphasized to achieve the desired effects on such fabric (**Figure 5**).

5. The method of stitching

The type of sewing or stitching as shown above is based on horizontal sewing that cuts the surface of the fabric neatly. Ideally, the distance of the stitched fabric

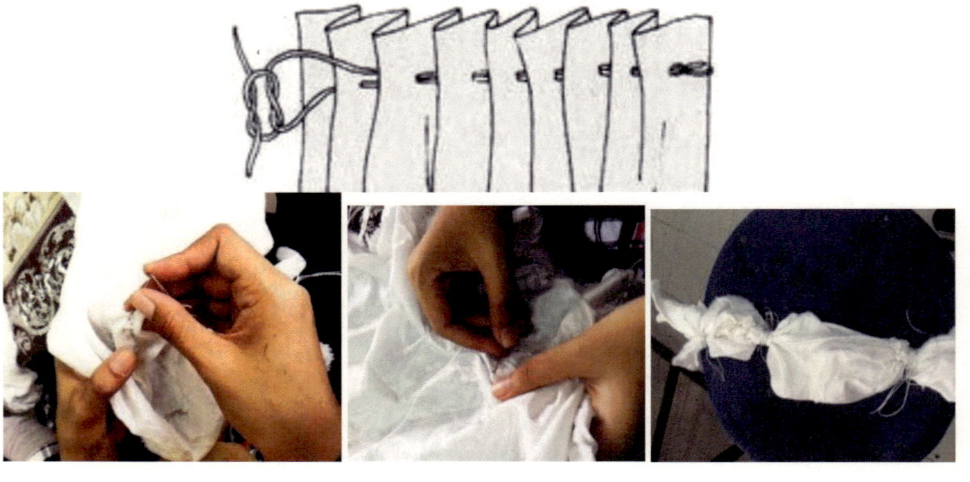

Figure 6.
The process of horizontal sewing performed on a fabric.

should be in the range between 1 cm and 2 cm. obviously, the direction of sewing that is straight and compact will create an amazingly appealing effect. In particular, the end of the cloth must be tied to achieve a better effect (**Figure 6**).

6. The tritik rayon-fabric patterns

As shown in **Figure 7**, the result of using the tritik technique on the Rayon fabrics showed stunning effects, visibly highlighting the effects of colors and stitching on such fabric. Evidently, the stitching the fabric horizontally did not in any way compromise the quality of the fabric. On the contrary, such a stitching method was able to project the undulation of the movement of colors together with the desired pattern on the fabric.

7. The type of satin fabrics

As contended by almost all practitioners, the satin fabric is regarded as the most elegant fabric compared to other types of fabrics, making it a high-class fashion material. This contention is not without reason, as this type of fabric has a surface is delicately soft and glossy, the characteristics that create stunning reflections under the light. In general, satin cloth consists of silk or Rayon, which makes its surface extremely soft. The drawback of this fabric, however, is that it needs constant care, given the delicate nature of its material, which is made up of the softest fibers. To date, satin fabrics have been widely used in many designer fashions throughout the world, notably in developed countries (**Figure 8**).

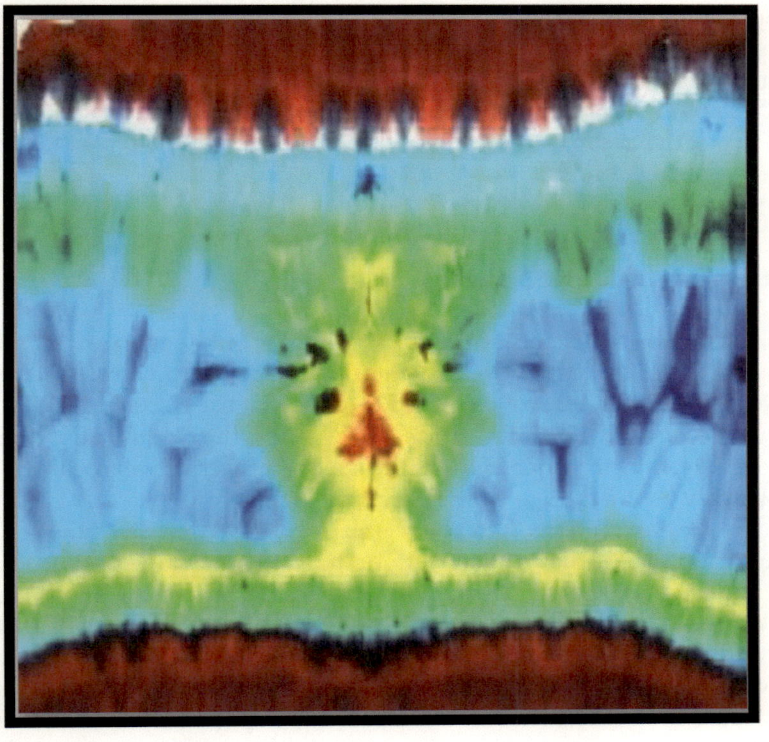

Figure 7.
Tritik technique on the rayon fabrics showed stunning effects, visibly highlighting the effects of colors and stitching on rayon fabric.

Figure 8.
Satin fabric.

8. The method of stitching

The appropriate configurations of such stitching for such fabrics are circular and horizontal. In this study, the configuration examined was based on circular sewing involving a single direction of movement, of which the closer the distance of the stitches the more attractive the effects on the surface of the fabric.

As shown in the **Figure 9**, spiral stitching based on the close distance among the stitches will create stunningly beautiful effects on the fabric. Furthermore, the edges of the cloth have to be permanently fastened by pulling the thread forcefully to create the desired effects.

9. The tritik satin-fabric patterns

Figure 10 shows the effects of the tritik technique on the surface of the satin fabric. Revealingly, it shows that a well-balanced use of colors can create spectacularly attractive and beautiful effects compared to those that use colors that are too bright or too dull. Through this practical studio-based study, the researchers examined the practice of the tritik technique in the batik-making process involving three types of fabrics, namely cotton fabric, rayon fabric, and satin fabric.

Based on the observations, it can be reasonably argued that each type of fabric has its own unique and beautiful tritik effects, despite using the same sewing or stitching configuration. Surely, such differences in the tritik effects lie in the properties of the fabrics, with each having different thickness and structure of fibers, which produce the unique texture of the fabrics. Clearly, the different types of fibers make some fabrics soft while others coarse, the impact of which will have a profound impact on the rate of absorption and the rate of evaporation of liquids that result in different effects on the patterns of the fabrics. Given such inherent differences, the selection of appropriate fabrics should be treated with caution – in fact, it should be treated as the basis – to help create specular and stunning patterns using the tritik technique.

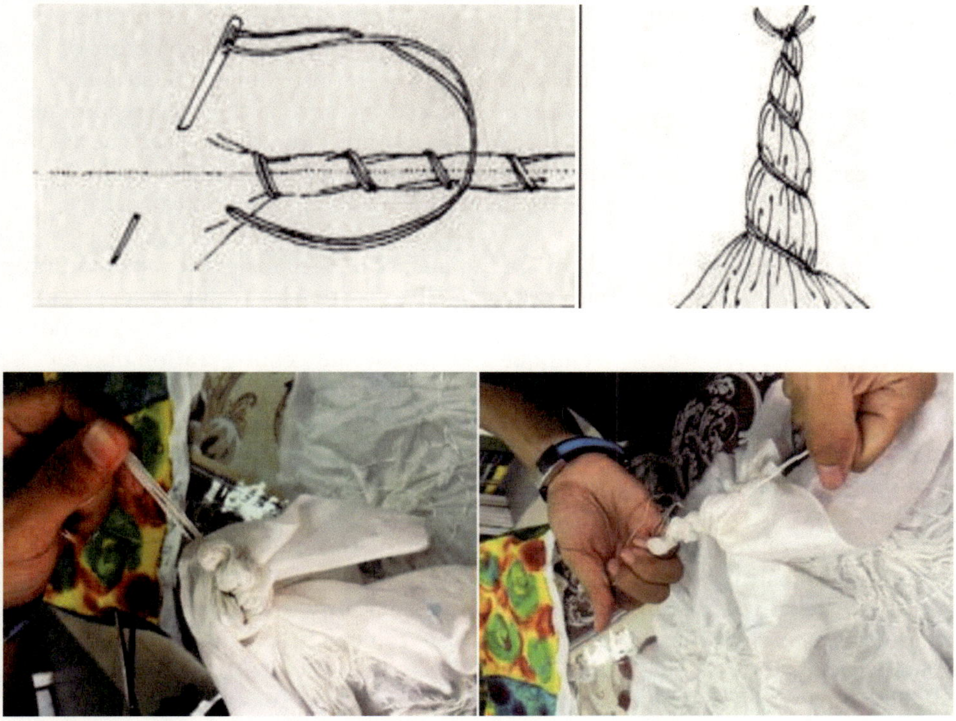

Figure 9.
Spiral stitching based on the close distance among the stitches will create stunningly beautiful effects on the fabric. Furthermore, the edges of the cloth have to be permanently fastened by pulling the thread forcefully to create the desired effects. The process of spiral stitching performed on a fabric.

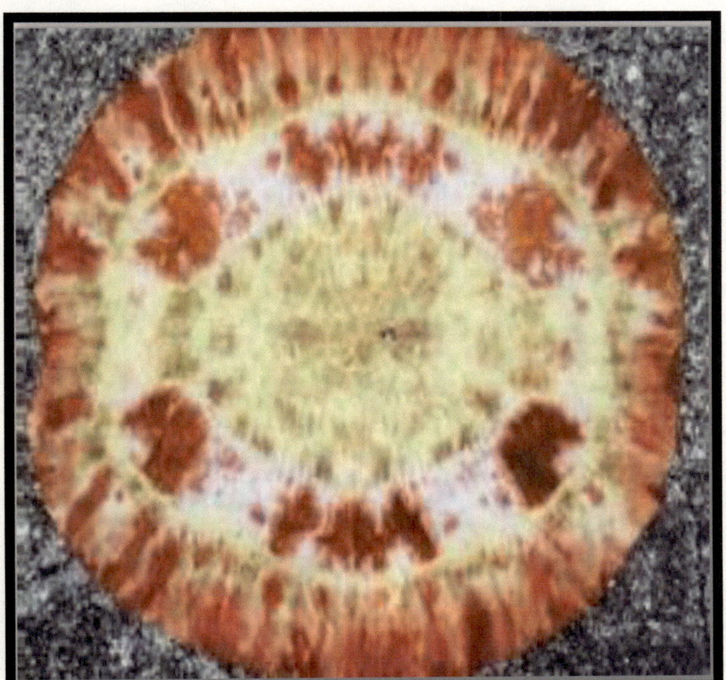

Figure 10.
Tritik technique on the surface of the satin fabric.

Moreover, the quality of stitching also depends on the sewing configuration that can help create beautiful effects by controlling the form or the structure of such a pattern. Also observed in this study was that pattern designs in various organic forms or shapes seemed to be the dominant pattern in the tritik technique to produce patterns with high aesthetical values. In addition, the distance between stitches can strongly influence the effects on the patterns made on the fabrics. Likewise, the strength of the knots is also important in creating such attractive patterns.

Evidently, the closer the stitching on the surface of the fabric, the more stunning the patterns will be. Similarly, the tighter the threads are tied, the more spectacular the tritik effects will be in producing beautiful, delicate lines of various sizes and quality. Undeniably, the tying technique and the stitching configuration play an important role in the tritik technique in creating beautiful, unique patterns. In terms of the use of coloring materials, the tritik technique heavily relies on relevant colors to create the desired tritik patterns on fabrics. In fact, such a technique emphasizes well-balanced and judicious use of colors, given that the tritik pattern entails the tone of colors that is neither too strong not too weak.

Clearly, a well-balanced use of colors in the tritik technique can produce patterns that harmoniously blend the chosen color to produce pleasing effects, highlighting a spectacular contrast of colors that enrich the beauty the batik fabric. In this regard, the mixing of Naphthol color and Diazo salt can help produce a color tone that represents the color of the earth's soil. Thus, it cannot be overstated that the coloring effect is an important element in designing beautiful, intricate patterns on the surface of fabrics, which can be carried out by experimenting with colors and sodium silicate. The effect of tritic techniques on fabrics has indirectly created new

Figure 11.
Type of fabrics: Cotton. Technique: Dipping Tritik. Medium: Naphthol color. Soaking duration (in sodium): 6 hours.

patterns with very unique organic and abstract shapes. The effects of color patterning the shapes on the surface of the fabric is one of the characteristic privileges tritik technique that can provide confirmation of the identity of batik fabrics are processed. The followings figure showcase the pattern designs of various fabrics created by the tritik technique (**Figure 11**).

10. Tradition technique *vs* global trending

The experimentation of the tritik technique in designing patterns is a new learning process that effectively has helped create a new, diverse technique in batik textile industry. Specifically, practitioners can use this unique technique, which is slowly being forgotten, to manipulate the method of sewing or stitching threads on the surface of fabrics, which, in principle, the experimentation with ways to create beautiful pieces of fashions with colorful pattern designs (**Figure 12**).

As demonstrated, the effects of decorative arrangements created by the tritik technique is both refreshingly amazing and attractively mesmerizing, with the surface of fabrics infused with design elements and principles that give rise to high aesthetical values of the fabric materials. In addition, both the intended effects and the unintended effects resulting from the application of colors in the tritik technique can help create the desired forms, shapes, lines and spaces on the fabric materials. Furthermore, exploring the techniques and integrating the knowledge and skills pertaining to synthetic coloring materials can pave a way for the improvement in the learning of pattern designs.

According to a study conducted by Bintan Titisari, Kahfiati Kahdar and Intan Rizky Mutiaz in writing an article entitled Development of Dye Sewing Techniques (Tritik) with patterns geometris [5] suggests a very significant finding on how the application of Dye Sewing techniques (Tritik) can be implied in the fashion world. The effect of the use of geometric patterns on political techniques will produce motifs with the effects of direction, depth, and movement (optical illusion) by using the composition of balance, rhythm and harmony. In addition to the presence of effects optical illusions that give the impression of depth, direction and motion, they can be used to create dimensions and illusions in fashion products. The effect of Sewing Techniques (Tritik) from this traditional heritage can also be adapted using

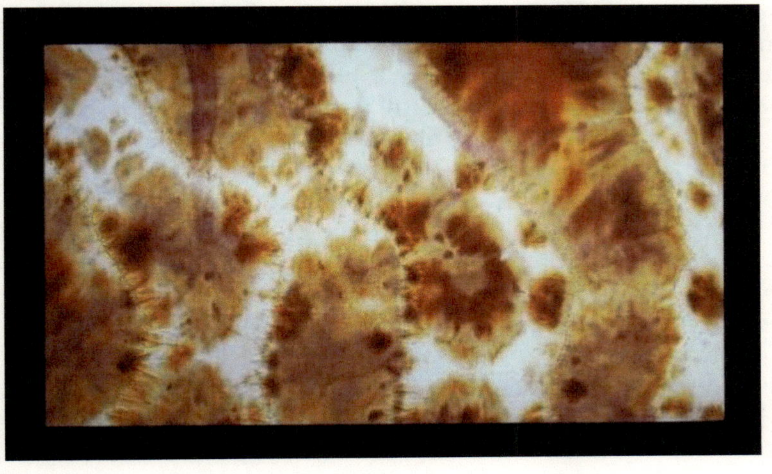

Figure 12.
Type of fabrics: Satin. Technique: Brush-swiping Tritik. Medium: Remazol color. Soaking duration (in sodium): No soaking involved.

Figure 13.
Type of fabrics: Rayon. Technique: Swiping and dipping Tritik. Medium: Naphthol color. Soaking duration (in sodium): No soaking involved. Year 2016.

the latest technology with the help of computer applications and industrial-scale sewing machine technology that can make new contributions in textile technology, for example, geometric patterns using vector graphics editor can be used as preliminary data for development in the CAM (Computer Aided Manufacture) program (**Figure 13**).

The effect of the Sewing Pattern design (Tritik) can also be commercialized in the Fashion industry design where the illusion effect of this geometric design gives a soft finish to the fabric and further highlights the design to visualize the camouflage effect (see **Figures 14–17**). The result of the tritik technique adapted from this traditional technique is an alternative effect that can be designed on the surface of batik fabric. Traditional techniques from hand sewing skills can further highlight the value of the beauty of decorative patterns on batik fabrics.

On the international scene, batik has already taken its place in the contemporary fashion industry. Now the fabric is not only used for traditional clothing, but has also found its way to applications such as haute couture as well as being used in accessories such as handbags [7]. Many popular figures have walked the red carpet proudly wearing batik, from Bill Gates, Nelson Mandela to Barack Obama, and from Beyoncé Knowles to Jessica Alba. The international fashion scene has seen batik designers introduce batik to the world through the mixing of fabrics with modern designs and production methods. For example, Malaysian fashion designer Fern Chua presented handmade batik designs to the world stage through the British Council's global campaign. Highlighting the theme of Crafting Futures, the campaign also brought together fashion and craft designers from around the world to explore and build the future of batik's potential globally. The works of others from the world's batik designers, and many more have also supported batik on the international stage.

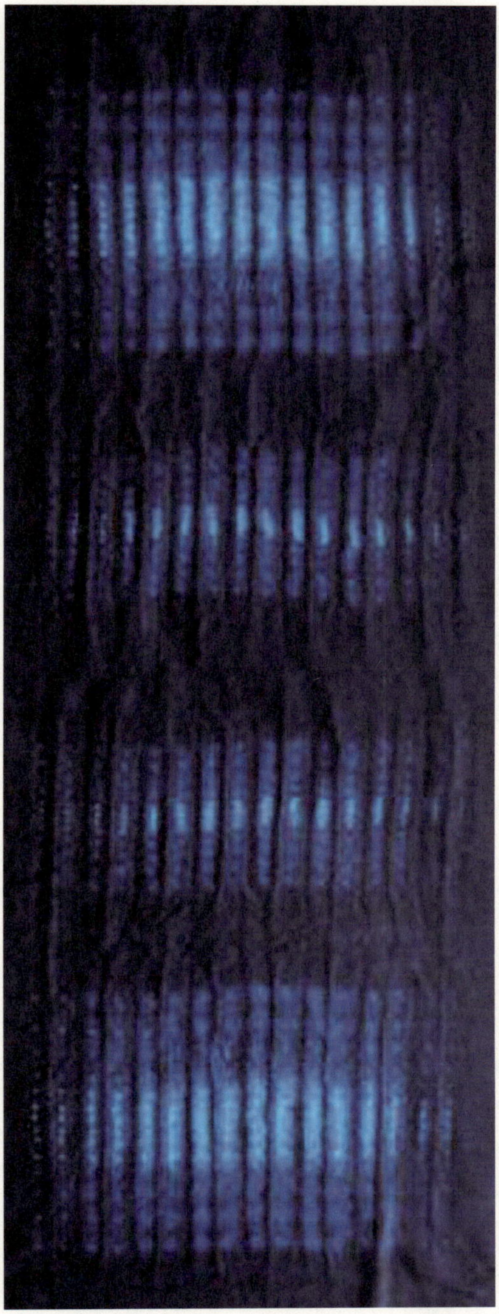

Figure 14.
Tritik techniques that can be used as an illusion pattern design for the fashion industry. Photo credit to Titisari et al. [6]

These advances have also influenced well -known designers from other countries to include batik in their design collections. Notably, Belgian-American designer Diane von Furstenberg's batik dress worn by Duchess of Cambridge Kate Middleton; while Angelina Jolie was seen wearing a batik dress by US designer Nicole Miller. Other international designers who also feature batik in their collections include Dries van Noten from Belgium, Ek Throngprassert Thailand, and Milo

Figure 15.
Fashion design that adapts sewing techniques (tritik) in Malaysia.

Figure 16.
Fashion design that adapts sewing techniques (tritik) by SEYMOUR. Photo credit to BLOG DESIGN BY LABINA @ PLEXICOD.

Figure 17.
Fashion design that adapts sewing techniques (tritik) by Humbang Shibori x Purana at JFW 2019. Photo credit to (Fimela.com/Nurwahyunan).

Milavica from Italy. In addition, one of the oldest fashion schools in Italy, Koefia, not only incorporates batik fashion in its curriculum, but also parades its stylish designs on the catwalk. Therefore, the practitioners of batik fashions can capitalize on the effects of the tritik technique to help them create spectacularly stunning and beautiful pattern designs on the surface of the fabrics of batik textile in global. To help realize this aim, it becomes the imperative of the stakeholders and practitioners to rejuvenate such a technique that is capable of creating immaculate and unique pattern designs with high aesthetical values.

Author details

Harozila Ramli* and Tajul Shuhaizam Said
Sultan Idris Education University (UPSI), Perak, Malaysia

*Address all correspondence to: harozila@fskik.upsi.edu.my

References

[1] Raffles, Thomas Stamford, (1817) The History of Java. London: Parbury and Allen

[2] Skeat, W.W. (1902) "Silk and Cotton Dyeing by Malay". The MBRAS,1,123-127

[3] Winsteadt, R.O. (1925). " Malay Industries Part1: Arts and Craft" in R.J Wilkinson (Ed). Papers on Malay Subjects. Kuala Lumpur: Government Press

[4] Harozila Ramli, Tajul shuhaizam & Siti Salwa (2019). The Beauty of Tritik Technique in Creating Batik-textile Pattern Designs, Journal of Advanced Research in Dynamical and Control Systems 11(05-special issues):1105

[5] Hestri Wulansari (2005). *Tesis PhD: Perancangan Teknik Tritik Dengan Penambahan Struktur Tenun Sebagai Pelengkap Busana.* Surakarta: Universitas Sebelas Maret Surakarta.

[6] Titisari, B., Kahdar, K & Mutiaz, I. R. Pengembangan Teknik Jahit Celup (Tritik) dengan Pola Geometris. ITB J. Visual, Art & Design 6 (2), 131 (2014).

[7] Ira Dhyani Indira (2009). *Batik Ceria*. Jakarta: Puspa Swara

Section 3

Testing and Analysis

Chapter 7

Smart Textiles Testing: A Roadmap to Standardized Test Methods for Safety and Quality-Control

Ikra Iftekhar Shuvo, Justine Decaens, Dominic Lachapelle and Patricia I. Dolez

Abstract

Test methods for smart or electronic textiles (e-textiles) are critical to ensure product safety and industrial quality control. This paper starts with a review of three key aspects: (i) commercial e-textile products/technologies, (ii) safety and quality control issues observed or foreseen, and (iii) relevant standards published or in preparation worldwide. A total of twenty-two standards on smart textiles – by CEN TC 248/WG 31, IEC TC 124, ASTM D13.50, and AATCC RA111 technical committees – were identified; they cover five categories of e-textile applications: electrical, thermal, mechanical, optical, and physical environment. Based on the number of e-textile products currently commercially available and issues in terms of safety, efficiency, and durability, there is a critical need for test methods for thermal applications, as well as to a lesser degree, for energy harvesting and chemical and biological applications. The results of this study can be used as a roadmap for the development of new standardized test methods for safety & quality control of smart textiles.

Keywords: smart textiles, wearable electronics, test methods, quality control, safety, efficiency, durability, electronic textiles (e-textiles)

1. Introduction

The smart/electronic textile market has recently exploded, mostly driven by personal healthcare. The term "smart textiles" refers to the "smart functionality" of a product, whereas "electronic textiles" (e-textiles) refer to the "hardware and/or technology" that is responsible for the smart functionality [1]. The market size of smart textiles already reached USD 4.72 billion in 2020 with Asia-Pacific countries leading the chart followed by Americas and Europe [2]. Vista Medical Ltd. (Canada), Myant (Canada), Interactive Wear (Germany), Schoeller Textiles (Switzerland), Intelligent Clothing (England), Google (US), International Fashion Machines (US), Textronics (US), Gentherm Incorporated (US), and Sensoria (US) are the major key players in the smart textile industry.

The convergence between textile substrates and conformable electronics like embedded sensors or actuators has given rise to wearable smart/e-textiles. E-textiles can augment the level of protection, comfort, and physiological performance of humans, with applications in many industries, including medicine, protective clothing, military, and automotive. A few authors have analyzed these current and potential applications. For instance, Honarvar and Latifi described the components, structures, and major application areas of smart e-textiles, including ambulatory measurements for patients with cardiovascular diseases, nonwovens for electromagnetic interference (EMC protection) for security, protective GPS-suits for military, bleeding sensor threads for surgeons, and flexible electronic keypads for dialing phone numbers [3]. Ismar et al. explored the use of e-textiles for futuristic clothes [4]. Dolez et al. analyzed the potential of smart textile technologies for occupational health and safety (OH&S) [5]. Finally, Stoppa et al. described different biomedical smart textile projects conducted within the European Commission's 6th and 7th framework programs: WEALTHY, MyHeart, BIOTEX, PROTEX, STELLA, OFSETH, CONTEXT, WearIT, and PLACE-it [6]. This convergence between clothing and electronics could pose some critical challenges for regulatory bodies, including US Food and Drug Administration (FDA), Health Canada, and National Institute for Occupational Safety and Health (NIOSH). Appropriate quality control methods are a critical tool for them to ensure that e-textiles do not to endanger users' health, safety, and privacy among others.

The lack of standardization of e-textiles is also considered one of the primary restraining factors for industrial growth. Even though the e-textile industries have generally been keen on designing products with improved safety and performance features, their efforts may not have met market expectations due to the current lack of dedicated standardized test methods. The two main disciplines at the root of e-textiles - textiles and electronics - are so much at odds with each other that dedicated standardization methods for smart/e-textiles are critical. However, progress in this area is lagging behind in comparison to the rapid pace of technological innovation. In this chapter, we will highlight critical challenges and provide some suggestions for the development of standardized test methods for smart/e-textiles.

2. Overview of smart/e-textile products and major barriers to market entry

As consumer electronics are marching towards the era of the Internet of Things (IoT), so are smart/e-textiles. Gradually, conformable electronics are embedded within textiles of various configurations to offer an on-body platform for pervasive computing, especially for healthcare and OH&S applications. Examples include a smart trouser for forest workers that can detect the proximity of chain saw and automatically turn it off [5]; industrial protective gloves that alert users of air toxicity by changing their color [5]; power vests to prevent unsafe movements of caregivers while lifting heavy weights [7]; and Myant's recent VOC (volatile organic compound) sensing facemasks to detect airborne infectious agents [8]. Also, smart textiles have been designed for protection against sexual assaults, with the SHE (Society Harnessing Equipment) anti-rape lingerie that can deliver a 3800 kV shock [9]. On a lighter note, Microsoft patented a smart cloth that alerts a user of an incoming text message or daily activity reminders, by generating a mild electric shock to the body [10].

A survey of technologies, solutions, and products based on smart textiles and flexible materials was done in 2017 [5]. The different technologies, solutions, and products in terms of sensors and actuators identified were grouped into seven categories based on the input signal or stimulus for the sensors and the output signal for actuators. These categories are thermal, mechanical, chemical/biological, electrical, physical environment, optical, and power. **Figure 1** shows the distribution of

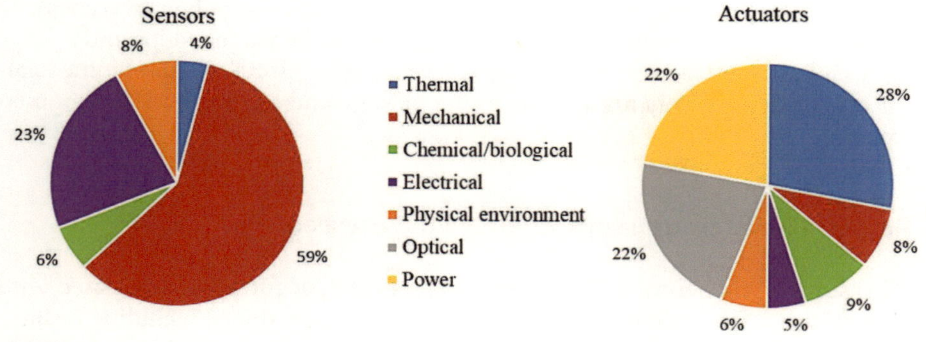

Figure 1.
Distribution of technologies, solutions, and products relevant to smart textiles and flexible materials as a function of the stimulus for sensors (left) and output signal for actuators (right) [5].

the technologies, solutions, and products based on smart textiles and flexible materials identified by the researchers in these different categories. In the case of sensors, the dominant category is associated with a mechanical stimulus, with 59% of the sensors. Electrical sensors account for 23%. In the case of the actuators, thermal, optical, and power outputs represent each about a quarter of the technologies, solutions, and products identified.

However, before the mass adoption of smart/e-textile products is possible, some burning questions need to be addressed: for instance, what is a safe electrical shock, both for user alert and assailant deterrence? Could a malfunctioning smart garment prevent activating a safety emergency shut-off system? What about potential privacy issues associated with the data generated by e-textiles? In an attempt to standardize their assessment of wearable electronic product performance, a group of electrical engineers evaluated the safety performance of wearable energy harvesters based on the device failures and user-related hazards [11]. However, to date, no one has provided a response to the questions customers could legitimately ask for the different applications smart/e-textiles are aiming for.

For instance, according to experts, the lack of standardization and quality control poses the highest barriers to smart textiles entry into the healthcare market (**Figure 2**) [12]. Since wearable electronic components are often worn close to the body, special attention is required to prevent health hazards. There are also potential issues of efficiency associated with the interconnections between the different components. The lack of standardized processes for welding, soldering or glueing for instance can significantly reduce the performance, durability, esthetic, and

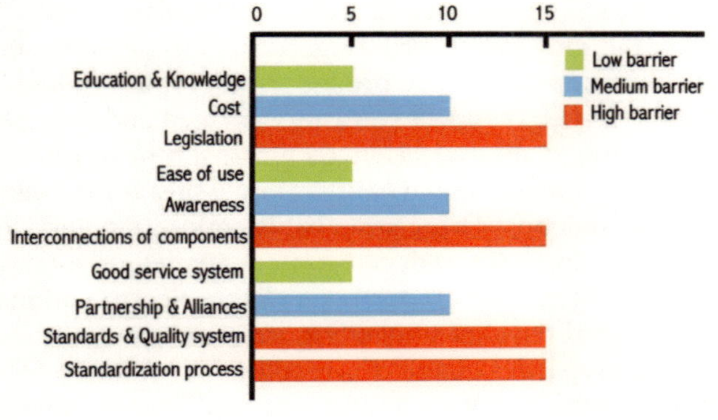

Figure 2.
Barriers to entrance of smart textiles in the healthcare market (based on data from [12]).

hand-feel of the product. Other barriers include product cost, public awareness, lack of education and knowledge [12]. Strategic industrial partnerships and multidisciplinary alliances among key players in textile, electrical, and biomedical engineering can positively impact the product design and test method development processes.

3. Review of different issues reported or foreseen

This section will describe different issues, reported or foreseen, associated with the durability, safety, and efficiency of smart/e-textile products, including health monitoring apparels, protective clothing, automotive actuator textiles, and textile-based physiological sensors.

3.1 Durability

Electrical elements embedded into textile structures to produce e-textiles include electronic circuits, electrodes and printed tracks. They must be extremely rugged, robust, and durable because of their regular exposure to mechanically demanding environments [13]. The issues of durability reported with e-textile products are discussed in the next sections.

3.1.1 During the manufacturing process

A first aspect of durability deals with the e-textile manufacturing process itself. For instance, conductive yarns embroidered on a textile may be damaged by three dominant forces: tension, bending, and shearing [14]. In particular, conductive fibers generally exhibit a low bending radius [15]. However, for flexible display applications, they would be typically subjected to bending radii lower than 1 mm. They also have to withstand friction stresses associated with the embroidery operations. In the case of weaving, fibers must possess the capacity to withstand bending radii as small as 160 μm and 20% tensile strains.

3.1.2 Effect of biomechanical stresses during wear

An apparel product is subjected to large biomechanical stresses during wear, including during donning and doffing. For example, a research conducted with Canadian combat clothing showed that maximum stresses of 2410 and 2900 N/m occur during squatting across the back seat of trousers and coveralls, respectively [16]. Other movements like when bending elbows (for sleeves), bending knees (for trouser legs) or bending over exerted significant stresses on combat clothing of Canadian Forces. If the smart/e-textile product is not robust enough to withstand such biomechanical stresses, they will be easily damaged and experience a loss in functionality like sensing, communication, data-transfer or power supply. Such problems of loss in functionality could cause safety issues for soldiers or first responders in the line of action. Using stretchable connection and electrode designs could allow accommodating body-induced stresses applied on e-textiles. **Figure 3** displays examples of strategies to produce stretchable electro-conductive textiles.

Researchers conducted tensile and bending resistance tests to assess the durability and elastic properties of smart/e-textiles. For example, PEDOT:PSS ((poly(3,4-ethylenedioxythiophene): poly(styrene sulfonate)) dyed cotton and silk yarns exhibited a tensile strength of 260 and 136 MPa with a conductivity of 12 S/cm compared to 305 and 157 MPa for the pristine (uncoated) cotton and silk yarns,

Figure 3.
Examples of stretchable electro-conductive textiles. (a) Elastic behavior of a conductive yarn (white on the pictures) under increasing deformation (i) undeformed, (ii) medium deformation, and (iii) high deformation. The conductivity was maintained even at high deformations. (b) Experimental set-up to measure the electrical resistance of a knitted electrode under stretch in a mechanical test frame.

respectively [17]. The conductive polymer coated silk yarns showed a robust electrical performance, displaying a reduction of conductivity of around 50% after 1000 bending cycles. Using a fabric test tester, Qui et al. measured the durability of a power generating textile fabric [18]; it did not exhibit any measured degradation after 10000 bending cycles of mechanical stimulation, showing superior durability. **Table 1** displays different test methods used by researchers to assess the durability of textile resistive heaters (**Table 1**).

3.1.3 Effect of surface phenomena in service

Besides biomechanical stresses, different surface phenomena such as wear, corrosion, chemical contamination could destroy the transmission functionalities of smart textile components like optical glass fibers [28]. For instance, **Figure 4** illustrates the effect of abrasion on a smart/e-textile webbing (white on the left image (a)) than includes conductive yarns. The multimeter on the right image (b) records the electrical resistance after successive series of abrasion cycles.

To simulate wear behavior, different mechanical tests can be conducted on textiles, for example to measure their abrasion resistance [29]; a lower abrasion resistance would potentially indicate a poor durability of the electrical functionality for conductive tracks on smart textiles. Recent work on a graphene-coated aramid fabric reported a resistance of up to 150 abrading cycles before the complete loss of electrical conductivity [30]. The stability to wear of a power generating textile fabric was analyzed after prolonged use of up to 15 days [18]. The fabric was successful at lighting up an array of LEDs under different dynamic conditions: raising hands, shaking clothes, and human running.

Durability against environmental degradation is another critical factor for smart textiles. For instance, silver-plated textile electrodes may lose their functionality if exposed to air for a longer period because metals are prone to atmospheric corrosion, including silver [31]. When silver is exposed to atmospheric pollution, the surface tarnishes due to a reaction between silver and reduced sulfur compounds in the ambient air [32]. As a result, a dark layer of Ag_2S (silver sulphide) is formed over the silver plating. Sulfur releasing bacteria could also be present in our washing machines, which may lead to a secondary sulfidation of textile silver electrodes [33].

Conductive elements	Test condition	Form factor	Durability test method	Study
Carbon nanotube (CNT) ink	—	Printed element	Tensile strength	[19]
Silver filament	80, 100, and 120 ° C in oven for 264 h	Yarn	Tensile strength	
Silver yarn	65% RH and 20 °C as per EN ISO 2062:2009	Plain, rib, and interlock fabric	Stretchability	[20]
Stainless steel yarn	—	Plain and interlock fabric	Stretchability	[21]
Copper nanowire -polyurethane film	—	Nylon glove with the printed film	Stretchability	[22]
LIG (Laser Induced Graphene) on polyimide film	—	LIG film in contact with copper tape and Ag-paint	Bending test	[23]
Composite ink (graphene-tourmaline- polyurethane)	—	Printed heater on woven cotton wrist band	Abrasion resistance	[24]
CNT-polypyrrole polymer	—	Polymer coated cotton yarn	Bending test	[25]
Multiwalled CNT	—	Coated cotton woven fabric	Bending test	[26]
Carbonized modal knit encapsulated with Ecoflex silicone rubber		Weft knitted fabric	Bending test	[27]

Table 1.
Methods used to assess the durability of resistive heating textiles.

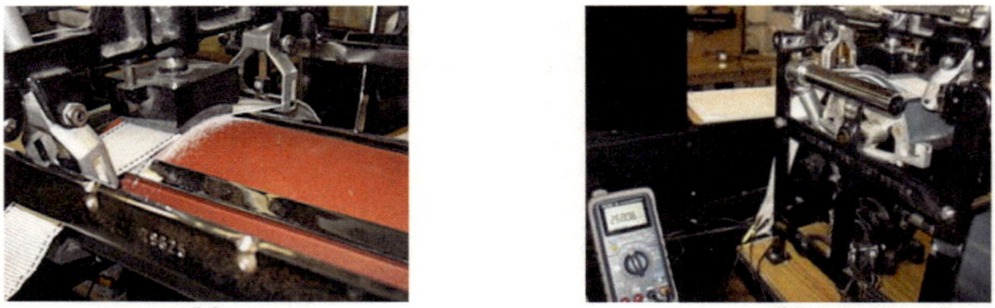

(a) (b)

Figure 4.
Abrasion testing on a white webbing with conductive yarns (a). A multimeter measured the change in resistance after successive series of abrasion cycles (b).

Salt from body sweat during workouts or from seawater in marine applications may also corrode metallic elements of smart textiles [34].

3.1.4 Thermal resistance

Resistance to heat is a critical factor for electro-thermal e-textiles, for example heating textiles. The heating components and the material in contact have to be able to sustain the heat generated with in operation without losing their conductivity, strength and other performance, and without getting on fire or melting. For instance, Liu et al. characterized the impact of heat exposure on Ag fabric heaters [35]. The study reviewed the heating performance of three different knitted fabric heaters, viz., plain single jersey (PSF), ribbed stitch (RSF), and interlock knit (ILK), fabricated with silver plating compound yarns (SPCYs) and polyester staple fiber spun yarns (PSFSYs). After 264 h of prolonged heating of the SPCYs in an oven at three different temperatures, 80, 100, and 120 °C, the electrical resistance of the SPCYs were evaluated. The resistance of the heater increased by ~17% and ~ 75% after the 264 h aging period at 80 and 100 °C, respectively. After 24 h of aging at 120 °C, the resistance exceeded the measuring range of the multimeter. In addition, the textile structure used may affect the thermal resistance of the system. For example, it was reported that a fabric woven with Ag-coated nylon and cotton yarn powered at 15 V exhibited different degradation temperatures depending on the weave structures: 69.8 °C for plain weave, 80.6 °C for twill weave, and 103.5 °C for sateen [36].

3.1.5 Resistance to washing

Washability is a massive barrier to successful commercialization and widespread adoption of e-textiles. It is a critical concern for e-textile users as the washing and drying processes subject the product to damaging conditions and could eventually destroy the connectivity between the electronic components or the electronic components themselves [37]. Chemical stress (detergent, surfactants), thermal stress (washing/drying temperature), solvent (water), and mechanical stress (e.g., friction, abrasion, flexion, hydro-dynamic pressure, garment twist) are the four dominant forces that could damage electronic components during washing cycles. A protective layer is typically used to protect the smart textile or its electronic components from getting damaged or exposed during laundry. Polyurethane (PU) is largely used as a waterproofing encapsulation layer for e-textiles [38]. It has the great advantage of being flexible and stretchable and can accommodate the stretch of the fabric underneath. Polypropylene thin films have also be laminated to provide protection to metallized polymer films on e-textiles against repetitive washing and abrasion [39]. However, the encapsulation may not be durable. For example, the extremity of metal wires encapsulated in an e-textile product could damage the encapsulation layer due to their intrinsic rigidity and configuration geometry. **Figure 5** shows how the extremity of soldered metal wires pierced through a PU lamination after repeated washing cycles.

(a) (b)

Figure 5.
Soldered wires in a smart/e-textile product piercing through the wash-resistant polyurethane encapsulation layer.

(a) (b)

Figure 6.
Resistive heating blanket: (a) electronic control module with wiring for power supply; (b) connection between electronic module and the blanket (identified with the red box). The rigid connector is sewn to the blanket with a single row of stitches (shown in the inset).

Concerns also exist for non-textile rigid elements that are sewn to e-textiles for instance. The laundering process may damage them if they are fragile. They may also hit the flexible conductive interconnects when the textile is tumbled during washing or drying. **Figure 6** displays the example of a resistive heating blanket. The plastic connector between the non-washable electronic modules and the blanket could get damaged when exposed to the mechanical stresses and the elevated temperatures associated with laundering and tumble drying.

Another issue is associated with the tendency of some textile fibers to absorb water. As water swells the hydrophilic fibers, it influences their physical properties [40]. It can also reduce the flexural strength, modulus, strength, hardness, and fracture toughness at the textile fiber-polymer matrix interface [41]. Even hydrophobic fibers can transport moisture by capillary action. The water can reach the different e-textile components and damage them. Electronic modules and batteries are the most sensitive to water ingress, which can instantly and permanently damage them [42, 43]. Proper encapsulation is once again the solution when complete unplugging is not possible [37]. Alternatively, a water-free, air-based laundry system has been designed for smart garments [44].

Product lifetime is very important for consumers. A typical 100% cotton t-shirt provides serviceability for at least 20 washes [45]. Hence, the expectation of consumers is no less for smart textiles, especially due to their high price tag. OMsignal claimed that their smart t-shirt, designed for tracking heart and respiration rate, could undergo 50 wash cycles [46]. However, the company no longer exists. Karaguzel et al. designed a silver ink screen-printed nonwoven electro-textile circuit that could resist up to 25 wash cycles [47]. Cho et al. reported no change in conductivity of an rGO-coated meta-aramid woven fabrics after ten 6-min washing cycles at 40 °C using a Laundero-meter [30]. Similarly, intrinsic conductive polymers like PEDOT:PSS were used to exhaust-dye silk yarns and showed no change in electrical conductivity for up to 4 washing cycles [17]. Laminated and metallized textile yarn electrodes sustained successfully 20 domestic washing cycles according to EN ISO 6330 [39].

Researchers also used prolong washing to demonstrate the stability-to-laundering of smart textiles. Qui et al. observed a constant electrical output (voltage: ~110 V, current: 2 μA) for their piezoelectric energy harvesting fabric based on biomechanical body movements after up to 2 h of continuous washing [18]; only a minor degradation in the output (voltage: ~106 V, current: 1.9 μA) was recorded after 12 h of prolonged washing. The impact of powder detergent (containing conventional chlorine-based bleaching agents), liquid detergent, and sodium percarbonate (an unconventional stain remover based on oxygen bleach) was compared when testing the resistance of Ag-plated nylon electrodes to 30 washing cycles [42]. Detergents with bleaching agents were reported to be more damaging to the Ag-electrodes, as the bleaching agents oxidize the Ag layer, making the conductive layer vulnerable to mechanical rubs and progressive wash cycles. The researchers recommended using liquid detergents, free of any form of bleaching agents, for e-textiles. Recently, researchers from the University of Toronto and the University of Waterloo developed two different electrocardiogram (ECG) electrodes made of silver-coated and carbon-suffused nylon yarns [48]. Although silver-coated ECG electrodes resisted well up to 35 washes, the carbon yarns yielded a longer lifespan and maintained a reasonable signal quality for the ECG biosignals.

3.2 Safety

Safety is the biggest concern for e-textile users because of the fear of electric shocks from embedded electronics. Even though the embedded electronics are responsible for the smart behavior of e-textiles, the safety of the product should not be compromised by the presence of electronic components.

3.2.1 Electric shocks and shorts

Embedded electronics in e-textiles may suffer from short-circuits or mechanical failures, e.g. due to body sweat or ambient moisture, similarly to what is observed for electronic devices in marine environments [49]. Such malfunctioning can cause serious health hazards or fire accidents. For instance, a recall was issue for the Omni-Heat electric jackets of the company Columbia [50]. A manufacturing defect was detected in the heating component of the wrist cuff, which could create an electric short and lead to burn injuries. The electrical insulation of conductive components can be achieved by surrounding the conductive components with an electrically insulating layer, for instance through core spinning, using a tubular intarsia knitting, or by encapsulation in a water-resistant polymer for instance [51].

3.2.2 Exposure to high temperatures

Burns due to exposure to high temperatures is a serious safety concern for users of heating textiles. Skin temperature is around 34-35 °C although it differs slightly between different regions of the human body while the core temperature of the human body is maintained at around 37 °C [52, 53]. **Figure 7** illustrates the effect of exposure of the human skin to different temperatures [54–56]. While the burning pain threshold is at 43 °C, extended exposure of the skin to 45 °C can lead to 2nd and 3rd degree burns. Temperature overshooting or the malfunctioning of heating textiles could cause severe burn injuries, in particular for people with impaired sensations. For instance, a 26-year-old male patient with paraplegia suffered from a hip burn due to a heated car seat while driving a 2004 Jeep Cherokee for 30 minutes [57]. While the patient was unaware that the car seat was preprogrammed to a high setting (~41 °C), it was later revealed that the seat heater malfunctioned and exceeded 41 °C.

Similar unfortunate cases include a 42-year-old post-traumatic paraplegic patient in Germany who required several reconstructive surgeries as a result of burns caused by a heated car seat [58], a 54 old paraplegic patient driving a 1999 Chrysler Town & Country minivan who suffered from blisters in the rear and upper thigh [59], and a 50-year-old diabetic and paraplegic woman who suffered from a

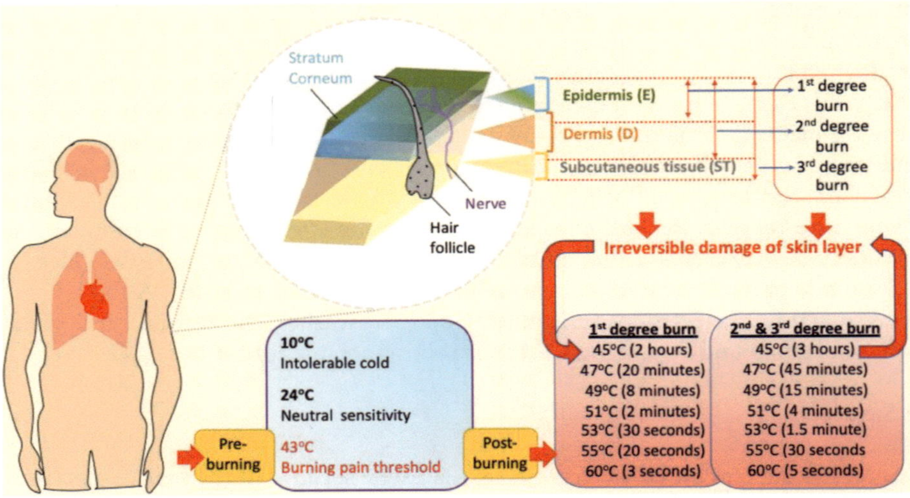

Figure 7.
Resistance of the human skin layers to low and elevated temperatures. Data retrieved from [54–56] (artwork by author).

partial-thickness burn on her medial buttocks [60]. Canada has cold winters and people use heated car seats during the winter; however, most of the heated seats do not display the temperature reached during operation. **Figure 8** displays the example of a North American 2020 full-sized sedan car with its heated car seats and different levels of heat settings; no indication of the actual temperature reached by the heated seats is available. As shown in **Figure 7,** an overheating at 50 °C may cause a 2nd or 3rd degree burn within 4 minutes. Beside medical patients, a temperature overshoot may also increase the risk of fire in the case of apparel articles with poor fire retardancy.

Similar smart heating technologies are also used by diabetic patients. Diabetic patients often suffer from nerve damages, termed as neuropathy, which involves sensory or motor impairment of small and large fibers of the body muscles [61]. The weakness of feet nerves is the most common type of diabetic neuropathy affects. Foot ulcers, sharp or burning pains in feet, and numbness of toes are other neuropathy symptoms [62]. To keep neuropathic pain under a manageable level, patients often undertake different physical therapies, including heat therapies by heating pads (**Figure 9a**) [63]. Since diabetic patients may suffer localized feet numbness and these heating pads are in direct contact with the skin, any temperature overshoot may cause serious skin burn injuries. Unfortunately, different heating textiles like heating blankets, mattress pads or throws (**Figure 9b**) are sold to consumers without proper instructions or clear indications. For instance, the heat regulator does not indicate the level of temperature it generates for its different heat settings (high, medium, low) (**Figures 9c**). Such an approach could harm sensitive skins as

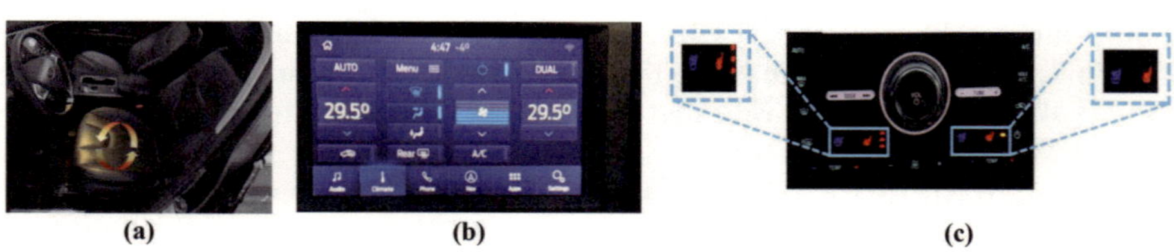

Figure 8.
A full-sized sedan with its heated seat (a) and dashboard control modules with set temperatures (b). The car seat heat indicator is in tally marks (with no reference of actual temperature) (c).

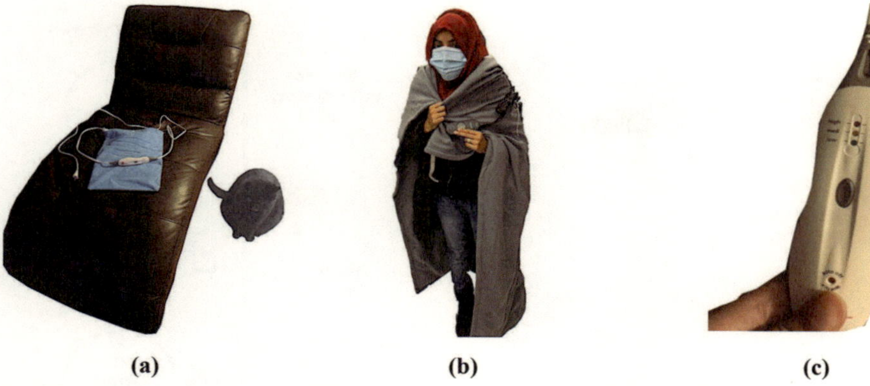

Figure 9.
Examples of heating textile products. (a) a heating pad; (b) a 120 V (A.C.)/60 Hz/ 115 W electronic heating throw made of 100% polyester fiber; (c) heating control module without numerical indication of temperature levels for the three different heat settings (high, medium, low).

the heat tolerance level differs from person to person and patients with diabetes or paraplegia suffer from reduced or impaired sensitive body organs. Operation manuals also miss indications about the temperatures at the different heat settings and warnings about the dangers of prolonged heating times.

In addition, the accumulation of heat over time could also make the users of different electronic wearables feel uncomfortable [64]. Such issue is particularly critical in the case of joule heating textiles where it could lead to burns for the user or instance of fires. Due to the accumulation of heat in the textile over the successive heating/cooling cycles for instance, the temperature may keep on increasing even if the power input remains constant – a phenomenon which was marked in smart nylon gloves embedded with a polyurethane-copper nanowire (PU-CuNW) resistive heating element [22]. Kim et al. reported a temperature increase of 5 °C, from 85 °C to 90 °C, during a 10 h prolong heating period of the smart nylon gloves [22]. This temperature increase over time could potentially be associated with the ~7% increase they observed in the resistance of the conductive element of the PU-CuNW-nylon glove after the 10 h heating period. Thermal inertia may also lead to issues of overheating as the temperature experienced may exceed the set value, which can be associated with a phenomenon of overshooting.

3.2.3 Battery ignition and fire hazards

Recently, the US Homeland Defense and Security Information Analysis Center described the need for integrating multiple energy harvesting textiles on US military protective clothing [65]. Indeed, most wearable electronic systems need to be powered to be able to function. Strategies for textile-based energy harvesting are generally based on triboelectric (based on the friction between pieces of garments during body motion) [66], piezoelectric (from deformation during body motion [67], thermoelectric (using body heat) [68], and photovoltaic (from solar energy) power generation [69]. Recent scientific works also showed the potential of producing biochemical energy from body sweat using textile-based biofuel cell systems [70]. An overview of different energy harvesting textile platforms is illustrated in **Figure 10** for an application for dismounted soldiers.

In addition to energy harvesting, on-body batteries or supercapacitors are needed to store the energy from these energy harvesting fabrics and/or provide some power supply autonomy to the wearable clothing system [71]. However, these integrated batteries could suffer from battery ignition. One such incident was

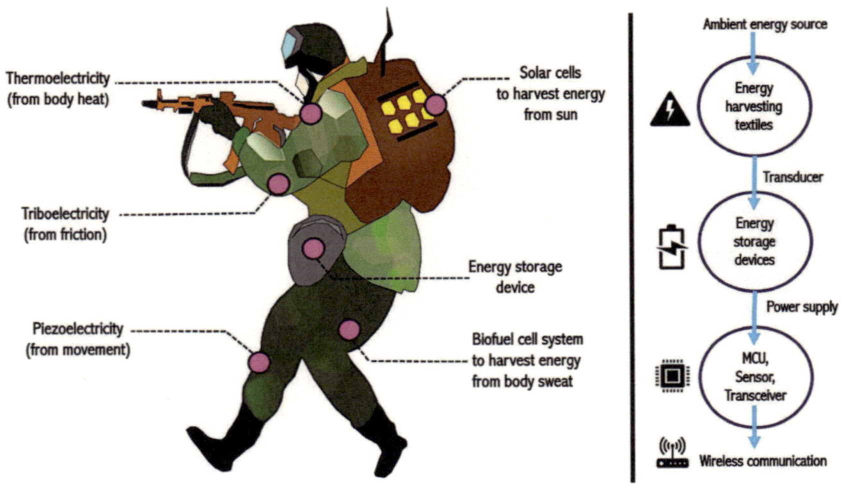

Figure 10.
An overview of energy harvesting systems for autonomous and self-charging protective military clothing with a simplified block diagram for wireless communication platform.

reported by the Department of Police in Arkansas (USA); a smart jacket caught on fire due to the ignition of its built-in battery [72]. Another example concerns the Omni-Heat electric jacket models of the company Columbia [50]. A recall resulted from defective batteries that could overheat and ignite the jacket. Similar incidents could have dramatic consequences, in particular in the military where an increasing number of e-textile systems are being encountered. For instance, in the US, the Future Force Warrior, Scorpion, and Land Warrior programs take advantage of copper and tinsel wire-based textile USB, radiating conductor, and electro-textile cables among others for improved flexibility and real-time information technology in military protective clothing systems [73].

3.2.4 Unstable connectivity

The reliance on smart textiles in case of emergencies is a growing trend for the biomedical, OH&S, and transportation industries. Any inconsistency or flaw in the interconnecting conductive tracks may render the emergency smart textiles dysfunctional, with potential dramatic consequences. Moreover, if the textile antennas or wireless communication system suffer any disruption in the communication protocol, it will make the user vulnerable to life-threatening situations. For example, Smart Enjoy Interact Light (SEIL) backpacks are manufactured for cyclists to avoid traffic accidents by displaying built-in LED lights or by expressing images in real-time [74]. The bag allows the user to show traffic signals like left/right, stop and emergency signs using a wireless controller. Any flaw in the conductive tracks or quality issues with PCBs (printed circuit boards) may disrupt the direct signal transduction, thereby putting the cyclists in danger. Other examples of smart textiles employed for health monitoring and disease prevention by early detection include the Vivago WristCare to monitor and transmit data on a person's health condition 24 hours a day – with benefits beyond the traditional push-button alarm; MARSIAN smart gloves to monitor and wirelessly transmit ANS (nonconscious) activities and real-time physiological (skin microcirculation, respiration rate, etc.) data; SenseWear body armband for measuring physiological parameters (motion, temperature, skin electrical conductance); VTAM biomedical t-shirt for teleassistance in medicine to monitor shock, fall, respiration, temperature, and location; and Vivometric's LifeShirt for ambulatory and plethysmographic respiration monitoring [75].

3.3 Efficiency

Thus far, the current chapter has discussed several aspects associated with the durability and safety of smart/e-textiles. Efficiency covers aspects such as the actuating performance against applied stimulus level or the quality of the biosignal detection in physiological applications. These aspects are also critical for the satisfaction of the smart/e-textile user.

3.3.1 Response time

Response time can be defined as the delay between the input, i.e., the activation by the stimulus, and the output of the smart/e-textiles. In the case of joule heating textiles, the response time can be determined from the time–temperature curves. Researchers have used different parameters to characterize the response time of heating textiles. For instance, R_{90} refers to the time required to reach 90% of the steady state temperature [24]. Xiao et al. reported a decrease in the R_{90} of a heating e-textile based on a carbon black nanoparticle-PU (polyurethane) composite film as the applied voltage was increased [76]. Another parameter used by researchers is the heat time constant (H_{TIME}) [25]. This is also known as response time constant (τ) [77]. The parameter τ characterizes the system's inertia [77]. It is defined as the time required to reach 63.2% of the maximum value, in this case the maximum temperature, according to the following equation (see Eq. (1)):

$$1 - e^{-1} = 1 - 0.3679 = 0.632 \; or \; 63.2\% \tag{1}$$

One solution developed by researchers to improve the reaction/response time of the carbon-based conductive materials is to take advantage of different metal fillers. For example, Ag nanowires were added to graphene oxide to prevent lattice defects during the reduction to rGO [78].

3.3.2 Power efficiency

As many smart/e-textiles require power to operate, power efficiency is critical to maximize the wearability of the device. For joule heating textiles, researchers generally express the maximum temperature reached as a function of the applied power density to characterize the heating performance of the heating system, for example flexible graphene heaters for wearable electronics [79]. Work on thermoregulatory devices for cooling and heating applications, stretchable knit heating cotton gloves, and stretchable smart textile heaters based on copper nanowires have relied on heat flux density measurements to quantify the resistive heating performance [22, 24, 25, 80]. However, power efficiency is still a weakness for products currently on the market [5].

3.3.3 Uniformity of actuation

The uniformity of actuation is a critical parameter when considering heating textiles. For example, Hao et al. characterized the uniformity of the heating performance of a cotton woven fabric spray-coated with a graphene nanosheet conductive mixture by showing the temperature distribution from four different perspectives: (a) in the horizontal direction over a span of 4 cm (the length of the heater), (b) in the vertical direction over a span of 2 cm (the width of the heater), (c) observed from the top (sprayed face of the fabric), and (d) the bottom (non-sprayed face) [24]. They also compared the heat distribution in the flat and bent (180°) conditions. The 2D plane surface temperature distribution was uniform in both the horizontal and vertical directions during the 10–60s heating phase, which confirmed the uniform distribution of the conductive coating. However, a small temperature gradient was observed in the periphery of the heating fabric; the authors attributed it to heat loss by convection. The comparison between the sprayed and non-sprayed faces of the fabric showed a 3.5 °C difference, with the sprayed face exhibiting a temperature of 83.8 °C. No significant effect was noticed in the heat distribution when the flexible heater was bent by 180°. As another example of temperature distribution inhomogeneity, **Figure 11** displays the temperature measurement of two different heating fabrics: (i) a nonwoven heater (R1) and (ii) a fabric with heating wires (R2). Different patterns of spatial heat distribution are observed with both types of heating textile structures.

3.3.4 Repeatability/stability of the actuation level

A similar approach was undertaken by several researchers to evaluate the thermal stability of electrothermal textiles during repeated heating–cooling cycles of different amplitudes. The test would involve a series of stepwise or periodic or cyclic applied voltages, with the resulting temperature changes being recorded [76]. Some researchers also used specific actuation patterns. For example, Sun et al. characterized his segregated carbon Nanotube/thermoplastic polyurethane (s-CNT/TPU) heater with three different types of heating–cooling cyclic patterns [19]: (a) ten on/off periodic cycles at 6 V, (b) three cycles of 1.5–3-4.5-6 V step increase followed by an off period, and (c) five on/off periodic cycles at increasing then decreasing voltages (3–4.5-6-4.5-3 V). In general, two types of approaches have been observed among researchers investigating the efficiency of

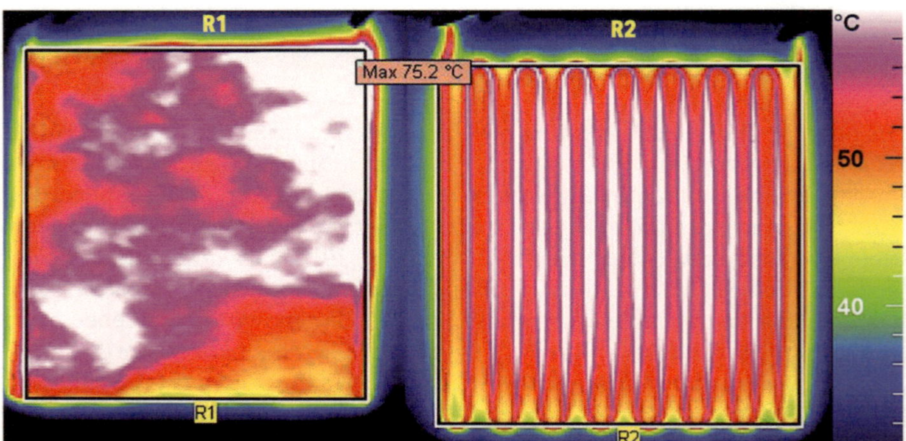

Specimen	Mean	Min	Max	Std. Dev	Eps
R1	56.3°C	40.2°C	75.2°C	4.68°C	1.00
R2	54.8°C	36.8°C	68.3°C	4.84°C	1.00

Figure 11.
Comparison between the infrared temperature measurement of heating textiles using a conductive nonwoven structure (R1) and a conductive wire (R2).

wearable heaters: (a) cyclic heating–cooling tests at a fixed voltage, and (b) repetitions of the continuous profile of variable voltages.

3.3.5 Quality of biosignal measured

Smart/e-textiles for biomedical applications often incorporate textile sensors or electrodes. The efficiency of these devices depends on the quality of the biosignals recorded. Dry textile sensors suffer from high contact impedance between the skin and the electrodes [81]. This results in high signal distortion and level of noise, lowering the overall efficiency of the biomedical devices. To overcome this challenge, researchers have integrated a water reservoir to continuously dispense moisture vapor to a Ag/Ti-coated polyester yarn embroidered electrode and lower the motion artifacts [82]. However, this system still does not offer a long-term solution as the reservoir dries out after a few hours, disrupting the signal measurement protocols, and thereby, the product efficiency [83]. Ultimately, the efficiency in the linearity of the output signals will have to be improved by reducing the impact of temperature, mechanical vibrations, ambient relative humidity, and other atmospheric factors [84].

3.3.6 Negative impact of moisture

Moisture reduces the performance of all types of batteries, including textile batteries or batteries integrated into smart textiles [43]. Moisture may also cause chemical and physical interferences in the control module of e-textiles, reducing its efficiency before a total failure occurs [85]. Besides the possibility of electric shocks or complete signal loss from corrosion, marine e-textiles could also experience decreased efficiency when exposed to the salt of seawater. As soon as the saltwater propagates the localized corrosion process of textile electrodes or conductive interconnects, it could affect the overall signal quality, lowering the transduction efficiency [49].

3.4 Other issues reported and concerns with the use of smart textiles

3.4.1 Longevity of the power supply source

For the consumer satisfaction, the longevity of the system supplying power to the e-textile, either a battery or an energy harvesting component, is critical. Unfortunately, the same situation experienced in the mobile phone sector will potentially be observed with e-textiles, in particular with batteries and chargers. Components may even reach obsolescence faster due to the combination of specific life cycle factors associated with both the electronics and textile sectors [86].

3.4.2 Maintenance and repairs

Fault detection and maintenance are another critical aspect of e-textiles. Due to their seamless integration into smart textiles, routine maintenance of electronic components can be extremely difficult. Also, any attempt to repair of the defective components may permanently damage the smart textile products.

3.4.3 Electronic component and software upgrades

In an effort towards real-time data analytics, smart textiles provide a platform for portable computing for the consumers, for instance for biosignal and physiological data collection. Any difficulty to update the electronic components, firmware, networking protocols, and software could seriously jeopardize the lifetime of the e-textile product.

3.4.4 E-waste and legislation

E-waste already raises a major challenge. With e-textiles, the situation becomes even worse as they are more integrated, have a shorter life span, and will be more likely disposed of with their batteries [86]. In addition, if people own one cell phone, they have several tee shirts in their wardrobe. E-textiles may lead to contamination of other materials' recycling processes as well the increased release of toxic substances. Hence, proper standardization and appropriate regulations are needed for the safe disposal of this new generation of electronics.

4. Test methods: current state of knowledge and future needs

Several national and international standardization organizations have been working over the last 10 years towards the development of standards for smart/e-textiles. This includes the European Committee for Standardization (CEN) with technical committee CEN TC 248/WG 31, the International Electrotechnical Commission (IEC) with technical committee IEC TC 124, ASTM International with technical committee ASTM D13.50, the International Organization for Standardization with technical committee ISO/TC 38/WG 32, and the American Association of Textile Chemists & Colorists (AATCC) with technical committee AATCC RA111. Several of them have published and/or are working on the development of test methods for smart/e-textiles. A total of 18 published/in-development standard test methods are listed in **Table 2**. They are organized according to the classification shown in **Figure 1**. Four documents relative to terminology are also included in the table.

The distribution of existing sensor and actuator-based textile technologies, solutions, and products by category of input/output signal (**Figure 1**) can be compared with the standard test methods (published and in development) identified (**Figure 12**). While most of the test method development efforts for e-textiles are in the electrical category, which accounts for 55% of the total test methods published and in development, technologies, solutions, and products in the electrical category only represent 28% and 5% of the sensors and actuators, respectively. For their part, mechanical test standards only represent 11% of the total, whereas technologies, solutions, and products in the mechanical category comprise 59% of the sensor-based smart/e-textiles. Also, very few standard test methods exist for thermal, optical and physical environmental aspects of e-textiles, while commercial products in these categories account for a large part of products/technologies in the market. No standards are available yet for power/energy harvesting and chemical/biological e-textiles, while some related products already exist on the market. This situation has led several researchers and research institutions to develop their own test methods [107]. It must be mentioned that test methods characterizing the electrical function were included in the electrical category while they may also, in a certain extent, apply to other categories of smart/e-textiles.

Based on the number of commercial e-textile products currently available and issues reported in terms of safety, efficiency, and durability, there is thus a critical need for test methods for thermal applications, as well as to a lesser degree, for energy (power) harvesting and chemical and biological applications. For this purpose, a trifactor model of performance assessment is illustrated in **Figure 13**.

Test method	Document
Electrical (Total of 10 test method standards, 1 published and 9 in development)	
ASTM WK61479- Durability of textile electrodes exposed to perspiration (in development)	[87]
ASTM WK61480- Durability of textile electrodes after laundering (in development)	[88]
AATCC RA111(a)- Electrical resistance of electronically integrated textiles (in development)*	[89]
AATCC RA111(b)- Electrical resistance changes after home laundering (in development)*	[90]
CEN EN 16812:2016- Linear electrical resistance of conductive tracks*	[91]
IEC 63203–204-1- Washable durability for leisure and sportswear e-textile system (in development)*	[92]
IEC 63203–201-3- Electrical resistance of conductive textiles under simulated microclimate (in development)*	[93]
IEC 63203–250-1- Snap button connectors (in development)*	[94]
IEC 63203–201-1- Basic properties of conductive yarns (in development)*	[95]
IEC 63203–201-2- Basic properties of conductive fabric and insulation materials (in development)*	[96]
Thermal (Total of 4 test standards, 1 published and 3 in development)	
CEN EN 16806–1:2016- PCM - Heat storage and release capacity	[97]
CEN EN 16806–2 PCM- Heat transfer using a dynamic method (in development)	[97]
CEN EN 16806–3 PCM- Determination of the heat transfer between the user and the product (in development)	[97]
IEC 63203–406-1- Measuring skin contact temperature (in development)	[98]
Mechanical (Total of 2 test standards in development)	
IEC 63203–401-1 - Stretchable resistive strain sensor (in development)	[99]
IEC 63203–402-1 – Finger movements in glove-type motion sensors (in development)	[100]
Physical environment (Total of 1 test standard in development)	
IEC 63203–402-2 - Fitness wearables – step counting (in development)	[101]
Optical (Total of 1 test standard in development)	
IEC 63203–301-1 - Electrochromic films for wearable equipment (in development)	[102]
Others (Total of 4 test standards, 2 published and 2 in development)	
ASTM D8248–19- Standard terminology for smart textiles	[103]
ASTM WK61478- New terminology for smart textiles (in development)	[104]
CEN 16298 - Definitions, categorization, applications and standardization needs	[105]
IEC 63203–101-1 – Terminology (in development)	[106]

**Also applies to other categories of products/technologies.*

Table 2.
Standards (existing and in development) test methods for smart textiles.

The product assessment should also take into account the product features, longevity, benefits/cost ratio, and the user experience.

Applying this trifactor model, we have identified the need for more than thirty standard test methods in the specific case of thermal e-textile products (**Tables 3–5**). They include a full-sleeve resistive heating jacket, battery-powered resistive heating boots, resistive heating car seats, battery-powered resistive heating gloves, a

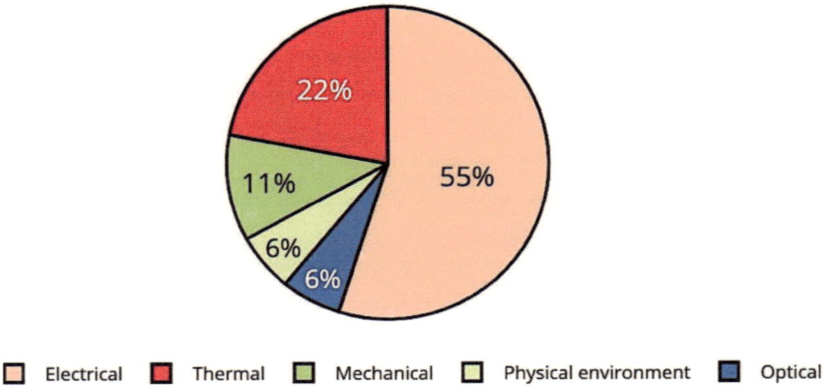

Figure 12.
Existing/in-development standardized test methods for smart textiles as of December 2020.

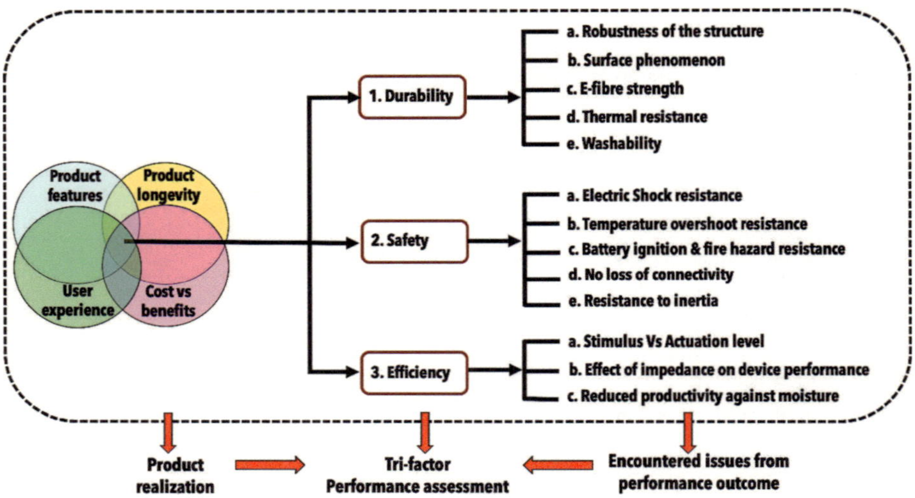

Figure 13.
Tri-factor framework for assessing the performance of smart/e-textiles.

Performance evaluation	Applicable standards
Efficiency of the overall functional/protective clothing system	Not available
Efficiency of the heat transfer system between textile and user (at the component level)	Not available
Efficiency of the induction-charging system	Not available

Table 3.
Test methods needed to evaluate the efficiency of smart/e-textiles.

reflective heating jacket, an air-exchange heating face mask, a cooling vest using water circulation and a Peltier module, and a thermo-regulated jacket with phase change materials. The few standard test methods already published and in development are also included in the tables when relevant: in several cases, the standard would not apply to the case of thermal e-textiles used as an example here. In the

Performance evaluation	Applicable standards
Efficiency of protection against shorts or open circuit, leading to shocks or fire hazards	IEC 63203–201-2 is in development, but it may not apply to high resistance conductive fabrics used for antistatic or heater purposes
Efficiency of controls (i.e., power limit) to avoid over-heating, leading to skin burn or damage	IEC 63203–406-1 ED1 is in development but it appears to be limited to wearable electronic devices
Impact of prolonged heating exposure on the skin and the surrounding environment	Not available
Overshooting of temperature difference between set temperature and experienced temperature by the skin	Not available

Table 4.
Test methods needed to evaluate the safety of smart/e-textiles (include test methods published and in development when relevant).

Performance evaluation	Applicable standards
Electrical resistance of the heater/resistive material to cleaning (washing/ laundering, dry-cleaning, drying)[a,d]	ASTM WK61480 (draft) AATCC RA11 (draft) IEC PN 63203–201-2 (draft)
Electrical resistance of the heating element to exposure of perspiration (from different parts of the body)[a,b,d]	ASTM WK61479 (draft) IEC PN 63203–201-2 (draft)
Electrical resistance of the heating element when subjected to mechanical stresses (tension /compression/ bending / fatigue/ abrasion/cutting /tearing / bodyweight)[a,b,c,d]	IEC PN 63203–201-2 is in development, but it does not appear to cover the aspects of abrasion, cutting, tearing, and fatigue
Electrical resistance of conductive parts to steaming or ironing (after laundering)[a]	Not available
Electrical resistance of the heating element to extreme weather conditions (e.g., rain and snow)[a,b,c,d]	Not available
Electrical resistance of the heating elements after exposure to severe use conditions (hot/cold/high humidity)[a,b,c,d]	Not available
Electrical resistance of the heating elements after exposure to different kinds of liquid (water, coffee, soft drinks)[c,d]	Not available
Electrical resistance of conductive track to cleaning[a,b,c,d]	ASTM WK61480 (draft) AATCC RA11 (draft) IEC PN 63203–201-2 (draft)
Electrical resistance of fasteners (e.g., switch, snaps, power supply) to cleaning[a]	Not available
Electrical resistance of fasteners to power supply to repetitive connection/disconnection for cleaning, i.e., fatigue[a,d]	Not available
Electrical resistance of fasteners to steaming/ironing[a]	Not available
Electrical resistance of fasteners to the power supply to exposure of perspiration (e.g., corrosion)[a]	Not available
Resistance of reflective thermal heating pattern to cleaning (washing/ laundering, dry-cleaning, drying)[e]	Not available
Resistance of reflective thermal heating pattern to body abrasion[e]	Not available

Performance evaluation	Applicable standards
Resistance of reflective thermal heating pattern to perspiration[e]	Not available
Preservation of thermal heat reflection of liner fabric over time i.e., aging behavior[e]	Not available
Resistance of antimicrobial property of ventilator to cleaning (washing/ laundering/ dry-cleaning/ drying)[f]	Not available
Resistance of structural integrity of the ventilator against external compression and abrasion[f]	Not available
Efficiency of the heat recovery of the ventilation system from the exhaled breath[f]	Not available
Efficiency of the transformation mechanism of cold inhaled air into warm air inside the ventilator[f]	Not available
Resistance of structural integrity of the bladder/reservoir to compression and abrasion (with zipper track while detaching)[g]	Not available
Heat storage and release capacity of phase change material (PCM)[h]	CEN EN 16806–1 (Part- 1)
Resistance of PCM and coatings (many contain binders) to washing/ laundering/ dry-cleaning/ drying[h]	Not available
Resistance of PCM and coatings to abrasion[h]	Not available
Resistance of PCM and coatings to perspiration[h]	Not available
Determination of cooling or heat transfer of the PCM (coated or portable packs) technology[h]	CEN EN 16806–1 (Part- 2)
Resistance of PCM and coating to steaming and ironing[h]	Not available
Efficiency of PCM (coated or portable packs) technology over the course of the time i.e., aging behavior (weather conditions)[h]	Not available
Efficiency of PCM (coated or portable packs) technology to fatigue[h]	Not available

[a] *Resistive heating jacket.*
[b] *Resistive heating boot.*
[c] *Resistive heating car seat.*
[d] *Resistive heating gloves.*
[e] *Reflective heating jacket.*
[f] *Air-exchange heating face mask.*
[g] *Cooling vest using water circulation and a Peltier module.*
[h] *Thermo-regulated jacket with phase change material.*

Table 5.
Test methods needed to evaluate the durability of smart/e-textiles (include test methods published and in development when relevant).

case of the durability assessment, the analysis considered the specificities of the application corresponding to the product under consideration.

5. Conclusion

After a brief overview of smart/e-textile products and major barriers to market entry, this chapter discussed different issues reported as well as foreseeable challenges that may result in injuries for instance, with electric shocks, skin burns and fires. Aspects related to the user's satisfaction, for instance in terms of the product

DOI: http://dx.doi.org/10.5772/intechopen.96500

longevity and the ability to maintain/repair it, were also covered. In particular, different conditions such as biomechanical stresses applied during use, ambient moisture, and laundering may reduce the life expectancy of the smart textile due to a damage of the conductive interconnects or a reduced actuation, for instance. As the world moves towards an increased adoption of smart e-textiles, such unwanted outcomes can put the lives of healthcare patients, first responders, and soldiers, for instance, at risk.

Due to the lack of dedicated standard test methods, manufacturers of e-textiles are limited in their attempt to control the quality of their products; as a result, they are unable to scale up and have their innovative e-textile technologies and products reach their full potential. It is clear that the issues reported in terms of safety, durability, and efficiency of e-textiles can be mitigated and eliminated through appropriate quality control using standard test methods. Currently, only 18 standard test methods published and in development by CEN, IEC, ASTM, and AATCC technical committees relevant to smart/e-textiles were identified. In some categories of e-textiles, e.g., thermal, chemical, biological, and energy harvesting, few or no test methods exist while several products are already on the market.

Using a trifactor model of performance assessment based on safety, efficiency, and durability, more than 30 standard test methods were identified for thermal e-textiles by considering a series of existing technologies/products: a full-sleeve resistive heating jacket, battery-powered resistive heating boots, resistive heating car seats, battery-powered resistive heating gloves, a reflective heating jacket, an air-exchange heating face mask, a cooling vest using water circulation and a Peltier module, and a thermo-regulated jacket with phase change materials. The development of such product-oriented test methods and their adoption by the manufacturing industries, will facilitate the design process towards a safer, more efficient, and durable smart/e-textile world. Adopting a collaborative and multidisciplinary approach, involving textile, materials, biomedical, and electrical engineers as well as relevant national and international standardization technical committees in textiles and electronics, is key to achieving this.

Acknowledgements

The authors disclosed receipt of the following financial support for the research: MITACS Canada and CTT Group, Canada.

Author details

Ikra Iftekhar Shuvo[1]*, Justine Decaens[2], Dominic Lachapelle[2] and Patricia I. Dolez[1]

1 Department of Human Ecology, University of Alberta, Edmonton, AB, Canada

2 CTT Group, St-Hyacinthe, QC, Canada

*Address all correspondence to: ikra@ualberta.ca; pdolez@ualberta.ca

References

[1] Köhler AR (2013) Challenges for eco-design of emerging technologies: The case of electronic textiles. Mater Des 51: 51–60. https://doi.org/10.1016/j.matdes.2013.04.012

[2] MarketsAndMarkets (2015) Smart Textiles Market by Type (Passive, Active, Ultra-smart), Function (Sensing, Energy Harvesting, Luminescence & Aesthetics, Thermo-electricity), Industry (Healthcare, Sports & Fitness, Fashion, Military, Automotive), & Geography - Global Forecast to 20

[3] Honarvar MG, Latifi M (2017) Overview of wearable electronics and smart textiles. J Text Inst 108:631–652. https://doi.org/10.1080/00405000.2016.1177870

[4] Ismar E, Kurşun Bahadir S, Kalaoglu F, Koncar V (2020) Futuristic Clothes: Electronic Textiles and Wearable Technologies. Glob Challenges 4:1900092. https://doi.org/10.1002/gch2.201900092

[5] Dolez PI, Decaens J, Buns T, et al (2020) Applications of smart textiles in occupational health and safety. IOP Conf Ser Mater Sci Eng 827:012014. https://doi.org/10.1088/1757-899X/827/1/012014

[6] Stoppa M, Chiolerio A (2014) Wearable Electronics and Smart Textiles: A Critical Review. Sensors 14: 11957–11992. https://doi.org/10.3390/s140711957

[7] Weigelt G (2015) Power vest-Research News. In: Fraunhofer. https://www.fraunhofer.de/en/press/research-news/2015/March/power-vest.html. Accessed 15 Nov 2020

[8] Myant Inc (2020) Myant unveils connected ppe concepts, creating new ways to assess health and performance as part of the skiin interconnected system of biometric garments. In: Myant. https://myant.ca/. Accessed 15 Nov 2020

[9] Graham F (2013) Wearable technology: The bra designed to shock attackers. In: BBC News. https://www.bbc.com/news/business-22110443. Accessed 15 Nov 2020

[10] Sky News (2015) Microsoft May Send Shock Alerts Through Clothes. In: Sky News UK. https://news.sky.com/. Accessed 15 Nov 2020

[11] Hisham AAB, Saad SA, Ahmad MRR, et al (2017) Review of Safety Evaluation of Thermal Wearable Power Harvesting Device. Int J Integr Eng 9:

[12] Schwarz A, Van Langenhove L, Guermonprez P, Deguillemont D (2010) A Roadmap on Smart Textiles. CRC Press

[13] Cherenack K, van Pieterson L (2012) Smart textiles: Challenges and opportunities. J Appl Phys 112:091301. https://doi.org/10.1063/1.4742728

[14] Kok M, de Vries H, Pacheco K, van Heck G (2015) Failure modes of conducting yarns in electronic-textile applications. Text Res J 85:1749–1760. https://doi.org/10.1177/0040517515573405

[15] Cherenack K, Zysset C, Kinkeldei T, et al (2010) Woven Electronic Fibers with Sensing and Display Functions for Smart Textiles. Adv Mater 22:5178–5182. https://doi.org/10.1002/adma.201002159

[16] Crow RM, Dewar MM (1986) Stresses in Clothing as Related to Seam Strength. Text Res J 56:467–473. https://doi.org/10.1177/004051758605600801

[17] Ryan JD, Mengistie DA, Gabrielsson R, et al (2017) Machine-Washable PEDOT:PSS Dyed Silk Yarns for Electronic Textiles. ACS Appl Mater

Interfaces 9:9045–9050. https://doi.org/10.1021/acsami.7b00530

[18] Qiu Q, Zhu M, Li Z, et al (2019) Highly flexible, breathable, tailorable and washable power generation fabrics for wearable electronics. Nano Energy 58:750–758. https://doi.org/10.1016/j.nanoen.2019.02.010

[19] Sun W-J, Xu L, Jia L-C, et al (2019) Highly conductive and stretchable carbon nanotube/thermoplastic polyurethane composite for wearable heater. Compos Sci Technol 181:107695. https://doi.org/10.1016/j.compscitech.2019.107695

[20] Hamdani S, Potluri P, Fernando A (2013) Thermo-Mechanical Behavior of Textile Heating Fabric Based on Silver Coated Polymeric Yarn. Materials (Basel) 6:1072–1089. https://doi.org/10.3390/ma6031072

[21] Hamdani STA, Fernando A, Maqsood M (2016) Thermo-mechanical behavior of stainless steel knitted structures. Heat Mass Transf 52:1861–1870. https://doi.org/10.1007/s00231-015-1707-z

[22] Kim D, Bang J, Lee W, et al (2020) Highly stretchable and oxidation-resistive Cu nanowire heater for replication of the feeling of heat in a virtual world. J Mater Chem A 8:8281–8291. https://doi.org/10.1039/D0TA00380H

[23] Bobinger MR, Romero FJ, Salinas-Castillo A, et al (2019) Flexible and robust laser-induced graphene heaters photothermally scribed on bare polyimide substrates. Carbon N Y 144: 116–126. https://doi.org/10.1016/j.carbon.2018.12.010

[24] Hao Y, Tian M, Zhao H, et al (2018) High Efficiency Electrothermal Graphene/Tourmaline Composite Fabric Joule Heater with Durable Abrasion Resistance via a Spray Coating Route. Ind Eng Chem Res 57:13437–13448. https://doi.org/10.1021/acs.iecr.8b03628

[25] Lima RMAP, Alcaraz-Espinoza JJ, da Silva FAG, de Oliveira HP (2018) Multifunctional Wearable Electronic Textiles Using Cotton Fibers with Polypyrrole and Carbon Nanotubes. ACS Appl Mater Interfaces 10:13783–13795. https://doi.org/10.1021/acsami.8b04695

[26] Rahman MJ, Mieno T (2015) Conductive Cotton Textile from Safely Functionalized Carbon Nanotubes. J Nanomater 2015:1–10. https://doi.org/10.1155/2015/978484

[27] Zhang M, Wang C, Liang X, et al (2017) Weft-Knitted Fabric for a Highly Stretchable and Low-Voltage Wearable Heater. Adv Electron Mater 3:1700193. https://doi.org/10.1002/aelm.201700193

[28] Bosowski P, Hoerr M, Mecnika V, et al (2015) Design and manufacture of textile-based sensors. In: Electronic Textiles. Elsevier, pp 75–107

[29] zdil N, Zelik G, Spren G (2012) Analysis of Abrasion Characteristics in Textiles. In: Abrasion Resistance of Materials. InTech

[30] Cho C, Elias A, Batcheller J, et al (2019) Electrical conduction of reduced graphene oxide coated meta-aramid textile and its evolution under aging conditions. J Ind Text 152808371986938. https://doi.org/10.1177/1528083719869387

[31] Matikainen A, Nuutinen T, Itkonen T, et al (2016) Atmospheric oxidation and carbon contamination of silver and its effect on surface-enhanced Raman spectroscopy (SERS). Sci Rep 6: 37192. https://doi.org/10.1038/srep37192

[32] Palomar T, Ramírez Barat B, García E, Cano E (2016) A comparative

study of cleaning methods for tarnished silver. J Cult Herit 17:20–26. https://doi.org/10.1016/j.culher.2015.07.012

[33] Denawaka CJ, Fowlis IA, Dean JR (2016) Source, impact and removal of malodour from soiled clothing. J Chromatogr A 1438:216–225. https://doi.org/10.1016/j.chroma.2016.02.037

[34] MacLeod ID, Stambolov T (1987) The Corrosion and Conservation of Metallic Antiquities and Works of Art. Stud Conserv 32:138. https://doi.org/10.2307/1506218

[35] Liu H, Li J, Chen L, et al (2016) Thermal-electronic behaviors investigation of knitted heating fabrics based on silver plating compound yarns. Text Res J 86:1398–1412. https://doi.org/10.1177/0040517515612359

[36] Sezgin H, Bahadir SK, Boke YE, Kalaoglu F (2016) Thermal analysis of e-textile structures using full-factorial experimental design method. J Ind Text 45:752–764. https://doi.org/10.1177/1528083714540699

[37] Tao X, Koncar V, Huang T-H, et al (2017) How to Make Reliable, Washable, and Wearable Textronic Devices. Sensors 17:673. https://doi.org/10.3390/s17040673

[38] Yang K, Torah R, Wei Y, et al (2013) Waterproof and durable screen printed silver conductive tracks on textiles. Text Res J 83:2023–2031. https://doi.org/10.1177/0040517513490063

[39] Altaş S, Yılmaz E, Adman N (2020) Improving the repetitive washing and abrasion resistance properties of fabrics produced with metallized yarns. J Ind Text 152808372094296. https://doi.org/10.1177/1528083720942961

[40] Booth JE (1964) Principles of textile testing: an introduction to physical methods of testing textile fibres, yarns, and fabrics. Chemical Pub. Co.

[41] Alomayri T, Assaedi H, Shaikh FUA, Low IM (2014) Effect of water absorption on the mechanical properties of cotton fabric-reinforced geopolymer composites. J Asian Ceram Soc 2:223–230. https://doi.org/10.1016/j.jascer.2014.05.005

[42] Gaubert V, Gidik H, Bodart N, Koncar V (2020) Investigating the Impact of Washing Cycles on Silver-Plated Textile Electrodes: A Complete Study. Sensors 20:1739. https://doi.org/10.3390/s20061739

[43] Nayak R, Wang L, Padhye R (2015) Electronic textiles for military personnel. In: Electronic Textiles. Elsevier, pp 239–256

[44] LONGINOTTI-BUITONI G, BOVIO D (2020) Laundry system for smart garments

[45] Badgett J (2019) An Evaluation of the Quality of Men's 100% Cotton Jersey Knit T-Shirts. J Text Sci Fash Technol 3:. https://doi.org/10.33552/JTSFT.2019.03.000557

[46] Metz R (2014) This Shirt Is (Really) Sensitive. In: MIT Technol. Rev. https://www.technologyreview.com/s/527136/this-shirt-is-really-sensitive/

[47] Karaguzel B, Merritt CR, Kang T, et al (2009) Flexible, durable printed electrical circuits. J Text Inst 100:1–9. https://doi.org/10.1080/00405000802390147

[48] Eskandarian L, Lam E, Rupnow C, et al (2020) Robust and Multifunctional Conductive Yarns for Biomedical Textile Computing. ACS Appl Electron Mater 2:1554–1566. https://doi.org/10.1021/acsaelm.0c00171

[49] Mariello M, Guido F, Mastronardi VM, et al (2019) Reliability of Protective Coatings for Flexible Piezoelectric Transducers in Aqueous Environments. Micromachines 10:739. https://doi.org/10.3390/mi10110739

[50] Murph D (2013) Columbia recalls Omni-Heat electric jackets due to burn hazard (or, for working too well). Engadget

[51] Castano LM, Flatau AB (2014) Smart fabric sensors and e-textile technologies: a review. Smart Mater Struct 23:053001. https://doi.org/10.1088/0964-1726/23/5/053001

[52] Starner T, Maguire Y (1998) A heat dissipation tutorial for wearable computers. In: Digest of Papers. Second International Symposium on Wearable Computers (Cat. No.98EX215). IEEE Comput. Soc, pp 140–148

[53] Kroemer AD, Kroemer KH. (2016) Office ergonomics. CRC Press, Second edition. | Boca Raton [Florida] : Taylor & Francis, CRC

[54] Gagge AP, Gonzalez RR (2011) Mechanisms of heat exchange: biophysics and physiology. In: Comprehensive Physiology. John Wiley & Sons, Inc., Hoboken, NJ, USA

[55] Iftimia N, Ferguson RD, Mujat M, et al (2013) Combined reflectance confocal microscopy/optical coherence tomography imaging for skin burn assessment. Biomed Opt Express 4:680. https://doi.org/10.1364/BOE.4.000680

[56] Moritz AR, Henriques FC (1947) Studies of thermal injury: ii. the relative importance of time and surface temperature in the causation of cutaneous burns. Am J Pathol 23:695–720

[57] Benjamin C, Gittler M, Lee R (2011) Burn from car seat heater in a man with paraplegia: case report. J Spinal Cord Med 34:332–334. https://doi.org/10.1179/2045772311Y.0000000005

[58] Demir E, O'Dey D m., Fuchs PC, et al (2006) Die Autositzheizung — eine potenzielle Verbrennungsgefahr für den Querschnittsgelähmten. Nervenarzt 77: 201–203. https://doi.org/10.1007/s00115-005-1960-3

[59] Vogel B (2008) Escaping the Hot Seat. New Mobil.

[60] Rakowski KRM, Sivathasan N, Sivathasan N (2011) Pain in your buttocks? Check your heated car seat isn't burning you. Spinal Cord 49:672–672. https://doi.org/10.1038/sc.2010.115

[61] Bansal V (2006) Diabetic neuropathy. Postgrad Med J 82:95–100. https://doi.org/10.1136/pgmj.2005.036137

[62] Healthline (2018) Find the Right Diabetic Socks. In: Healthline. https://www.healthline.com/health/find-right-diabetic-socks

[63] Tufts Medical Center (2020) Peripheral Neuropathy. In: Tufts Med. Cent. https://hhma.org/healthadvisor/aha-perineur-crs/#:~:text=Put moist heat on the,any type of hot pad.

[64] Mbise E, Dias T, Hurley W (2015) Design and manufacture of heated textiles. In: Electronic Textiles. Elsevier, pp 117–132

[65] Swanner J, Bito J, Nichols G, et al (2017) Integrating Multiple Energy Harvesting Systems for Department of Defense Applications. In: EESAT Conference - Evolution & Revolution

[66] Choi AY, Lee CJ, Park J, et al (2017) Corrugated Textile based Triboelectric Generator for Wearable Energy Harvesting. Sci Rep 7:45583. https://doi.org/10.1038/srep45583

[67] Matsouka D, Vassiliadis S, Bayramol DV (2018) Piezoelectric textile fibres for wearable energy harvesting systems. Mater Res Express 5:065508. https://doi.org/10.1088/2053-1591/aac928

[68] Sun T, Zhou B, Zheng Q, et al (2020) Stretchable fabric generates

electric power from woven thermoelectric fibers. Nat Commun 11: 572. https://doi.org/10.1038/s41467-020-14399-6

[69] Nocito C, Koncar V (2016) Flexible photovoltaic cells embedded into textile structures. In: Smart Textiles and their Applications. Elsevier, pp 401–422

[70] Lv J, Jeerapan I, Tehrani F, et al (2018) Sweat-based wearable energy harvesting-storage hybrid textile devices. Energy Environ Sci 11:3431–3442. https://doi.org/10.1039/C8EE02792G

[71] Kim S, Vyas R, Bito J, et al (2014) Ambient RF Energy-Harvesting Technologies for Self-Sustainable Standalone Wireless Sensor Platforms. Proc IEEE 102:1649–1666. https://doi.org/10.1109/JPROC.2014.2357031

[72] KFSM-TV (2017) Springdale Woman's Heated Jacket Bursts Into Flames After Battery Ignites. 5 News

[73] Carole W, Justyna T, Patricia W, et al (2005) Development of Electronic Textiles to Transport Data and Power in Future U.S. Military Protective Clothing Systems. J ASTM Int 2:

[74] Chun J, Lee M (2016) Developing a SEIL (Smart Enjoy Interact Light) bag utilizing LED display. Int J Cloth Sci Technol 28:. https://doi.org/10.1108/IJCST-02-2015-0026

[75] Axisa F, Schmitt PM, Gehin C, et al (2005) Flexible Technologies and Smart Clothing for Citizen Medicine, Home Healthcare, and Disease Prevention. IEEE Trans Inf Technol Biomed 9:325–336. https://doi.org/10.1109/TITB.2005.854505

[76] Xiao Z, Sheng C, Xia Y, et al (2019) Electrical heating behavior of flexible thermoplastic polyurethane/Super-P nanoparticle composite films for advanced wearable heaters. J Ind Eng Chem 71:293–300. https://doi.org/10.1016/j.jiec.2018.11.038

[77] Skogestad S (2008) Chemical and Energy Process Engineering. CRC Press

[78] Lai Y-T, Tai N-H (2015) One-Step Process for High-Performance, Adhesive, Flexible Transparent Conductive Films Based on p-Type Reduced Graphene Oxides and Silver Nanowires. ACS Appl Mater Interfaces 7:18553–18559. https://doi.org/10.1021/acsami.5b04875

[79] Huang Y, Tao L-Q, Yu J, et al (2020) Improved Performance of Flexible Graphene Heater Based on Repeated Laser Writing. IEEE Electron Device Lett 41:501–504. https://doi.org/10.1109/LED.2020.2965585

[80] Hong S, Gu Y, Seo JK, et al (2019) Wearable thermoelectrics for personalized thermoregulation. Sci Adv 5:eaaw0536. https://doi.org/10.1126/sciadv.aaw0536

[81] Pani D, Dessi A, Saenz-Cogollo JF, et al (2016) Fully Textile, PEDOT:PSS Based Electrodes for Wearable ECG Monitoring Systems. IEEE Trans Biomed Eng 63:540–549. https://doi.org/10.1109/TBME.2015.2465936

[82] Weder M, Hegemann D, Amberg M, et al (2015) Embroidered Electrode with Silver/Titanium Coating for Long-Term ECG Monitoring. Sensors 15:1750–1759. https://doi.org/10.3390/s150101750

[83] Soroudi A, Hernández N, Wipenmyr J, Nierstrasz V (2019) Surface modification of textile electrodes to improve electrocardiography signals in wearable smart garment. J Mater Sci Mater Electron 30:16666–16675. https://doi.org/10.1007/s10854-019-02047-9

[84] Nasiri S, Khosravani MR (2020) Progress and challenges in fabrication of

wearable sensors for health monitoring. Sensors Actuators A Phys 312:112105. https://doi.org/10.1016/j.sna.2020.112105

[85] Suh M (2015) Wearable sensors for athletes. In: Electronic Textiles. Elsevier, pp 257–273

[86] Köhler AR, Hilty LM, Bakker C (2011) Prospective Impacts of Electronic Textiles on Recycling and Disposal. J Ind Ecol 15:496–511. https://doi.org/10.1111/j.1530-9290.2011.00358.x

[87] Draft ASTM WK61479 (2018) New test method for durability of smart garment textile electrodes exposed to perspiration

[88] Draft ASTM WK61480 (2018) New test method for durability of smart garment textile electrodes after laundering

[89] Draft AATCC RA111(a) (2019) Electrical resistance of electronically-integrated textiles

[90] Draft AATCC RA111(b) (2019) Electrical resistance changes of electronically-integrated textiles after home laundering

[91] CEN EN 16812 (2016) Textiles and textile products - electrically conductive textiles - determination of the linear electrical resistance of conductive tracks

[92] Draft IEC 63203–204-1 (2019) Wearable electronic devices and technologies – part 204–1: electronic textile – washable durability test method for leisure and sportswear e-textile system

[93] Draft IEC 63203–201-3 (2019) Wearable electronic devices and technologies - Part 201–3: Electronic Textile - Determination of electrical resistance of conductive textiles under simulated microclimate

[94] Draft IEC 63203–250-1 (2018) Wearable electronic devices and technologies – part 250–1: electronic textile –snap button connectors between e-textile and detachable electronic devices

[95] Draft IEC 63203–201-1 (2018) Wearable electronic devices and technologies – part 201–1: electronic textile – measurement methods for basic properties of conductive yarns

[96] Draft IEC 63203–201-2 Wearable electronic devices and technologies – part 201–2: electronic textile – measurement methods for basic properties of conductive fabric and insulation materials

[97] CEN EN 16806–1 (2016) Textiles and textile products - textiles containing phase change materials (pcm) - part 1: determination of the heat storage and release capacity

[98] Draft IEC 63203–406-1 (2019) Wearable electronic devices and technologies - Part 406–1: Test methods of on-body wearable electronic devices for measuring skin contact temperature

[99] Draft IEC 63203–401-1 (2019) Wearable electronic devices and technologies - Part 401–1: Devices and Systems – Functional elements - Evaluation method of the stretchable resistive strain sensor

[100] Draft IEC 63203–402-1 (2019) Wearable electronic devices and technologies – part 402–1: devices and systems – accessory – test methods of glove-type motion sensors for measuring finger movements

[101] Draft IEC 63203–402-2 (2019) Wearable electronic devices and technologies – part 402–2: performance measurement of fitness wearables – step counting

[102] Draft IEC 63203–301-1 (2019) Wearable electronic devices and technologies - Part 301–1: Test method

of electrochromic films for wearable equipment

[103] ASTM D8248 (2019) Standard Terminology for smart textiles

[104] Draft ASTM WK61478 (2017) New terminology for smart textiles

[105] CEN 16298 (2011) Textiles and textile products - smart textiles - definitions, categorization, applications and standardization needs

[106] Draft IEC 63203–101-1 (2019) Wearable electronic devices and technologies – Part 101–1: Terminology

[107] Decaens J, Vermeersch O (2018) Specific testing for smart textiles. In: Advanced Characterization and Testing of Textiles. Elsevier, pp 351–374

Chapter 8

Modeling and Validating Analytic Relations for Electromagnetic Shielding Effectiveness of Fabrics with Conductive Yarns

Ion Razvan Radulescu, Lilioara Surdu, Emilia Visileanu, Cristian Morari and Marian Costea

Abstract

Electromagnetic (EM) radiation may be harmful for human's health and for functioning of electronic equipment. The field of Electromagnetic Compatibility approaches various solutions to tackle this problem, while shielding of the radiation is one of the main solutions. Since the development of spinning technology for producing conductive yarns for fabrics, textile electromagnetic shields have become a valuable alternative to metallic shields. Their main advantages are given by the flexibility, the low weight and the good mechanical resistance, as well as by the possibility to precisely design the shield. The scientific literature includes several analytic relations for estimating the electromagnetic shielding effectiveness (EMSE), in case of woven fabrics with conductive yarns, which may be modeled as a grid of electric conductors. This book chapter tackles three different analytic models for estimating EMSE, which are useful to predict this functionality in the design phase of fabrics. The analytic relations are subsequently comparatively validated by EMSE measurements via TEM cell equipment of two woven fabrics with conductive yarns out of stainless steel and silver with a grid of 4 mm. Results of validated analytic relations are used for the approach of designing textile shields with regard to final application requirements.

Keywords: fabrics, yarns, stainless steel, silver, shielding effectiveness, modeling, validating

1. Introduction

Textile materials have reached in the last two decades a lot of additional functionalities in connection to new application fields [1]. Keywords such as technical textiles or smart textiles are increasingly gaining popularity among end users since various products are already available on the market. One important domain of technical textiles are fabrics destined for the construction sector – BUILDTECH, having electromagnetic (EM) shielding properties [2].

The research field of EM shielding fabrics combines at least two major disciplines, namely textile science and electromagnetic compatibility. As such, this chapter tackles interdisciplinary research with a lot of potential applications. In order to understand why EM shielding functionalities are relevant for textile materials, one has to consider the huge amount of EM radiation of our environment these days, caused by telecommunication or other electric energy sources [3].

Several studies prove that EM radiation is harmful for human beings [4]. Non-ionizing radiation produce a heating of the cellular tissues of the human body with negative impact on the health and may produce cancer [5]. Although the scientific studies have not done a strict correlation between radiation and deterioration of human's health, it is evident that a causality does exist [6].

Another problem are undesired radiation sources which cause interference with electronic devices. EM Interferences (EMI) are a deep topic of electromagnetic compatibility (EMC) science and one of the most applied solutions is shielding [7]. By shielding of interferences between electronic devices a proper functioning is ensured [8].

Shielding of EM radiation has been done traditionally by use of metallic plates. Due to their outstanding electric conductivity and the formation of Eddy currents, which produce an opposite EM field, the incident EM field is attenuated. The Shielding Effectiveness (EMSE) may be considered as basic quantity defining this attenuation. One of the most applied relation for measurement of EMSE is (1) [9]:

$$\begin{aligned} \boldsymbol{EMSE} &= \mathbf{10 \log_{10}} \frac{\boldsymbol{Power\ of\ the\ incident\ field}}{\boldsymbol{Power\ of\ the\ transmitted\ field}} = \\ &= \boldsymbol{Reflection} + \boldsymbol{Absorption} + \boldsymbol{Multiple\ Reflections} \end{aligned} \tag{1}$$

However, although metallic plates have most effective EMSE properties due to their excellent electric conductivity properties, advanced technical textile materials are also used for such applications. The evolution of spinning technologies [10, 11] and the possibility to produce electric conductive fibers and yarns to be inserted into the fabrics structure, have made possible the manufacturing of textile electromagnetic shields. Either being inserted as yarns within the woven or knitted fabrics or as fibers within nonwoven fabrics, these novel textile materials render electric conductivity properties. The fabric structure with inserted conductive yarns may be designed according to weaving/knitting principles of textile science, having various yarn counts, fabric densities, weaves or float repeats [12].

When compared to metallic plates, electrical conductive textiles are flexible, lightweight and have a good mechanical resistance, and have as well the possibility to precisely design the shield for a certain application. Fabrics are currently preferred in many shielding applications over metallic plates, for the lower costs and the adaption in shape and size to the protected area. The insertion of conductive yarns into the fabric structure may yield various geometric patterns. One of the most common pattern is the grid pattern, which is obtained by inserting conductive yarns in warp and weft system of a woven fabric. Although such conductive woven fabrics do not reach the EMSE of metallic plates, they are used due to their special, mentioned properties.

Modeling of EMSE for electrical conductive materials was a task given by the necessity to estimate the material's properties in the design phase. Manufacturing EM shields without a prior estimation of their EMSE properties requires subsequent physical determinations and an ongoing loop of manufacturing and measurement. This undesired loop may be overcome by modeling of the desired property, in order to be able to estimate the property's values without manufacturing in view of the end application requirements. As such, important resources of time, materials, energy and men power are saved.

Modeling may be done by two main principles [13]:

- Phenomenological modeling: tackles the system as black box and uses experimental plans and statistics to get insight of experimental data;

- Mechanistic modeling: tackles the system as white box and describes by means of analytic relations their internal mechanism.

A simple mechanistic method to model the shielding effectiveness achieved through conductive plane shields was provided by Schelkunoff, and was named impedance method [7]. This method performs an analogy of the discontinuity of shield impedance with a lossy section within long bifilar transmission lines. The impedance method is considered to estimate sufficiently precise shielding effectiveness, by using following geometric and electric parameters: thickness of the shield, electric conductivity and magnetic permeability of the shield, over a specified frequency domain. Shielding is considered to occur due to three mechanisms: reflection loss of the wave due to mismatch of impedance between air and shield (electric conductivity and magnetic permeability are parameters with high sensitivity), absorption loss due to heat loss of the EM wave within the shield (thickness of shield is parameter with high sensitivity) and multiple reflection correction term due to re-reflection of the EM wave within the shield (1). In case of shields with grid structure such as the woven fabrics with conductive yarns, adaptations of the impedance method were accomplished, by introducing additional parameters of the fabric structure, such as the distance between conductive yarns or the conductive yarn's diameter. For modeling of EMSE by woven fabrics is a growing field of research, up-to-date contributions may be found in [14, 15]. A model based on the circuit method for near field electromagnetic waves was described in [16]. The circuit method was developed by Kaden and is considered next to the impedance method, main mechanistic model to estimate EMSE for electromagnetic shields, based on geometric and electric parameters [7].

This book chapter tackles modeling of EMSE for grid structures of woven fabrics with inserted conductive yarns, comparatively by mechanistic models of different analytic relations [17–20], with the purpose of being able to estimate EMSE in the design phase of the textile shielding products.

2. Materials and methods

This subchapter describes the analytic relations for estimation of the EMSE and the fabric samples with inserted conductive yarns and their properties, destined for validation.

2.1 Electric and geometric parameters for fabrics

Figure 1 presents the main geometric parameters of woven textile structures:

h - fabric thickness [m].

a - distance between conductive yarns (ratio between conductive and nonconductive yarns-float repeat) [m].

d – optical diameter of the conductive yarn [m].

The following electric parameters apply for the textile materials:

σ_y – Electric conductivity of the metallic yarns [S/m].

σ_f – Electric conductivity of the fabrics [S/m].

ε – Electric permittivity of the metallic yarns [F/m]

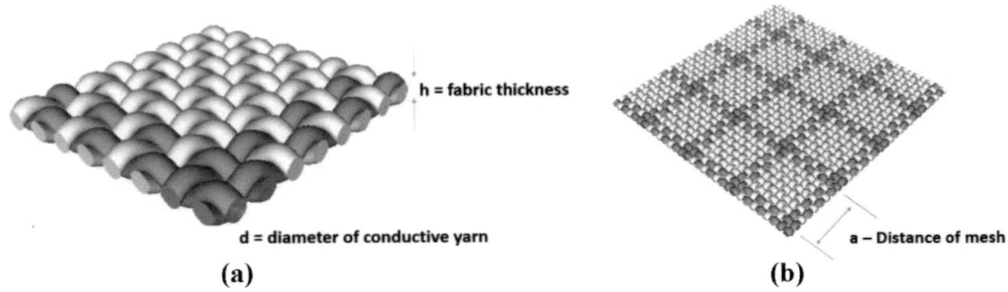

Figure 1.
(a) and (b) – Fabric structures and geometric parameters.

$\varepsilon = \varepsilon_0 \varepsilon_r$, where:
ε_0 – Electric permittivity of vacuum $\varepsilon_0 = 1/(36\pi{*}10^9)$ F/m.
ε_r – Relative electric permittivity [1].
μ_y – Magnetic permeability of the metallic yarns [H/m].
μ_f – Magnetic permeability of the metallic fabrics [H/m]
$\mu = \mu_0 \mu_r$, where:
μ_0 – Magnetic permeability of vacuum $\mu_0 = 4\pi{*}10^{-7}$ H/m.
μ_r – Relative magnetic permeability [1].
f – Frequency of electromagnetic (EM) field [Hz].
ω – Angular frequency of EM field $\omega = 2\pi f$ [rad/s].
δ_y – Skin depth of the conductive yarn [m], expressed as:

$$\delta_y = \frac{1}{\sqrt{\pi f \mu_y \sigma_y}} \qquad (2)$$

δ_f – Skin depth of the fabric [m], expressed as:

$$\delta_f = \frac{1}{\sqrt{\pi f \mu_f \sigma_f}} \qquad (3)$$

2.2 Materials: the fabrics and their properties

Two types of fabrics with inserted conductive yarns made of stainless steel (F1) and silver (F2) were manufactured for validating the analytic relations of EMSE. The fabrics have similar structures, plain weave and a distance between conductive yarns of 5 mm, while F1 is based on stainless steel Bekinox BK50/2 yarns and F2 is based on silver Statex yarns. Cotton yarns of Nm50/2 were set as basic yarns. Both fabrics have been manufactured on rapier weaving looms, by inserting the conductive yarns in warp and weft system, with float repeat 6:2. **Table 1** presents their designed structure and their physical-mechanical and electric properties.

Figures 2 and **3** present pictures of the weaving loom and the warp beam at the company SC Majutex SRL (www.majutex.ro), used for manufacturing the textile woven fabrics F1 and F2.

* The electric conductivity of the yarns was computed considering the linear electric resistance of the yarns and the cross-section of the yarns, based on the measured optical diameter – relation (4).

		F1	F2
Conductive yarn (raw material / linear density)		Stainless steel and cotton spun yarn, Nm 50/2	Silver coated PA, 140x2 dtex
Linear electric resistance [Ω/m] R_l		3500	30
Electric conductivity yarns σ_y [S/m] *		7700	927000
Relative magnetic permeability yarns μr_y **		7.36	1
Basic yarn (raw material / linear density)		100% Cotton Nm 50/2	
Optical diameter yarn d [μm]		222	214
Weave		Plain weave	
Float repeat Basic: conductive yarn	Warp	6:2	6:2
	Weft	6:2	5:2
Fabric thickness h [mm]		0.55	0.49
Specific mass [g/m^2]		143	118
Fabric density [yarns/10 cm]	Warp	180	168
	Weft	170	150
Distance between conductive yarns a [mm]		5	5
Electric conductivity of fabrics σ_f [S/m] ***		41.54	589.0
Relative magnetic permeability of fabrics μr_f **		2.33	1

Table 1.
Textile structures and physical-mechanical & electric properties.

Figure 2.
Weaving loom.

$$\sigma_y = \frac{l}{R_l \cdot A} \tag{4}$$

Where:
σ_y – electric conductivity of the yarns [S/m].
R_l – linear electric resistance [Ω] for l = 1 m.
A – Cross-section of the yarn [m^2] with d optical diameter of yarn in **Table 1**.

** The relative magnetic permeability of the composite yarn respectively fabric was computed considering the relation presented in [21].

$$\mu_R = M_{de} MM_{de} \mu_{MM} \quad (5)$$

Where:
M_{de} - the equivalent percentage of bulk material from total volume;
MM_{de} – the equivalent percentage of the magnetic material from bulk material;
μ_{MM} – relative magnetic permeability of the magnetic material;
A relative magnetic permeability of stainless steel was set to μ_{MM} = 40, according to [22], while M_{de} and MM_{de} were computed according to the textile structure of yarn and fabric.

Ferromagnetic properties were computed for fabric F1 with inserted Bekinox stainless steel yarns, since fabric F2 with inserted Statex silver yarns has no ferromagnetic properties ($\mu_r \cong 1$).

The Bekinox BK50/2 stainless steel yarn is a spun yarn with 80% content of cotton fibers and 20% content of Bekinox VS stainless steel fibers, thus $MM_{de} = 0.20$.

For fabric F1, the cover factor of the woven fabric was considered for computing the parameter $M_{de} = 0.96$ (in one of our previous research studies [23]) and the ratio between conductive and basic yarns for computing the parameter $MM_{de} = 0.33$ (**Table 2**).

***The electric conductivity of fabrics was computed based on the measurement of the electric resistance of a piece of fabric (**Figure 4**) and relation (6).

The electric resistance R was measured via digital Ohmmeter and thickness h was measured via digital caliper for (6).

$$\sigma = \frac{1}{R} \cdot \frac{L}{l \cdot h} \quad (6)$$

Figure 3.
Warp beam.

μ_{MM} = 40 (stainless steel [23])	Stainless steel yarn	F1 fabric
M_{de}	100%	96%
MM_{de}	20%	33%
→Relative magnetic permeability μ_r	μ_{ry} 7.36	μ_{rf} = 2.33

Table 2.
Computation of relative magnetic permeability.

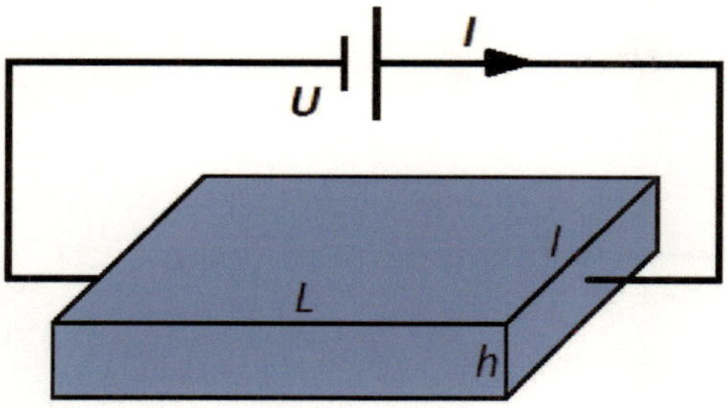

Figure 4.
Experimental setup for measurement of fabric's electric conductivity

2.3 Measured EMSE of the woven fabric samples

EMSE may be measured according to two main principles with following standardized methods:

- Principle of TEM cell measurement system (standards ASTM ES-07 and ASTM D-4935);
- Principle of emitting and receiving antenna system (standards IEEE 299 and IEEE 299.1 – shielded enclosures with max length of 2 meter).

EMSE of both fabrics F1 and F2 was measured via a TEM cell measurement system, including signal generator, amplifier and oscilloscope, according to standard ASTM ES-07. **Figure 5a**) presents the scheme of a TEM cell and the washer-shaped textile sample, while **Figure 5b**) presents a picture of the TEM cell.

Figure 6 presents the diagram of EMSE to frequency on logarithmic scale with measured values for F1 and F2 in the frequency domain 0.1–1000 MHz.

The samples were coated on the edge with silver paint, in order to ensure a good electric conductivity at connection points to the TEM cell (**Figure 7**).

These experimental measurements of the fabrics with inserted conductive yarns of stainless steel (F1) and silver (F2), were done in order to validate two analytic relations for estimation of EMSE.

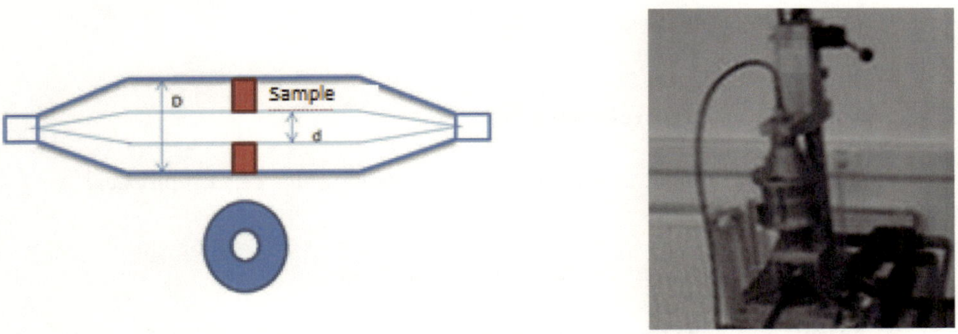

(a) **(b)**

Figure 5.
(a) and (b) TEM cell for EMSE measurement according to standard ASTM ES-07.

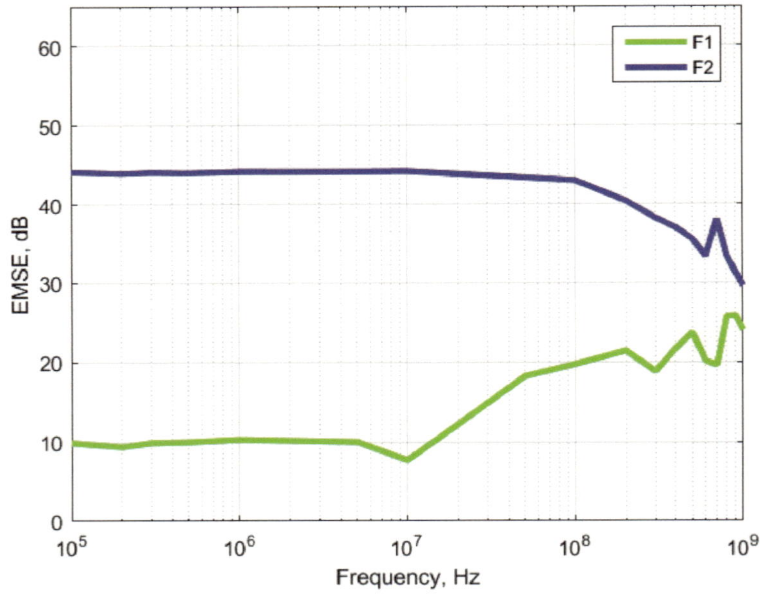

Figure 6.
EMSE measured for F1 and F2.

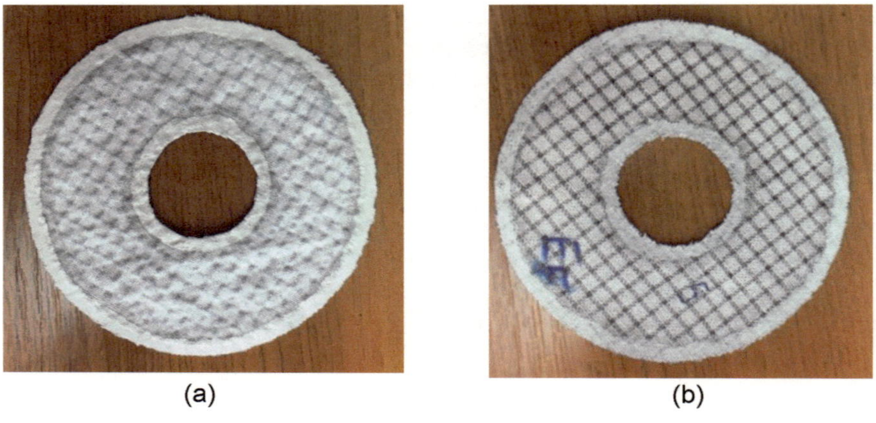

Figure 7.
(a) Sample F1 for TEM cell. (b) Sample F2 for TEM cell.

2.4 Methods: the analytic relations

The analytic relations of modeling EMSE are valid under certain physical premises:

- The incoming field is an electromagnetic far field;
- Both warp and weft inserted conductive yarns contribute to EMSE;
- The impedance method (Schelkunoff) [7], which performs an analogy of the discontinuity of shield impedance with a lossy section within a long bifilar

transmission lines, represents the basic source of all adapted relations for EMSE modeling;

- The validation of EMSE analytic relations may be done by measurements of EMSE via Transverse Electromagnetic (TEM) cell system.

2.4.1 Impedance method with correction factors for meshed materials

For flexible, meshed materials like woven fabrics with inserted conductive yarns, the shielding effectiveness in given by Eq. (7) [17]:

$$\boldsymbol{EMSE} = \boldsymbol{A_a} + \boldsymbol{R_a} + \boldsymbol{B_a} + \boldsymbol{K_1} + \boldsymbol{K_2} + \boldsymbol{K_3} \tag{7}$$

Where:
A_a = attenuation introduced by a particular discontinuity, dB.
R_a = aperture single reflection loss, dB.
B_a = multiple reflection correction term, dB.
K_1 = correction term to account for the number of like discontinuities, dB.
K_2 = low-frequency correction term to account for skin depth, dB.
K_3 = correction term to account for coupling between adjacent holes, dB.

Term $\mathbf{A_a}$. A premise for this relation is that the frequency of the incident wave is below the cut-off frequency, given by: $\boldsymbol{f_c} = \boldsymbol{c}/\boldsymbol{\lambda_c}$, with the cut-off wavelength 2.0 times of the maximum rectangular opening. With $\boldsymbol{a} = \mathbf{5\ mm}$ and $\boldsymbol{\lambda_c} = \mathbf{2}\,\boldsymbol{a} = \mathbf{10\ mm}$, it results a cut-off frequency of $\boldsymbol{f_c} = \mathbf{30\ GHz}$. Our frequency domain limit being 1 GHz, this premise is fulfilled.

$$\boldsymbol{A_a} = 27.3\left(\frac{\boldsymbol{h}}{\boldsymbol{a}}\right)[\mathbf{dB}] \tag{8}$$

Where:
h = depth of opening (fabric thickness) [cm].
a = width of the rectangular opening perpendicular to E-field (distance between conductive yarns) [cm].

Term $\mathbf{R_a}$. The aperture single reflection loss term depends upon both the impedance of the incident wave and the shape of the aperture.

$$\boldsymbol{R_a} = \mathbf{20}\log_{\mathbf{10}}\left(\frac{\mathbf{1} + \mathbf{4}\boldsymbol{K}^{\mathbf{2}}}{\mathbf{4}\boldsymbol{K}}\right)[\mathbf{dB}] \tag{9}$$

Where

$$\boldsymbol{K} = \mathbf{j6.69} \cdot \mathbf{10}^{-\mathbf{5}}\boldsymbol{f} \cdot \boldsymbol{a} \tag{10}$$

for rectangular apertures and plane waves,
f = frequency in MHz and.
a, the same significance as above and also expressed in [cm].

Term $\mathbf{B_a}$. The multiple reflection term is given by relation (11):

$$\boldsymbol{B_a} = \mathbf{20}\log_{\mathbf{10}}\left(\mathbf{1} - \frac{(\boldsymbol{K}-\mathbf{1})^{\mathbf{2}}}{(\boldsymbol{K}+\mathbf{1})^{\mathbf{2}}}\mathbf{10}^{-\frac{\boldsymbol{A_a}}{\mathbf{10}}}\right)[\mathbf{dB}] \tag{11}$$

$$\text{valid for}: \boldsymbol{A_a} < \mathbf{15\ dB} \tag{12}$$

For sample F1 and F2 $\boldsymbol{a} = \mathbf{5\ mm}$ and with $\boldsymbol{h_1} = \mathbf{0.55\ mm}$, respectively $\boldsymbol{h_2} = \mathbf{0.49\ mm}$, it result $\boldsymbol{A_a} \approx \mathbf{3.003\ dB}$ and the premise (12) is fulfilled.

Term K$_1$. For a source distance from the shield that is large compared with the aperture spacing, the correction term for the number of discontinuities is given by relation (13):

$$K_1 = -10 \log_{10}(S \cdot n)\ [\mathrm{dB}] \tag{13}$$

Where:
S = area of each hole (sq cm).
n = number of holes/sq. cm.

For fabric samples F1 and F2: $\boldsymbol{S} = \mathbf{0.25\ cm^2}$ and $\boldsymbol{n} = \mathbf{4/sq\ cm}$, thus $\boldsymbol{K_1} = \mathbf{0}$ may be neglected. Moreover the term K1 can be ignored for sources close to the shield, which is the case of the TEM cell.

Term K$_2$. The skin depth correction term is introduced for the reduction of EMSE when the skin depth becomes comparable to the screening wire diameter or the dimension between apertures. An empirical relation was developed for the skin depth correction term (14, 15):

$$K_2 = -20 \log_{10}\left(1 + 35p^{-2.3}\right) \tag{14}$$

Where:

$$p = d/\delta_y \tag{15}$$

As a computation example, the skin depth of silver yarns of fabric sample F2 – with a higher electric conductivity of the yarns than F1, is given by relation (16):

$$\delta_y = \frac{63}{\sqrt{f}}\ [\mathrm{mm}] \tag{16}$$

For frequency $\boldsymbol{f} = \mathbf{1\ MHz}$ we obtain $\boldsymbol{\delta_y} = \mathbf{63 \cdot 10^{-3} mm}$.

The term K2 is the single correction factor of the analytic relation sum which encounters the electric parameter of the yarns (electric conductivity and magnetic permeability), within the relation of skin depth. It is thus a factor with high sensitivity on the overall EMSE relation. The electric parameters were considered for the conductive yarn (not for the fabric), since the ratio $\boldsymbol{p} = \boldsymbol{d}/\boldsymbol{\delta_y}$ is a property of the yarn.

Term K$_3$. Attenuation is relatively high when apertures are tight and the depth of the openings is small compared to the aperture width. This is the result of coupling between adjacent holes and especially important for small openings. The relation is given by (18):

$$K_3 = 20 \log_{10}(\coth A_a/8.686)[\mathrm{dB}] \tag{17}$$

2.4.2 Impedance method with correction between foil and grid

The paper [18] and the handbook [19] propose an analytic relation for EMSE, based on impedance method, with correction between the fabric as a foil and the fabric as a grid. Since at low frequencies, the wavelength λ is much larger than the distance of the aperture opening *a*, the incident radiation sees the fabric as a foil (thin electric materials in relation to the skin depth). At higher frequencies, when the wavelength is comparable with the aperture opening, the grid structure of the fabric becomes relevant (thick electric materials). As such, an exponential function, depending on the frequency f, the dimension of the aperture a and the constant C, was proposed to balance the two relations of EMSE for foil and grid structure - Eq. (18).

$$EMSE = \exp\left(-Ca\sqrt{f}\right)(EMSE)_{foil} + \left[1 - \exp\left(-Ca\sqrt{f}\right)\right](EMSE)_{grid} \quad (18)$$

Where:
a = Distance between conductive yarns [mm].
f = frequency of EM field [MHz].
C = constant.

Analytic relation (18) was firstly introduced for metalized textiles by [18]. The two analytic relations for $(EMSE)_{foil}$ – the shielding effectiveness of a metallic foil of the same thickness (h) as the fabric and $(EMSE)_{grid}$ – the shielding effectiveness of an aperture (of size $L \times D$), subjected to a plane wave radiation are given by Eqs. (19, 20) and are valid for metallized textiles:

$$(EMSE)_{foil} = 20\log_{10}\left\{\left[\exp\left(\frac{h}{\delta}\right)\right]\left[\frac{(1+k)^2}{4k}\right]\left[1 - \frac{(k-1)^2}{(k+1)^2}\exp\left(-\frac{2h}{\delta}\right)\right]\right\} \quad (19)$$

$$(EMSE)_{grid} = 100 - 20\log_{10}(a \cdot f) + 20\log_{10}[1 + \ln(a/s)] + 30D/a \quad (20)$$

where:
$k = Z_0/Z_m$ (ratio of wave impedance to shield impedance).
h – thickness of the fabric [m].
δ – skin depth of the material [m].
a – maximum distance between conductive yarns [mm].
s – minimum distance between conductive yarns [mm].
D – depth of the aperture [mm].
f - frequency [MHz].
C = 0.129 (constant).

The constant C is derived by equaling the two relations $(EMSE)_{foil} = (EMSE)_{grid}$ and considering the frequency of three skin depths 3δ, as the point below current is negligible in a foil, (95% of the current flows within 3δ) [18].

The Eqs. (18–20) were derived by contribution of [20] in order to be applied for fabrics with inserted conductive yarns. Following conditions apply for the proposed Eqs. (19–20) to be valid for fabrics with inserted conductive yarns:

a. $\sigma \gg \omega\varepsilon$

This condition applies to conductive materials for which $\omega\varepsilon$ is negligible as compared to σ (in metals the condition is practically always fulfilled). This leads to a simplification of *EMSE* relations through the simplification of the shield material impedance definition:

$$Z_m = \sqrt{\frac{j\omega\mu}{\sigma + j\omega\varepsilon}} \approx \sqrt{\frac{j\omega\mu}{\sigma}} \quad (21)$$

Sample F1 has an electrical conductivity σ_{f1} = 41.5 S/m and the electrical conductivity of sample F2 is σ_{f2} = 589 S/m. Since it is considered that the fabrics have the electric permittivity of vacuum ($\varepsilon_0 = 1/36\pi^*10^9$ F/m), the term $\omega\varepsilon$ has a value of 0.055 at a frequency of 1 GHz. Therefore, this condition is fulfilled for both samples in the analyzed frequency range (100 kHz–1 GHz), since the difference between the two terms is of three orders of magnitude for sample F1 (41.5 > > 0.055) and four orders of magnitude for sample F2 (589 > > 0.055).

b. $Z_m \ll Z_0$

This condition corresponds to a situation when there is a substantial mismatch between the wave impedance and shield impedance. When this condition is fulfilled, *EMSE* relations greatly simplify as the ratio between the wave impedance and the shield impedance becomes much greater than 1 ($k > > 1$). Thus, Eq. (19) becomes:

$$(EMSE)_{foil} = 20\log_{10}\left\{\left[\exp\left(\frac{h}{\delta}\right)\right]\left[\frac{k}{4}\right]\left[1 - \exp\left(-\frac{2h}{\delta}\right)\right]\right\} \tag{22}$$

By using the simplified version of Z_m, the following values are obtained for the impedance of the two textile samples: 0.21 Ω for F1 and 0.037 Ω for F2 at a frequency of 100 kHz, and 21 Ω for F1 and 3.66 Ω for F2 at 1 GHz. The lowest impedance ratio is $k \approx 18$ for sample F1 at 1 GHz which is still much greater than 1. $k = 103$ for sample F2 at 1 GHz. Therefore, we can say that this condition is fulfilled for both samples in the analyzed frequency range (100 kHz–1 GHz). Note that sample F1, which has lower electric conductivity, fulfills this condition by two degrees of order only for the frequency domain 100 kHz-74.5 MHz. For higher frequencies, the EMSE relation is still valid for F1 but with a greater error, according to [20].

c. $h < 3\delta$ (Thin materials)

It refers to the fact that the thinner the material the higher the reflections from the second interface of the material and thus the re-reflection term in *EMSE* relations is more significant. The amplitude of the incident wave will decrease with about 95% at a distance of three skin depths (3δ) from the first interface inside the material. From the skin depth definition, $\boldsymbol{\delta = 1/\sqrt{\pi f \mu \sigma}}$, one can see that it is lower for materials with higher conductivity and at higher frequencies. For sample F2 (which has a higher conductivity than F1), $\delta \approx 0.66$ mm at 1 GHz and $3\delta \approx 1.98$ mm, which is lower than the fabric thickness ($h = 0.55$ mm) showing that the condition is fulfilled.

These conditions are needed to express relations (19-20) in case of fabrics with inserted conductive yarns according to [20] and are valid both for F1 and F2.

$$(EMSE)_{foil} = 168.14 + 20\log_{10}\left(\sqrt{\frac{\sigma_r}{f\mu_r}}\right) + 8.6859\frac{h}{\delta} + 20\log_{10}\left|1 - e^{-2h/\delta}e^{-j2\beta t}\left(\frac{Z_0 - Z_m}{Z_0 + Z_m}\right)^2\right| \tag{23}$$

where β is the phase constant.

$$(EMSE)_{grid} = 158.55 - 20\log_{10}(a \cdot f) - 20\log_{10}\sqrt{n} \tag{24}$$

$\boldsymbol{\sigma_r}$ was considered in Eq. (23) the relative electric conductivity of the fabric in relation to the electric conductivity of copper $\boldsymbol{\sigma_{Cu}} = \mathbf{5.8 \;\; 10^7 S/m}$. The Eq. (24) is valid for electrically thin materials and square apertures.

C is a constant for woven fabrics with conductive yarns computed in [20], in relation to the electric conductivity of the fabric and the number of apertures of the maximum length of the washer-shaped sample for the TEM cell **Table 3**:

$$C = 1.972 \cdot 10^{-5}\sqrt{\sigma_f\, n} \tag{25}$$

And

$$n = \frac{l_c}{a} \tag{26}$$

With:

n – number of apertures.

l_c – maximum length of fabric for washer-sized sample within TEM cell.

a – distance between conductive yarns.

n	σ_f	C
18	589	0.0005
18	41.54	0.002

Table 3.
Number of apertures and electric conductivity of fabric to compute constant C *according to [20].*

The parameter l_c is needed according to [9] in order to compute the number of apertures of the maximum linear distance. The number of apertures on maximum linear distance are only relevant for EMSE of multiple apertures [9]. *C* is a factor with high sensitivity.

However, relation (24) is valid under certain conditions [20]:

- the maximum number of apertures counts on a linear dimension
- distance between conductive yarns $a < \lambda/2$
- distance between conductive yarns $a >$ thickness h
- the material is thin ($h < 3\delta$) and the n apertures are equally sized

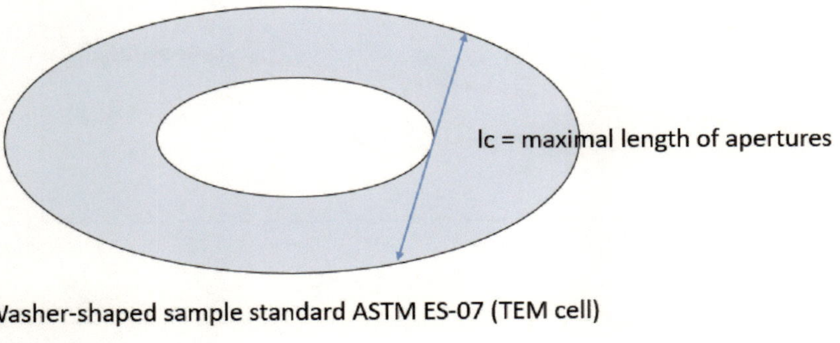

Figure 8.
The maximal distance of the sample for TEM cell meant to compute the number of apertures.

Figure 8 presents the scheme of a washer-shaped sample of TEM cell, with the longest distance, for counting the maximal number of apertures.

The measurements show the values: lc = 75 mm and n = 18 for F1 and F2. These values were used to compute constant C of Eq. (25), in order to be used in general equation of EMSE (18), with Eq. (23) for $(EMSE)_{foil}$ and Eq. (24) for $(EMSE)_{grid}$.

3. Validation of the two analytic relations by the two samples

Both the analytic relation of impedance method with correction factors (described at 2.4.1), and the analytic relation with correction between foil and grid (described at 2.4.2), were applied for sample F1 and sample F2, by introducing the electric and geometric parameters of the samples and generating the related EMSE diagrams.

Figures 9 and **10** present the modeled and the measured values in case of the relation of impedance method with correction factors for F1 and F2.

Figures 9 and **10** show a good modeling of the impedance method with correction factors, both for the fabric with stainless steel yarns and the fabric with silver yarns. The analytic relation (7) shows for the fabric with stainless steel yarns a maximal difference to the measured values of about 10 dB at the frequency of 10 MHz and has for the frequency domain of 0.1–1 MHz and 100–1000 MHz, close values with maximal differences of 2–5 dB. Same analytic relation shows for the fabric with silver yarns a maximal difference to the measured values of 5 dB, especially in the end points of the frequency domain, namely 0.1 MHz and 1 GHz. **Figures 11** and **12** present the modeled and the measured values in case of the relation with correction between foil and grid for F1 and F2.

Figures 11 and **12** show a good estimation of EMSE by relation with correction factor between foil and grid, both for the fabric with stainless steel yarns and the fabric with silver yarns. The analytic relation (18) shows a mean difference to the

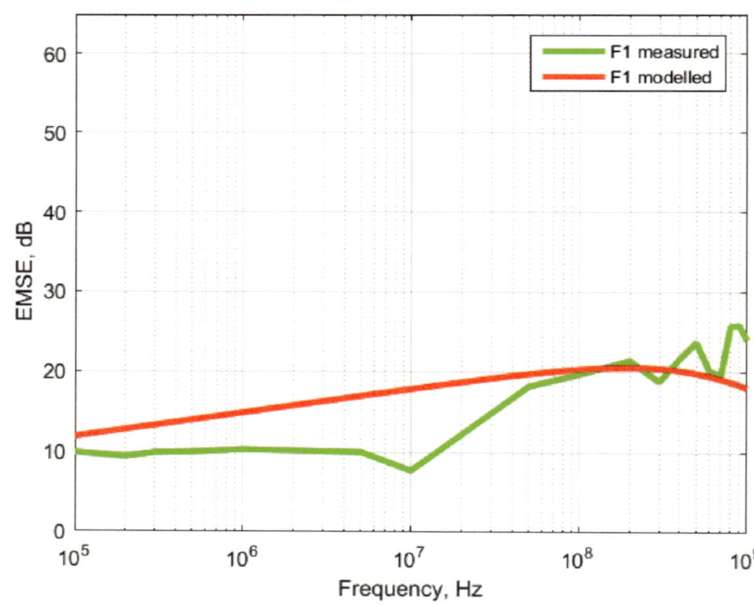

Figure 9.
F1 (stainless steel) – Modeled (7) and measured.

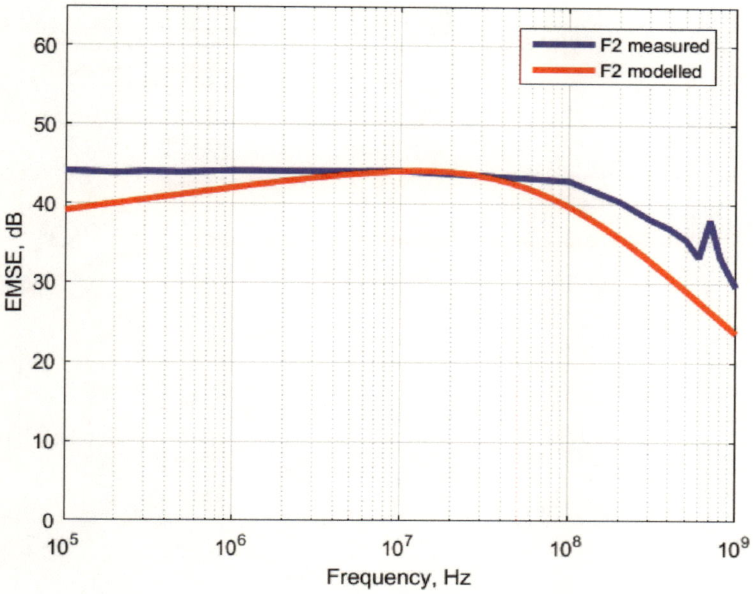

Figure 10.
F2 (silver) – Modeled (7) and measured.

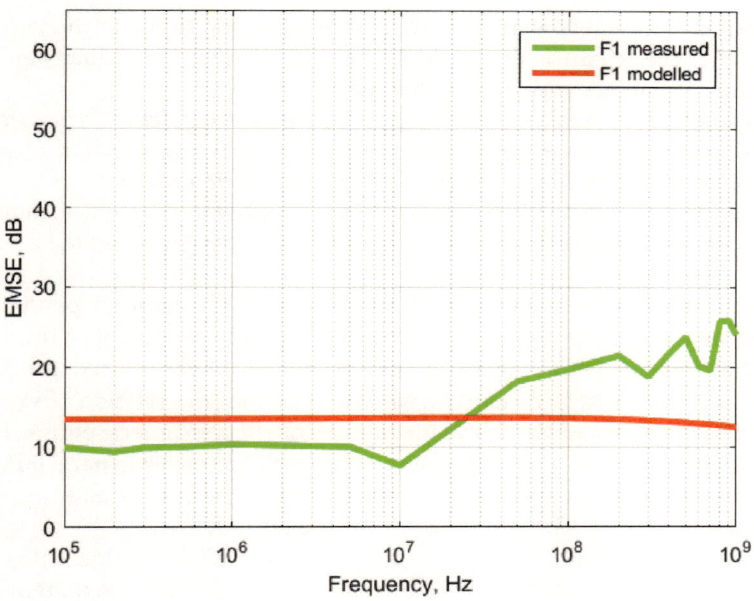

Figure 11.
F1 (stainless steel yarns) – modelled (18) and measured.

measured values of 5–10 dB on the frequency domain for F1 and a mean difference to the measured value of 9–10 dB on the frequency domain for F2.

The two methods for estimating EMSE in case of fabrics with inserted conductive yarns, relay on different principles of analytic relations. Both methods use as input parameters the electric and geometric parameters of the fabric. These electric

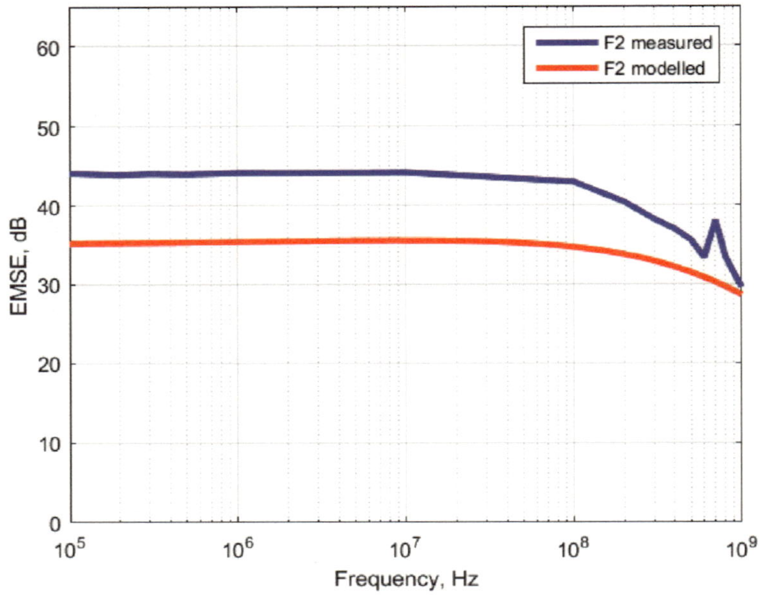

Figure 12.
F2 (silver) – modelled (18) and measured.

and geometric parameters were presented with same notations for the two methods, since the scientific literature tackles the methods individually and presents the parameters with various notations [17–20].

Two types of fabrics with inserted conductive yarns, having similar structure, but different raw materials for the conductive yarns, namely stainless steel and silver, were used to validate EMSE by the two methods, based on experimental data.

Some of the parameters of the two methods have greater significance in overall estimation of EMSE, some have less significance. This aspect is tackled by the sensitivity analysis.

As such, for the first method, all factors of relation (7) do not depend on frequency, neither on the electric parameters of the fabrics, except of the factor *K2*, which is expressed as ratio between yarn diameter $\boldsymbol{d}$ and yarn skin depth $\boldsymbol{\delta_y}$. *K2* is the single factor depicting the electric parameters, since the relation of skin depth for the conductive yarn of the fabric includes frequency, magnetic permeability and electric conductivity. By analyzing influence of each factor for general outcome of EMSE, it results that factor *K2* has the most significant influence and this is due its characterization of the electric parameters of the fabrics.

The second method aims to balance the relations of EMSE for the foil and for the grid, depending on frequency and the related thin or thick electric material. Main parameter with significant influence in this regard is the constant *C*. This constant depicts the balance between the EMSE of the foil and EMSE of the grid. Constant *C* was computed according to [20] depending on electric conductivity of the fabric and number of apertures on the longest distance of the washer-shaped fabric sample for the TEM cell (according to standard ASTM ES-07).

The geometric parameters of fabrics have for both methods an important significance too, however with less sensitivity than the electric parameters.

Another aspect to be tackled is the tolerance of the EMSE measurement via TEM cell system according to standard ASTM ES-07. The contour of the washer-shaped samples was coated with silver paste in order to ensure a good electric connection to the metallic parts of the TEM cell. Since EMSE was measured as ratio between the electric power values without and with material present and all measurement devices have they own tolerance (network analyzer, amplifier, TEM cell connection), the overall measurement error could be estimated at 3–4 dB.

As such, EMSE values were compared between models of estimation based on electric a geometric parameters of the fabric and experimental measurements. It can be stated that the validation approach has an indicative character. By comparing validation results of the two analytic relations (7) and (18) we may conclude that analytic relation (7) has closer values to experimental data. However, both relations have close values to the experimental values of EMSE and may be used to predict this special functionality of fabrics. Design of fabrics for electromagnetic shielding applications may be supported by both methods.

4. Specified requirements of electromagnetic shielding textiles

The designer of fabrics may relate on the document [FTTS-FA-003 v2], in order to evaluate what the needed requirements are for fabrics destined for EM shielding. This document states the limits of EMSE for fabrics of professional and general use (**Tables 4** and 5).

According to the specification, F2 (silver yarns) has for professional use a good EMSE in the frequency domain of 0.1–100 MHz and a moderate EMSE in the frequency domain of 100–1000 MHz. Moreover, the same fabric sample F2 has for general use an excellent EMSE in the frequency domain of 0.1–100 MHz and a very good EMSE in the frequency domain of 100–1000 MHz.

The sample F1 (stainless steel yarns) has professional use a fair EMSE in the frequency domain of 100–1000 MHz, while for general use the same sample has a very good EMSE. For general use in the frequency domain of 0.1–100 MHz, sample F1 has good EMSE.

These considerations prove that the achieved fabric sample may be successfully used as electromagnetic shields for various applications, with different grades of EMSE on the frequency domain.

Grade	5 Excellent	4 Very good	3 Good	2 Moderate	1 Fair
EMSE range	EMSE>60 dB	60 dB ≥ EMSE >50 dB	50 dB ≥ EMSE >40 dB	40 dB ≥ EMSE >30 dB	30 dB ≥ EMSE >20 dB

Table 4.
Class I – Textiles for professional use.

Grade	5 Excellent	4 Very good	3 Good	2 Moderate	1 Fair
EMSE range	EMSE>30 dB	30 dB ≥ EMSE >20 dB	20 dB ≥ EMSE >10 dB	10 dB ≥ EMSE >7 dB	10 dB ≥ EMSE >7 dB

Table 5.
Class II – Textiles for general use.

5. Discussion on structure–property relationship of EM shielding fabrics

The two proposed models have high relevance in product development for EM shielding applications. By having set the target EMSE value and the frequency of the application (**Tables 4** and **5**), one of the first choices is selection of the raw material for the conductive yarn (stainless steel, silver, copper etc.). The quantity (mass) of conductive yarn leads to a cost-effectiveness calculation for the shielding fabric. A second fabric parameter to be designed is the distance between conductive yarns a. A tighter distance between conductive yarns a, means a higher EMSE, according to both methods (7) and (18). On the other hand, too tight conductive yarns are critical, due to technological limitations on non-conventional weaving machines destined for insertion of conductive yarns with metallic content.

Other fabric parameters, such as weave, density and yarn count are third choice to design EM shields. Although, none of the presented models depicts these parameters, some other important properties of fabrics, such as mechanical resistance, flexibility and drape highly depend on these parameters.

Basically, the proposed models offer a quick overview of the functionality of shielding, especially useful when starting designing a fabric.

6. Conclusion

This book chapter tackles electromagnetic shielding as an important functionality to be achieved by textile materials. In order to be estimate this functionality, quantified by the electromagnetic shielding effectiveness (EMSE), the book chapter proposes two analytic models from the scientific literature [7, 20]. The approach of modeling EMSE offers to the textile designer relevant insight on the relation between fabric properties and shielding functionality. Most relevant electric and geometric parameters of woven fabric shields were considered within the analytic models, such as: fabric thickness, conductive yarn's diameter, distance between conductive yarns, electric conductivity and magnetic permeability of yarns and fabrics.

The two analytic relations were meant to be validated by experimentally measured woven fabrics with inserted conductive yarns out of stainless steel and silver. EMSE was measured for both these fabrics via TEM cell system, according to the standard ASTM ES-07. The experimental results show that both analytic relations have a good estimation of EMSE with a maximum difference on the frequency domain 0.1–1000 MHz of 5–8 dB. Moreover, the achieved fabrics show effective EMSE values according to the requirements specifications of textile shields for general and professional use.

In conclusion, the proposed analytic models do support the textile engineer in designing woven fabrics, to achieve desired shielding functionality. Some technological aspects regarding textile processing should be also considered, like limitations of inserting the metallic yarns on non-conventional weaving machines, achieving desired mechanical resistance properties by modifying yarns density of fabrics and overall cost-effectiveness of using expensive yarns with metallic content. This contribution was possible by interdisciplinary knowledge of textile field and electromagnetic compatibility field.

Author details

Ion Razvan Radulescu[1]*, Lilioara Surdu[1], Emilia Visileanu[1], Cristian Morari[2] and Marian Costea[3]

1 INCDTP, Bucharest, Romania

2 ICPE-CA, Bucharest, Romania

3 Faculty of Power Engineering, University Polytechnica Bucharest, Romania

*Address all correspondence to: razvan.radulescu@incdtp.ro

References

[1] Van Langenhove L. Smart textiles for protection: an overview. In: Chapman R. A., editor. Smart textiles for protection. Woodhead Publishing Limited, 2013. p. 3–33. ISBN 978–0–85709-762-0.

[2] Harlin A, Ferenets M. Introduction to conductive materials. In: Mattila H, editor. Intelligent textiles and clothing. Woodhead Publishing Limited in association with The Textile Institute, 2006. p.217–238. ISBN-10: 1–84569–162-8.

[3] Costea M. Methods and instruments to ensure electromagnetic immunity. AGIR Publishing House Bucharest, 2006, ISBN 973–720–041-1

[4] Syaza S.K.F., et al, Non-ionizing radiation as threat in daily life. Journal of Fundamental and Applied Sciences, 2017, 9(2S), p. 308–316, http://dx.doi.org/10.4314/jfas.v9i2s.21

[5] The Interphone study [Internet]. 2011. Available from: https://interphone.iarc.fr/ [Accessed: 2020-09-23]

[6] WHO - Radiation: Electromagnetic fields [Internet]. 2016. Available from: https://www.who.int/ news-room/q-a-detail/radiation-electromagnetic-fields [Accessed: 2020-09-23]

[7] Schwab A., Kuerner W. Electromagnetic compatibility. AGIR Publishing House Bucharest, 2013, ISBN 978–973–720-359-5

[8] Paul C. R., Chapter 1: Introduction to Electromagnetic compatibility. In: Introduction to Electromagnetic compatibility, Wiley Inter-science, 2006. ISBN-10: 0–471–75500-1

[9] Ott H. W. Chapter 6 Shielding. In: Electromagnetic compatibility engineering. Wiley Inter-science, 2009. p. 238. ISBN 978–0–470-18930-6

[10] BekinoxBK yarns [Internet], 2020, Available from: https://www.bekaert.com/-/media/Files/Download-Files/Basic-Materials/Textile/Bekaert-Bekinox-BK-yarns.pdf [Accessed: 2020-10-23]

[11] SilverStatex yarns [Internet], 2020, Available from: https://statex.de/en/shieldex-yarns/ [Accessed: 2020-10-23]

[12] Dodu A., The Handbook of the Textile Engineer. AGIR Publishing House Bucharest, 2005. ISBN 9738466857

[13] Velten K. Mathematical modelling and simulation. WILEY-VCF Verlag Germany:, 2009. ISBN: 978–3–527-40758-8

[14] Chandra, R. J., Shivamurthy, B., Kulkarni, S. D., & Kumar, M. S. (2019). Hybrid polymer composites for EMI shielding application-a review. Materials Research Express, 6(8), 082008.

[15] Yin, J., Ma, W., Huang, Z., Liu, K., & Leng, Z. (2020). A model for predicting electromagnetic shielding effectiveness of conductive fiber plain fabric. Journal of Industrial Textiles, 1528083720929356.

[16] Rădulescu, I. R., Surdu, L., Visileanu, E., Costea, M., PĂTRU, I., & Voicu, V. (2018). Modelling and testing the electromagnetic near field shielding effectiveness achieved by woven fabrics with conductive yarns. Industria Textila, 69(3), 169–176.

[17] Keiser B. Chapter 6: Shielding. In: Principles of Electromagnetic Compatibility, ARTECH House, 1985, Standard Book Number: 0–89006–065-7

[18] Henn, A.R.; Crib, R.M. Modeling the shielding effectiveness of metallized fabrics. In Proceedings of the International Symposium on

Electromagnetic Compatibility, Anaheim, CA, USA, August 1992.

[19] Neelakanta, P.S. Handbook of Electromagnetic Materials: Monolithic and Composite Versions and Their Applications. CRC Press: Boca Raton, USA, 1995; p. 447–490, ISBN 0–8493–2500-5.

[20] Neruda M., Vojtech L., Electromagnetic Shielding Effectiveness of Woven Fabrics with High Electrical Conductivity: Complete Derivation and Verification of Analytical Model, In: Materials, 2018, 11, 1657, https://doi.org/10.3390/ma11091657

[21] White D. A Handbook on Electromagnetic shielding materials and performance, USA: Don White Consultant, 1980.

[22] Permeability of stainless steel [Internet], 2018, Available from: http://www.ugitech.com/uploads/ugitransfer/docs/GED_1591.pdf [Accessed: 2020-10-23]

[23] Surdu, L., Visileanu, E., Rădulescu, I.R., Sandulache, I., Mitran, C., Mitu, B., Stancu, C., Ardeleanu, A., Research regarding the cover factor of magnetron sputtering plasma coated fabrics, In: Industria Textila, 2019, 70, 2, p. 154–159, http://doi.org/10.35530/IT.070.02.1618

Section 4

Application Areas

Chapter 9

Face Mask: A Novel Material for Protection against Bacteria/Virus

Thilagavathi Govindharajan and Viju Subramoniapllai

Abstract

Facemask is defined as a loose-fitting device which creates a physical barrier between the mouth and nose of the individual wearing mask and likely pollutants in the immediate environment. Evolution of severe viral respiratory infectious agents such as pandemic COVID-19, severe acute respiratory syndrome, pandemic influenza and avian influenza has driven the use of protective face masks by public and health workers. In this chapter, characteristics features and uses of different types of masks are discussed. Characteristics of various nonwoven technologies for manufacturing face masks are also discussed. Test methods and recent developments in face masks are briefly covered.

Keywords: Face Mask, Bacteria, Virus, Filtration

1. Introduction

Facemasks are in general used for reducing breathing exposure to airborne particles such as virus and bacteria that may be connected with a wide range of health effects [1]. Facemasks are considered to impede or reduce the spread of airborne particles that causes annoying health issues. Facemask is defined as a loose-fitting device which creates a physical barrier between the mouth and nose of the individual wearing mask and likely pollutants in the immediate environment [1]. Based on the use, face mask is commonly characterized as medical, isolation, dental and surgical masks. Usually a face mask is made-up of flat or pleated fabric with one to three layers, which in turn secured to the head with ear loops [1, 2]. An evolution of severe viral respiratory infectious agents such as pandemic COVID-19, severe acute respiratory syndrome, pandemic influenza and avian influenza has driven the use of protective face masks among public and health workers [3].

The face masks worn must prevent the infectious microbes such as viruses and bacteria from penetrating through the fabric structure. Hence it is recommended to produce fabric structures with pore sizes lesser than the microbe size. If the size of the microbes is known, then the fabrics can be manufactured according to the requirements. **Tables 1** and **2** shows the specific viruses and bacteria and associated diseases. From **Tables 1** and **2**, fabric producers can get an idea regarding the pore size required in face masks for filtering viruses and bacteria respectively [4]. This chapter discusses the different types of face masks, filtration mechanisms, manufacture and characteristic features of various face masks. Test methods and recent developments in face masks are also covered.

IntechOpen

Species	Size (mm)	Associated diseases
Hepatitis Virus	0.042–0.047 diameter	Hepatitis B
Adenovirus	0.07–0.09 diameter	Respiratory Infections
HIV	0.08–0.11 diameter	Acquired Immuno Deficiency Syndrome
Filoviruses	0.08 diameter	Ebola Virus
Bunyaviridae	0.08–0.12 diameter	Hanta Virus
Orthomyxoviridae	0.08–0.12 diameter	Influenza A, B, and C
Coronaviridae (SARS)	0.10–0.12 diameter	SARS
Varicella-Zoster Virus	0.11–0.12 diameter	Herpes
Cytomegalovirus	0.12–0.20 diameter	Pneumonia, Hepatitis, Retinitis

Table 1.
Size of some highly infectious disease viruses [4].

Species	Size (mm)	Associated diseases
Serratia marcescens	0.45 diameter	Extra-intestinal Infections, Nosocomial Infections
Pseudomonas aeruginosa	0.50–1.0 diameter	Endocarditis, Pneumonia, Osteomyelitis, Nosocomial Infections, Meningitis, Septicemia
Staphylococcus aureus	1.0 diameter	Pneumonia, Osteomyelitis, Pneumonia, Osteomyelitis
Mycobacterium tuberculosis	1.0–5.0 diameter	Tuberculosis
Bacillus anthracis	1.0–1.5 diameter	Anthrax Infection

Table 2.
Size of some disease-causing bacteria [4].

2. Different types of masks

Masks are fabricated based on the size of pores and particles desired to be filtered out, which is mainly determined by the health and medical professionals. Different types of masks as per the characteristics features and specific uses are discussed below.

2.1 Dust mask

Dust masks are in general flexible and these masks are developed to provide protect against dust, molds, pollens and other irritants. They normally not provide protection against any pathogens and therefore they should not be used for viral protection [3].

2.2 Single-use face mask

Single-use face masks are disposable and commonly used for single application. They are usually made from wood pulp tissue paper or single layer nonwoven fabric and are very thin. They are normally used for providing protection against larger dust particles, at construction sites and in other similar industries. It is not recommended to use such type of face masks for protection against covid-19 [3].

2.3 Surgical mask

Surgical mask is defined as a loose-fitting and disposable device that creates a physical barrier between mouth and nose of the wearer and the probable pollutants in the immediate environment [3]. Surgical mask is normally composed of 3 layers. The innermost layer contains an absorbent material which absorbs moisture while the wearer is breathing. The middle layer is made of melt blown nonwoven which work as a filter, and the outer layer repels liquids [5]. The surgical mask prevents the splashes, larger-particle droplets or sprays with a diameter above 100 mm. It also controls the spread of respiratory secretion and a person's saliva to others. The Severe Acute Respiratory Syndrome Coronavirus 2 (SARS-CoV-2 virus) is spherical, though slightly pleomorphic, has a diameter of 60–140 nm. These masks will not be able to prevent inhalation of very small particles existing in the air and therefore it does not offer full protection against pathogens [3].

2.4 N95 respirator

N95 respirators are normally non-oil resistant and also termed as electrets filter. The word N95 denotes that these types of face masks can filter at least 95% of aerosols at particle size 0.3 μm [6, 7]. It is reported N95 respirators sometimes do not offer adequate protection against the aerosol particles which are lesser than 300 nm. Hence protection given by certain N95 respirator masks can also drop below 95%, during high inhalation flow rates [6]. The N95 respirators of different industries has varied performance mainly based on the penetrating particle's size. The N95 respirator is made up of four main layers namely inner layer, support layer, filter layer, and layer mask filter layer from inside to outside of it. In addition, a ventilator fan is mounted on the outer layer of N95 to enhance the breathing performance. N95, N99 and N100 masks filter the corona virus effectively [6]. International regulation suggests that the surgical masks and N95 respirator are not advisable to be worn continuously for more than 4 hours and 8 hours respectively, if not it becomes soiled, damaged or contaminated.

2.5 P100 respirator

This is a sort of filtering face piece respirator. It is highly oil-proof and the filtration efficiency of the aerosol particles is 99.97%. From the comparisons made by the researchers regarding the filtration efficiency of N95 and P100 respirators, it was found that there was no considerable gap present in the permeability values before use. Whereas results after post exercise were more expedient for the P100 respirator, on the other hand, N95 failed the post-exercise criterion [6]. Moreover, owing to the likely effects on the breathing resistance, face seal, and moisture retention during the use and hard work, there is the threat of reshaping the face mask. P100 face masks could able to maintain their shape in humid and high temperature compared to the N95 [8].

2.6 Full face respirator

Full-face respirators are constructed using rigid plastic material which consists of both apparent part and central port part [9]. Such masks are used for the management of breathing problems and sleep troubles by providing respiration to patients [6]. Face covering part is made from flexible elastomeric material that fits well to cover the anatomy of face. Due to the elastic nature, straps aids in generating adequate force which gives good adherence of mask on to the wearer's face.

However, this arrangement not works well if the person rolls during sleep. In such situations, masks are removed from the wearer, which disconnects the sealing between mask and face of the wearer and thus leads to poor protection [10].

3. Mechanisms of filtration

The filtration mechanism of filter fabrics is mainly based on physical filtration mechanism namely interception, inertial impact, diffusion, gravitation, and electrostatic attraction [4]. During a particle filtration process, an interception takes place as the particle radius is equal to or greater than the fiber-particle distance (within 0.1 to 1 μm particle size) [4]. The particles are instantly arrested outside the face masks as soon as it comes in contact with and attaches to the fiber following the air streamline around the fiber. Inertia impact occurs when a particle size is bigger than 1 μm with a larger mass, which are not capable of going after the arc pathway of the air streamline crashes into the fiber. For the particles in high velocity, it becomes harder to penetrate through the pores of the mask sieve and to arrive at wearer. The diffusion mechanism [2, 3] makes the particles to diverge from the actual flow lines arbitrarily when they are close to fibers, mainly for particles sizes which are lesser than 0.1 μm. In the filter fabrics the electrostatic attraction occurs through the electrostatically charged mats [3]. By means of electrostatic driven adsorption the oppositely charged particles are attracted. The electrostatic attraction is efficient for capturing sub-micron particles with no increase in pressure drop [3]. On the other hand, the filtration state will alter when the fibers of the filter fabrics are within nanoscale. By means of charging techniques namely corona charging, tribocharging, and electrospinning the aerodynamic behavior of airflow around the filter can be made [4]. Nevertheless, the charges will decompose steadily during use or long-term storage.

4. Polymeric materials used in face masks

Fibers are the tiniest component of a textile structure and their characteristic features are mainly dependent on their physical and chemical properties. Fibers with irregular surfaces/irregular cross-sections have the capability to trap particles efficiently than the fibers with smooth surface/regular cross-sections. Cotton is characterized by convolutions, ideal for capturing small particles. Viscose rayon fibers have striations, which offer certain irregularity to the fibers surface. However, these striations are not as effective in inhibiting particle movements as in cotton fibers. Synthetic fibers such as polypropylene, polyester and nylon intrinsically possess smooth surface structure and round cross-sections that permit particles to certainly slide by the fiber. The smooth and round cross-section also creates capillary forces between fibers. Synthetic fibers with different and irregular cross-sections can be manufactured by using suitable spinnerets during fiber manufacture process and irregularity in the fiber structure can be made by means of texturisation process. Polyester fibers with longitudinal channels possess good wicking property. Fiber length is another factor influencing the fabric barrier properties. Longer fibers are not much effective in capturing small particles than shorter fibers [10].

Commonly used materials for face masks applications includes polypropylene, polyesters, polyamides polyphenylene oxide and polycarbonates. Trifluorochloroethylene is a kind of fluorinated polymers is also used for face mask applications. In general, hydrophobic and thermoplastic polymers are preferred. One of the main advantage of using polypropylene fiber in face mask is its nonabsorbent properties. Antibacterial face mask has been prepared from polypropylene by treating it with dimethyl-dioctadecyl-ammonium bromide. The face mask showed good resistance against bacteria [6]. Madsen et al. [11] analyzed the filtration efficacy of face masks made from polypropylene, polyester, glass fiber and cellulose material. The polypropylene showed highest filtration efficiency followed by polyester, glass and cellulose material. The filtration efficiencies of several commonly available fabrics from silk, cotton, flannel, chiffon, and synthetic fibers has been evaluated by Konda et al. [12]. It is observed that for particle sizes of <300 nm, the filtration efficiencies ranged from 5–80% whereas, for particle sizes >300 nm, filtration efficiencies ranged from 5–95%. It is also noted that the face mask produced from cotton-flannel, cotton-chiffon and cotton-silk showed more than 80% filtration efficiency for particle size less than 300 nm and also showed 90% filtration efficiency for particle size greater than 300 nm. Recently, micro fibers and nanofibers are gaining greater attention due to its unique features.

5. Technology of Manufacturing Face mask

The filtration efficiency of a face mask depends on several factors including fiber type, fiber's cross-sectional shape, technology of manufacture and web's structure. There are several methods to develop fibrous matt that can be used as filters in masks. Face masks are generally prepared from nonwoven fabrics as nonwoven fabrics offers superior filtration efficiencies than woven or knitted fabrics owing to their randomness and 3D structure which permits higher thickness which increases the particle travel distance. Layered nonwoven structures can also be used for the development of face masks. The key nonwoven technology for manufacturing face masks includes melt blown, spun bonding and electrospinning [3]. **Table 3** shows the characteristics of various nonwoven technologies for manufacturing face masks [3].

Factors	Melt blowing	Spun bonding	Electrospinning
Extrusion	A high-velocity air propels a molten thermoplastic resin from extruder	Rotating screw obliges the melt to extrude	The charged polymer jet is extruded via nozzle and drawn with the help of applied electrostatic force
Bonding	Self-bonding of fibrous web in the air.	Bonding by means of mechanical, chemical or thermal method	Thermal bonding used as a post-processing treatment can enhance the strength of fibers.
Die	Hot air converges with the fiber as it come out from the die	Hot air flow is at a cross flow to the out coming fiber	Positive charge is applied at extruder.
Fiber diameter	Ranges from 1 to 5 μm	Ranges from 15 to 50 μm	Ranges from 100 nm to 500 nm
Applications	Surgical face masks, gas filtration, liquid filtration, cartridge filters and clean room filters	Hygiene and healthcare textiles, disposable diapers, automotive industry, filtration, civil engineering, carpet backings packaging.	Tissue engineering, wound dressing, drug delivery and cosmetics, protective clothing, filters and smart textiles

Table 3.
Characteristics of various nonwoven technologies for manufacturing face masks [3].

5.1 Melt blown technology

In this technique, a high-velocity air blows a molten thermoplastic polymer from an extruder die with hundreds of tiny nozzles onto a conveyor or take-up screen to structure a fine fibrous and self-bonded web. The fibers produced via such method have relatively small diameters (1–5 mm) and thus, the pore size is much smaller, this makes them superior in filtration performance [13]. Hence, this technology is mostly used for the development of nonwoven fabrics for various filtration applications such as respirators, surgical face masks, liquid filters, cartridge filters and clean room filters. By introducing electrostatic charges and superabsorbent polymer in melt blown polymer fibers it is possible to improve the particle capturing efficiency, air moisture absorption capability and hygienic comfort [3].

Electrostatic-assisted melt blown fabric has been developed by applying electrostatic field directly to the meltblown spinning head [14]. The mean pore diameter of polypropylene microfibers formed using this technique reduced from 1.69 to 0.96 mm, this also resulted in a narrow fiber size distribution. The mean average pore size of conventional melt blown and electrostatic-assisted melt blown fabric were 33.415 mm and 29.285 mm, respectively [14]. The application of electrostatic force resulted in decreased pore size and mean fiber diameter for electrostatic-assisted melt blown fabric. It is also observed that electrostatic-assisted melt blown fabric had better filtration efficiency than conventional meltblown fabric [14]. For 0.3 mm particle size, the filtration efficiencies of conventional melt blown fabric and electrostatic-assisted melt blown fabric were 40.65% and 50.82%, respectively. Also, for 1 mm particle size, the filtration efficiencies were 73.98% and 86.44% and for 2.5 mm particle sizes, the filtration efficiencies were 95.35% and 98.96%, respectively [3, 14]. With reduced fiber diameter, the pore size also reduces and the even distribution of fibers per unit area increases this gives the particles additional chance to adhere to the nonwoven fabric [3]. Furthermore, electrostatic-assisted melt blown fabric shows good adsorption ability than conventional fabrics. Melt blown technology has also been used to develop N95 respirator mask filters and surgical mask filters. The melt blown fabrics are the most important filter of face masks that arrests the bacteria as well as microbes from entering or exiting the mask. Polypropylene is the most widely used material to produce surgical face mask. The porosity ranges from 75 to 95% and the basis weight of polypropylene web is around 5–1000 g/m^2. Polycarbonate, polystyrene, polyethylene, or polyester compositions are also used to prepare face masks [3].

5.2 Spun bonding technology

In this process, molten polymer is extruded onto a conveyor belt via spinneret. Extrusion is a process where the polymeric chips are generally fed into an extruder containing a rotating screw in a heated barrel in which polymer chips are mixed, then melted and finally pumped through a die, that gives the fibers a uniform thin profile and can also endure higher temperatures. Prepared fibers are then quenched by a cool air, and are collected on a conveyor belt and then are additionally bonded by means of thermal, mechanical or chemical means to form the spun bonded nonwoven fabric. The developed spun bonded fabric have random fibrous structure with a weight ranging from 5 to 800 g/m2, the thickness of the web range from 0.1 to 4.0 mm, and the diameter of the fiber vary from 1 to 50 mm [15]. When the fiber diameter decreases, the pore size of the nonwoven fabric decreases; resulting in more even allotment of fibers which increases the particle capturing efficacy. Polymers such as polyester, polypropylene, polyethylene, nylon, polyurethane, etc. are appropriate polymers for the spun bonding process [16]. Among different polymers, isotactic polypropylene is one of the most extensively used polymer for spunbonded nonwoven fabric manufacture as polypropylene is relatively economical and offers the highest yield (fiber per kilogram). The polypropylene based nonwoven products offers lowest density [3].

5.3 Electro spinning

Electro spinning is a method wherein a polymer based solution is discharged with the help of electric field. The polymeric fluid becomes finer as it passes through the electromagnetic field and deposits on a plate ensuing the production of nonwoven nano-fibrous web. The negative charge is supplied at the collector and positive charge is supplied at the nozzle making the fluid jet to acquire charge at the nozzle tip for the creation of Taylor cone. Charged polymeric fluid is extruded through the nozzle and are drawn by the electrostatic force [17]. As the jet speed up and thins in the electrical field, radial charge repulsion results in splitting the primary jet into numerous filaments, known as "splaying', thereby manufacturing polymeric nanofibers which are electrically charged which are collected at the collector. By means of the subsidiary jets formed, the mean fiber diameter is determined which ranges from 100 nm to 500 nm and the structure and properties of nanofibers is mainly affected by electrostatic forces and viscoelastic behavior of the polymer [18].

In order to examine the relationship between the fiber diameter and volume charge density, a model for the fluid jet inside the electric field has been developed by Fridrikh et al. [19]. The model predicts a terminal jet diameter, which is a result of balance between normal stresses owing to surface charge repulsion and surface tension. It can be found from electrical current, flow rate and the surface tension of the fluid. In another study polyacrylonitrile polymer was used to produce patterned nanofibrous membranes through electro spinning technique [20]. The bulged bubble template was used as collector for manufacturing the patterned membrane. They noted that though the pressure drop reduced in the range of 151.7 to 24.7 mm H_2O, the filtration efficiency reduced only in the range 99.94% to 96.33% when compared to nanofibrous filter [20].

In air filtration applications, reducing the pressure drop and maintaining filtration efficiency play an important role in patterned nanofibers manufacture. The results suggest that electrospinning could be a possible option to produce nanofibrous membranes which can be used in filters to manage hazardous ultrafine particles such as viruses. Electro spun nanofibrous nonwoven mats can be effectively used for the filtration of submicron and nanoparticles for enhanced health protection from different contaminants (e.g. coronavirus). Lackowski et al. [21] reported that these nonwoven mats encompass higher filtration efficiency for nano and submicron particles, which are superior than high efficiency particulate air filters having a filtration efficiency of 99.97% for the particle of size 0.3 micron. The filter created by an electrospinning process is generally charged and due to the presence of electrostatic force it has very high filtration efficiency. The commonly used polymeric materials includes polyvinyl chloride (PVC) and poly vinylidene fluoride (PVDF) [3].

6. Recent developments in face masks

6.1 3D-printed masks

Polypropylene due to its unique features such as easy processability, printability, recyclability, mechanical integrity and low cost, it is normally used for various technical and industrial applications [4, 6]. Alternatively, styrene-(ethylene-butylene)-styrene is a polymeric elastomer which has low processing temperature and low distortion during extrusion. Thus, the combination of polypropylene and styrene-(ethylene-butylene)-styrene can be used for the processability of 3D printed N95 face masks. Furthermore, it could manage the thermoplastic elastomeric ratio tailoring the elasticity and flexibility of the 3D model material to have a fitted face masks. Accordingly, 3D printing procedure is suitable to produce stable and biocompatible N95 masks that are similar to industrial brands [22]. Swennen et al. [23] fabricated a re-usable face mask by employing 3D printing procedure based on the materials and methods (3D imaging and 3D printing). In a 3D protective face mask there are two 3D-printed reusable polyamide composite components (a filter membrane support and a face mask) and two disposable components (filter membrane and head fixation band). The 3D modeling of the masks can be made quickly using computer-aided design. Cai et al. [24] developed a new technology for enhancing the wearing comfort and fit by using a three-dimensional laser scanning method. Acrylonitrile butadiene styrene plastic using the 3D printing method is employed to make the face seal prototypes.

6.2 The nostril filters

The use of nostril filters is an added innovative approach to protect individuals from airborne allergenic particles. The nasal filters located inside the nasal passages are thought to prevent airborne particles coming to the respiratory system. The conventional nasal filters are usually from non-woven web, woven nontoxic mesh, or porous filters. It reduces every day runny nose and sneezing by an average 12% and 45%, respectively [4]. Electrospinning technology enabled nasal filters are more effective in catching nanoparticles before entering into the host and also provides features such as flexibility and minimum pressure drop [25]. Nanofiber nasal filter (NNF) has been prepared by overlaying a carbon filter substrate onto electrospun nylon nanofibers [26]. The filtering efficiency of the filter is more than 90% for particles greater than 1 μm and 50% efficiency for particles less than 0.5 μm. These filters have immense potential in personal protective equipment against exposure to ultrafine particles [4].

6.3 Transparent mask

There is a communication difficulty between the deaf-mute patient and the doctor while wearing a mask. The filter fabric with high optical lucidity can be worn for personal protective equipment while making lip-reading available. On the other hand, there are some challenges in creating a transparent mask and maintaining the filtration efficiency [4]. Electrospinning technology was employed to develop patterned nanofiber air filters with high optical transparency and effective particulate matter 2.5 capture capability [27]. Developed patterned nanofibers showed high particulate matter 2.5 filtration efficiency of 99.99% and high porosity (>80%) with 69% transmittance. Liu et al. developed bilayer electrospun nanofibrous mat from poly (methylmethacrylate) and polydimethylsiloxane. They nonwoven mat showed particulate matter filtration efficiency of over 96% with high optical transmittance 86% [28]. The transparent nanofiber filter is reusable with the ability to retain high particulate matter removal efficiency even after five washing cycles. In a recently study, biodegradable face mask with a hierarchical structure and transparent look has been prepared by printing polylactic acid polymer struts on a polylactic acid nanofiber web by means of electrospinning and 3D printing technology [4, 28]. Hello Mask has been developed by researchers via electrospinning technology [29].

The transparent face mask contains very fine membranes with a pore size of around 100 nanometers, this offers efficient protection against pathogens.

7. Common tests used for respirators and surgical masks

The ASTM standards given by FDA, as the certified standard in the US. Standards ASTM F2100–11 (2011), indicates the performance necessities for the respirators and the face masks. The face masks are classified depending on the performance according to various testing namely breathability, bacterial filtration efficiency, flammability, fluid resistance etc. **Table 4** shows the ASTM F2100–11 levels of protection for face mask selection.

7.1 Bacterial filtration efficiency

Bacterial filtration efficiency (BFE) can be performed as per American society of testing and materials (ASTM) F2101 protocol. This test procedure determines the ability of the face masks to prevent the penetration of microorganisms generated through various activities namely sneezing, coughing and speech. As per this this test standard, the fabric face mask is placed between an aerosol chamber and a six-stage cascade impactor. The aerosol from *Staphylococcus aureus* is fetched into the chamber and using vacuum it is allowed to pass through the mask material. The air flow rate during testing is maintained at 28 L/min. For a minute duration,

ASTM F2100–11 Levels	Resistance to penetration by synthetic blood, minimum pressure in mm Hg for pass result	Differential pressure, mm H20/cm2 (Breathability)	Bacterial filtration efficiency	Sub-micron particulates filtration efficient at 0.1 micron	Flame spread
Level 1: Low barrier protection					
General use for short procedures and exams that do not involve aerosols, spray or fluids	80 mm Hg	< 4.0	≥ 95%	≥ 95%	Class1
Level 2: Moderate barrier protection					
For low to moderate levels of aerosols, spray and/or fluids	120 mm Hg	< 5.0	≥ 98%	≥ 98%	Class1
Level 3: Maximum barrier protection					
For heavy levels of aerosols, spray and/or fluids	160 mm Hg	< 5.0	≥ 98%	≥ 98%	Class1

Table 4.
ASTM F2100–11 levels of protection for face mask selection.

Staphylococcus aureus aerosol is passed to the nebulizer. Subsequently, cascade impactor and the air pressure is allowed to pass through the sample for two minutes. The concentration of the *Staphylococcus aureus* aerosol plays important role, hence it requires to be monitored and may be kept at 2200 ± 500 CFU per test. The mean value of diameter of the bacteria aerosol and the geometric standard deviation must be in the range of 3.0 ± 0.3 μm and 1.5 respectively.

$$BFE = 100\,\frac{(C - F)}{C} \tag{1}$$

where C and F indicates the amount of bacteria colonies present in the control and in the presence of the filter, respectively.

7.2 Particulate filtration efficiency

Particulate filtration efficiency method can be determined as per American society of testing and materials (ASTM) F2299 protocol, and denotes the quality of the surgical masks. The test methods measure the quality of the face masks for filtering the particles with various sizes. According to the FDA regulation certificate, the Particulate filtration efficiency test can be performed using the 0.1 μm polystyrene latex particles. The utilization of latex spheres provides an accurate test for determining a submicron efficiency performance. The polystyrene latex particles have been suspended in water, and the aerosols were produced using the particle generator, which is generally adaptable and can offer the favorable particles concentration. Particle counter downstream can be used to count the particles. The concentration of the aerosol can be attuned from 10,000 to 15,000 particles by feeding through the drying chamber by means of HEPA filtered air. As per the FDA protocols, the used particles are not charge neutralized.

7.3 Viral filtration efficiency

The viral filtration efficiency is normally measured for some face mask such as N95 and N99. In fact this test procedure is similar to ASTM F2101 followed for determining bacterial filtration efficiency. In this method, the bacteriophage ΦX174, contaminates the bacteria namely *Escherichia coli* is utilized as the experiment virus which is aerosolized to produce virus holding water droplets with approximate size of 3:0 ± 0:3 μm. In this method, the agar plates are first inoculated with *Escherichia coli. As a result*, the bacterial cells are lysed to produce plaques and hence parts in contact with the viral droplets become clear. The viral filtration efficiency can be determined by calculating the amount of bacteria colonies present in the control and in the presence of the filter as explained in bacterial filtration efficiency method.

7.4 Fluid resistance

This test method predicts the capability of the masks and respirators to lower the squirted synthetic blood or sprayed fluid which can go through the outer layer of the mask and transmit through the inner part by altering the pressure. As per ASTM F 1862, the penetration resistance capability of the medical face mask is calculated using the high-velocity synthetic blood, which is in contact with the surface layer of the face mask (in a particular time between 0 s and 2.5 s). Some factors such as viscosity polarity, the structure, surface tension, and the relative hydrophobicity or hydrophilicity of the face mask material have shown considerable effect on the penetration and the wetting of the body fluids. By regulating the surface tension of synthetic blood, the wetting properties of blood can be simulated which must be lower than the surface tension range for body fluids, blood excluding saliva approximately in the range of 0.042 Nm^{-1} to 0.060 Nm^{-1}.

7.5 Flame resistance

The hospitals contain various sources of the oxygen, heat, and fuel, the ASTM F2100–11 standards needs an assessment concerning the flame resistance for face masks. The used material for face masks should not contain any hazards for the consumer, and their flammability characteristics should not be high. The sample mask used for the flammability test should not be flamed or remain in flame after five seconds from burning. The flame spread test determines the time required for the flame to reach the sample in 5 inch distance. "Class 1" corresponds to group of the materials, which have standard flame resistance, and they are appropriate for the use in face masks and respirators.

7.6 Differential pressure (Delta-P)

The differential pressure test provides information regarding airflow resistance of the face masks and their breathability characteristics. While performing the test, air passes through the face mask in a controlled way, and different pressures are determined for the inner and outer layers of the face mask. The ratio of differential value to the surface area (cm^2) of the face mask is used to determine the breathability, where high Delta P values denotes a harder breath for the consumers. ΔP can be measured through following relationship.

$\Delta P = PM/4.9$, where PM denotes the mean value of the differential pressure of the face mask, in Pa. As per ASTM F2100–11, the minimum value for Delta P must be lower than 5.0 mm H_2O/cm^2 (or not more than 49 Pa). The Delta P values which are lesser than 0.2 or greater than 0.5 are not considered as standard values for the common surgical use.

8. Conclusions

Wearing a mask is a main factor to retard the spread of the virus. A summary of different types of face masks is discussed in relation to the structure and performance in filtering out the bacteria and viruses. Comparison of properties of nonwoven technology such as melt blown, spun bonding and electro spinning for manufacturing face masks has been discussed. Several researchers are currently making large efforts to innovatively develop 3D-printed respirators, transparent and nostril filters, responding to COVID-19. The progression of advanced masks and respirators will play a critical role in providing protection against COVID-19.

Author details

Thilagavathi Govindharajan* and Viju Subramoniapllai
Department of Textile Technology, PSG College of Technology, Coimbatore, Tamilnadu, India

*Address all correspondence to: thilagapsg@gmail.com

References

[1] Akduman, C., & Akcakoca Kumbasar, E.P. (2018). Nanofibers in face masks and respirators to provide better protection. IOP Conference Series: Materials Science and Engineering, 460,012013.

[2] Rengasamy, S., Shaffer, R., Williams, B., & Smit, S. (2017). A comparison of facemask and respirator filtration test methods. Journal of Occupational and Environmental Hygiene, 14(2), 92-103.

[3] Adanur, S., & Jayswal, A. (2020). Filtration mechanisms and manufacturing methods of face masks: An overview. Journal of Industrial Textiles, doi:10.1177/1528083720980169.

[4] Behera, B. K., & Arora, H. (2009). Surgical Gown: A Critical Review. Journal of Industrial Textiles, DOI: 10.1177/1528083708091251

[5] Derrick, J.L. & Gomersall, C.D. (2005). Protecting healthcare staff from severe acute respiratory syndrome: filtration capacity of multiple surgical masks. Journal of Hospital Infection, 59, 365-368.

[6] Tcharkhtchi, A., Abbasnezhad, N., ZarbiniSeydani, M., Zirak, N., Farzaneh, S., & Shirinbayan, M. (2021). An overview of filtration efficiency through the masks: Mechanisms of theaerosols penetration. Bioactive Materials, 6(1), 106-122.

[7] Rodriguez-Martinez, C. E., Sossa-Briceno, M.P., & Cortes-Luna, J.A. (2020). Decontaminationand reuse of N95 filtering facemask respirators: a systematic review of the literature, American Journal of Infection Control, 48(12), 1520-1522.

[8] Kim, J.H., Wu, T., Powell, J.B., Roberge, R.J. (2016). Physiologic and fit factor profiles ofN95 and P100 filtering facepiece respirators for use in hot, humid environments. American Journal of InfectionControl, 44 (2), 194-198.

[9] W.D. Ungar, T.L. Grimsley, B. Mishkin, Full Face Respirator Mask, Google Patents, 2010.

[10] Scott, R A. (2005). Textiles for protection, Woodhead publishing., Uk., 2005

[11] Madsen, P O., Madsen, R E. (1967) A study of disposable surgical masks, Am. J. Surg. 114(3) 431-435

[12] Konda, A., Prakash, A., Moss, A. (2020) Aerosol filtration efficiency of common fabrics used in respiratory cloth masks, ACS Nano, 14(5), 6339-6345

[13] Soltani, I., & Macosko, C.W. (2018). Influence of rheology and surface properties on morphologyof nano fibers derived from islands-in-the-sea meltblown nonwovens. Polymer, 145, 21-30.

[14] Pu, Y., Zheng, J., Chen, F., Long, Y., Wu, H., Li, Q., Yu, S., Wang, X., & Ning, X. (2018) Preparation of polypropylene micro and nanofibers byElectrostatic-Assisted melt blown and their application. Polymers, 959, 1-12.

[15] Midha, V., & Dakuri, A. (2017) Spun bonding technology and fabric properties: a review. Journal of Textile Engineering FashionTechnology,1, 126-133.

[16] Adanur, S. (1997). Paper machine clothing. USA: Technomic Publishing Co., Inc., 1997.

[17] Meng, K. (2017). Investigation on compound field of electrospinning and melt blowing for producing nanofibers. International Journal of Numerical Methods for Heat & Fluid Flow, 27, 282-286.

[18] Subbiah, T., Bhat G.S., Tock, R.W., Parameswaran, S., & Ramkumar, S.S. (2005). Electrospinning of nanofibers. Journal of AppliedPolymer Science, 96, 557-569.

[19] Fridrikh, S.V., Yu, J.H., Brenner, M. P., & Rutledge, G.C (2003). Controlling the fiber diameter during electrospinning. Physical Review Letter, 90, 144502.1-144502.4.

[20] Lou, L.H., Qin, X.H., & Zhang, H. (2017). Preparation and study of low-resistance polyacrylonitrile nano membranes for gas filtration. Textile ResearchJournal,87, 208-215.

[21] Lackowski, M., Krupa, A., & Jaworek, (2011). A. Nonwoven filtration mat production by electrospinning method. Journal of Physics. Conference Series,301, 1-4.

[22] 22.Ishack, S., & Lipner, S. R. (2020). Applications of 3D printing technology to address COVID 19related supply shortages, American Journal of Medicine, 133(7), 771-773.

[23] Swennen, G.R.,Pottel, L., & Haers, P.E. (2020). Custom-made 3D-printed face masks in case of pandemic crisis situations with a lack of commercially available FFP2/3 masks, International Journal of Oraland Maxillofacial Surgery, 49(5), 673-677.

[24] Cai, M., Li, H., Shen, S., Wang, Y., & Yang, Q. (2018). Customized design and 3D printing of face seal for an N95 filtering facepiece respirator. Journal of Occupational and Environmental Hygiene, 15 (3), 226-234.

[25] Kadam, V.V., Wang, L., &Padhye, R. (2018). Electrospunnanofibre materials to filter air pollutants – A review. Journal of Industrial Textiles, 47(8), 2253-2280.

[26] Han, T.T., Yang, L., Lee, K.B., &Mainelis, G. (2018).Design and Development of a Novel Nanofiber Nasal Filter (NNF) to Improve Respiratory Health. Aerosol and Air Quality Research, 18 (8), 2064-2076.

[27] Cao, J., Cheng, Z., Kang, L., Lin, M., & Han, L. (2020). Patterned nanofiber air filters with high optical transparency, robust mechanical strength, and effective $PM_{2.5}$ capture capability. RSC Advances. 10, 20155 - 20161.

[28] Liu, H., Huang, J., Mao, J., Chen, Z., Chen, G., & Lai, Y. (2019). Transparent Antibacterial Nanofiber Air Filters with Highly Efficient Moisture Resistance for Sustainable Particulate Matter Capture. iScience, 19, 214-223.

[29] G. Fortunato, Hello Mask, 2020. https://www.empa.ch/web/s604/schutzmaske

Chapter 10

Antimicrobial Functional Textiles

Jamiu Mosebolatan Jabar

Abstract

Most textile materials are potential substrates for microbial growth. In order to make textile materials suitable as functional materials, the microbial growth must be reduced to the barest minimum or quenched due to their undesirable effects; such as offensive odor, discoloration, degradation, mechanical strength reduction etc. Chemical finishing of textile materials (such as application of silver nanoparticles, quaternary ammonium compounds, chitosan, some synthetic and natural dyes to mention a few) is capable of imparting this functional property among others to the textiles. Although, mechanism of antimicrobial activities of treated textiles is yet to be clearly defined, but in most cases, antimicrobial action of treated textiles usually occurs through interaction of cation in antimicrobial agents with anionic charged microbial cell wall. Antimicrobial treated textiles are usually less prone to offensive odor, discoloration, deteriorating mechanical properties and make the consumers free of skin problems. In fact, they can be used as cheap materials for production of hospital gowns, hand gloves and face masks for containing microorganism borne diseases, such as the current Covid-19 pandemic.

Keywords: Antimicrobial, Cell wall, Chemical finishing, Covid-19, Offensive odor, Skin, Textiles

1. Introduction

Wide varieties of textile materials are available in the market. Each of them is meant for a particular purpose; ranging from covering nakedness, decoration to medical purpose [1]. Moisture content (MC) and moisture regain (MR) are some of the important properties to be considered for chosen these materials for intending purpose. Generally, natural fibers are of relatively high MC and MR when compared with man-made fibers [2]. Hence, they are more prone to microbial attack than their man-made counterparts, due to combination of fiber's moisture and warm environment (body temperature between 36 and 38°C) that favors microbial growth on textile substrates.

Therefore, many textile materials especially those from natural source are potential media for microbial growth in their unmodified form. The major demerits of microbial growth in textile materials are offensive odor, discoloration, staining and mechanical strength reduction. At times, microbial growth on clothing materials may cause dermal infection on the skin of the wearers [1]. These effects are more likely to occur in natural textile materials. They are disliked by textile manufacturers and therefore, need to be avoided to the barest maximum [3]. Several researchers have tried in one way or the other to work on textile fibers (especially the natural fibers) in order to reduce the rate at which they get attacked by microorganisms. Bhuiyan et al. [3] reported *Lawsonia inermis* L dyed jute fabric to have a

IntechOpen

good microbial inhibition and when surface modified with chitosan the percentage inhibition greatly improved. Hong [4] observed that tannin mordanted spent coffee grounds dyed wool fabric inhibited above 95% of microbial growth. Jabar et al. [1] reported that 1, 3-bis[(furan-2-l) methylene]thiourea functional dyed wool and cotton fabrics have excellent fungi inhibition.

This section will focus on after-treatments modification of natural fibers, since they are more associated with microbial attacks.

2. After-treatment modification of textile materials

The after-treatment of these fibers is meant to modify their surfaces for combating microorganisms when in contact with them. Although, some of the chemical finishing agents do not only modify the surface of textile materials but as well form chemical bond with textile matrix. The commonly used reagents for production of antimicrobial textile materials are silver compounds/composites and quaternary ammonium compounds [5]. High cost and environmental awareness on these compounds made researchers to be looking for low cost and environment friendly alternative from naturally derived materials. The reagents reported for this purpose are chitosan, natural and synthetic dyes [6].

How these reagents perform their roles as antimicrobial agents on textiles will be discussed in the below sub-sections.

2.1 Silver compounds/composites

Silver in form of chloride, nano-particle, organo-silver or composite is used as antimicrobial agent for textile material. The application of these antimicrobial agents to man-made textile materials (silk inclusive) can be through pad-dry, coating, spray, foam technique or can be done directly by their inclusion in spinning dope during spinning operation. Their application on natural textile materials is often done through exhaustion in aqueous solution or done during dyeing operation [7]. Addition of stabilizing agent (e.g. propylene glycol, sodium salt of alkyl sulfate, polyethylamine, polyethylpyrrolidine, dendrimers etc.) to any of these antimicrobial agents during aqueous application performs triple functions. Foremost, it prevents precipitation/agglomeration of the antimicrobial agent in solution. Polymeric stabilizing agents form silver–polymer giant molecules that enhance performance of antimicrobial agent in natural textile substrates and lastly, it reduces probability of Ag^+ bleeding out from textile substrates [2].

The way and manner silver performs anti-microbial function in textile substrates is yet to be clearly identified [2]. The general belief is that Ag^+ will be released slowly from silver–polymer complex in textile matrix or surface and diffused into the microbial cell(s) when in contact with micro-organisms. Diffusion of Ag^+ into the microbial cell's site hinders activities of microorganisms and finally destroys them. This action is usually a very fast action that occurs within few minutes of contact with microorganisms. Although, silver radical (Ag^*) may be formed along with Ag^+ released from silver–polymer complex or formed separately instead of the release of Ag^+. When in contact with microorganisms, Ag^* attacks microbial cell's protein structure, destroys it and kills the microorganisms. Although, silver antimicrobial agent applied to natural textile substrates through exhaustion are usually active for more than fifty cycles of laundry operation, but the slow release of Ag^+/Ag^* from silver–polymer complex results into reduction in quantity of available silver in the functional textiles [5]. The one applied through pad-dry, coating, spray or foam technique may no longer active after few laundry operations. These antimicrobial agents are used for production of cloths for active underwear garment, socks and sport wears. They are equally useful in production of antimicrobial cloths for medical applications, such as towels, wound dressing, mattresses spread sheets and pillow cases [7].

Silica matrix stabilizes silver nanoparticles (**Figure 1**) induced exceptional antimicrobial property onto textile substrates. Silver nanoparticles are weakly held within the silica matrix by physical forces and the release of Ag^+ for attacking microorganism is gradual over a long period of time. Application of antimicrobial film prepared from mixture of silver nanoparticles and fluoroalkyl siloxane (FAS) on textile substrates induces a better antimicrobial functional property on textiles [2].

It is essentially useful in production of clothing materials for medical applications and home textiles [3].

However, high cost of silver containing substances made researchers to be looking for low cost and effective alternatives as antimicrobial agents for inducing antimicrobial property onto the textile substrates [5].

2.2 Quaternary ammonium compounds

Quaternary ammonium compounds (**Figure 2**) are applied to textile substrates as antimicrobial agents through sol–gel chemistry.

Organic–inorganic gel network structure of quaternary ammonium compound can be applied in continuous liquid phase through pad-dry-cure process to form thin film of about 10 nm thickness on textile surface [2]. Prior to curing process between temperature of 140 and 170°C, the thin nano-composite of quaternary ammonium compound film links to textile substrate (e.g. cotton) through covalent bond formation (**Figure 3**) [6–7].

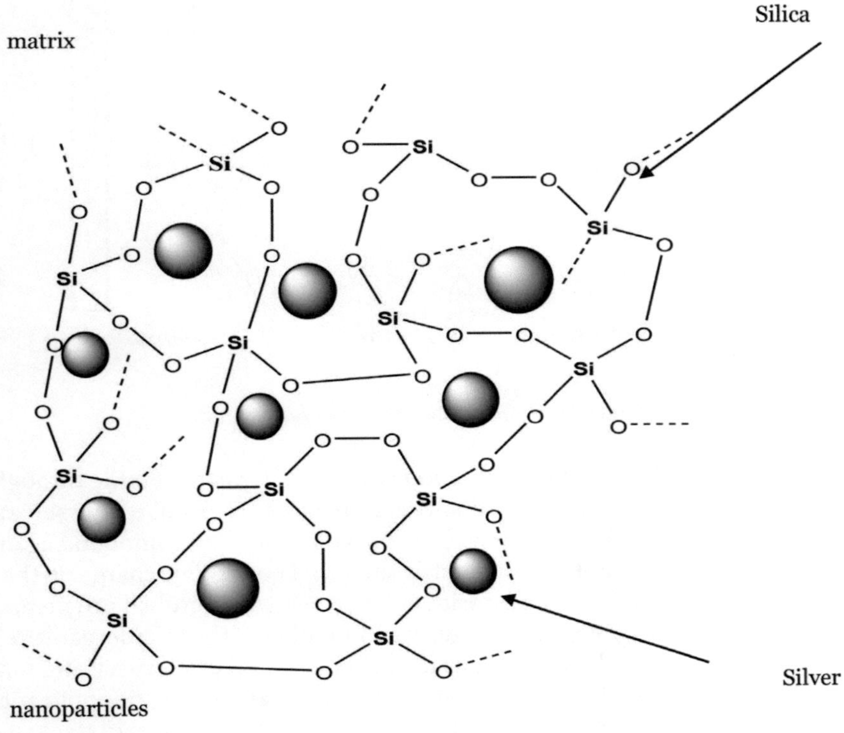

Figure 1.
Silver nanoparticles weakly held in silica matrix.

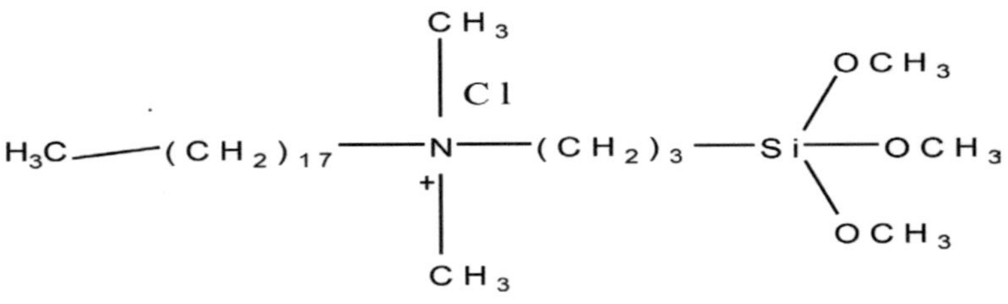

Figure 2.
Trimethylsilyl-propyldimethyloctadecyl ammonium chloride.

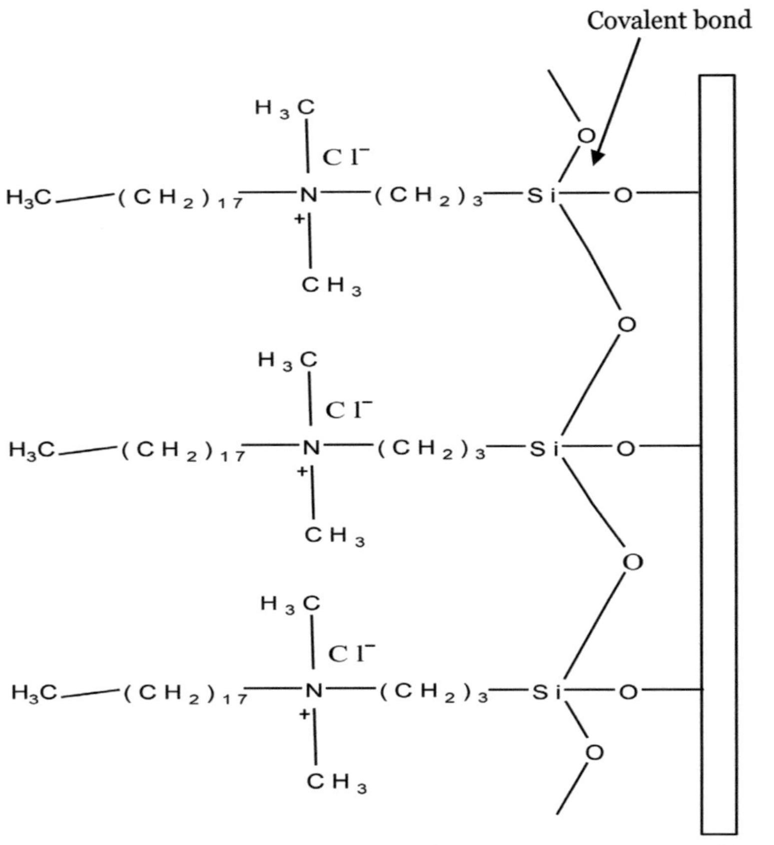

Figure 3.
Propyldimethyloctadecylammonium chloro silicone pad-dry-cured textile substrate.

The covalent bond formed imparts excellent durability to the textile through strong bond of attraction between the antimicrobial thin film and textile substrate. It also gives a control action of cationic quaternary ammonium compound against microorganism when in contact with textile surface. The cationic charge on the antimicrobial agent formed complex with anionic charged microbial cell membrane, thereby hindering microbial activity and finally kill the microorganism [2]. Just like in Ag^+ nanoparticles treated antimicrobial textiles, Quaternary ammonium compound treated cloths are equally useful for medical and domestic applications such as towels, wound dressing, mattresses spread sheets and pillow cases, mattresses and window blind.

Recent investigation has shown possibility of preparing antimicrobial precursor mixture from combination of quaternary ammonium compound and fluoroalkyl siloxane (**Figure 4**). Application of this antimicrobial mixture on textile substrates inhibits microbial growth very effectively. The unique features of this antimicrobial mixture besides the primary property are lowering tendency of adhesion of microorganism to textile substrates (by lowering fiber surface energy), importation of oleophobic and superhydrophobic properties to the textile fibers [2]. Therefore, it can be said that textile material treated with this mixture will possess multi-functional properties for use in production of out-door textiles (non-skin contact textiles), such as marquees, tarpaulins and awnings.

2.3 Chitosan

Chitosan (2-amino-2-deoxy- (1, 4)-*β*-D-glucopyranan) is a product of partial deacetylation reaction of chitin (**Figure 5**) obtained from shell of snails, crayfish, lobsters, cuttlefish, fungi cell walls, crabs and shrimps. It is most abundant natural polymer after cellulose. Chitosan is a polysaccharide with molecular structure similar to cellulose, apart from presence of nitrogen in its structure [6].

Chitosan has distinctive properties, such as antimicrobial activity, biodegradability, non-toxicity, solubility in both mineral and organic acids [7]. These properties account for its suitability as antimicrobial agent in textile production. Just like in Ag^+, the mechanism of its antimicrobial activities is not clearly understood. Although, it is generally believed that interaction of positively charged

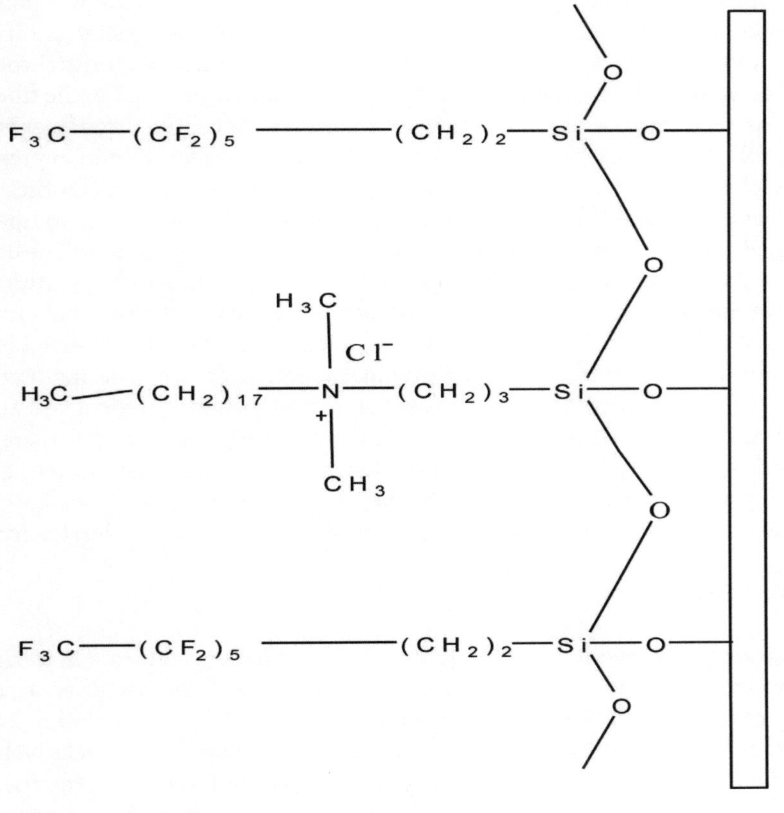

Figure 4.
Antimicrobial mixtures on cellulosic fiber.

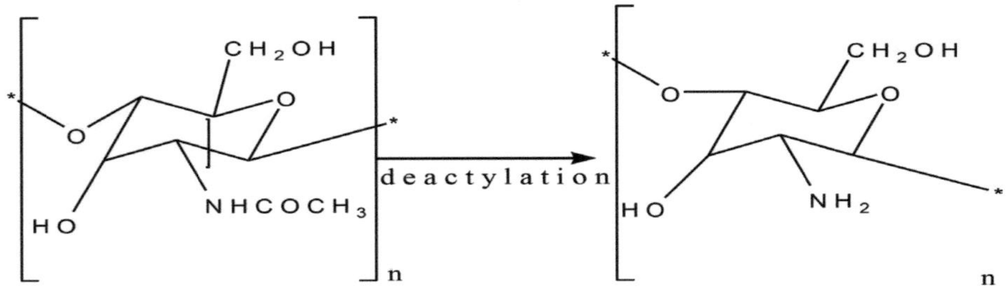

Figure 5.
Deacetylation of chitin.

chitosan amino group with negatively charged cell membrane of microorganism leads to inhibition of microbial growth and eventually results into their death [2]. Antimicrobial activity of chitosan can also be through binding of microbial DNA by chitosan, which results into microbial cell protein synthesis inhibition. Microorganisms' inability to biosynthesize protein in chitosan environment leads to their death [7].

Chitosan can be applied onto textile fibers through exhaustion process and made crosslinked giant molecule in fiber matrix through curing process (other possible mode of applications are pad-dry, pad-dry-steam and pad-batch) [5]. It is first dissolved in acidic medium (of a known concentration) by introduction of a known weight of biopolymer in glacial acetic acid (known volume and concentration) for 1 h (ripening time) and the mixture stirred mechanically for 2 h for complete dissolution of chitosan. If the curing process will be through radiation, a known quantity of photoinitiator will be added and required volume will be made through addition of distilled water [7]. But if curing process will be through thermal, required volume of distilled water will be added without addition of photoinitiator. Textile fiber will then be inserted into the prepared chitosan solution to be coated through exhaustion process, followed by drying at 70°C for about 1 h and cured for 30 min in presence of ultraviolet radiation (for photo chitosan cured fibers) or at 170°C (for thermally cured fibers-chitosan) [5]. During the curing process chitosan enters into fiber matrix in linear biopolymer form and undergoes crosslinking reaction. The curing process makes the treated fibers durable, even after several laundry operations [7].

Like in the case of quaternary ammonium compounds, functional antimicrobial property of mixture of chitosan and silver nanoparticles has been reported by researchers to be better than that of individual agent. Therefore, coating textile fibers with mixture of chitosan and silver nanoparticles imparts very good antimicrobial property onto them [4]. Chitosan or chitosan/Ag^+ coated textile materials are very good as surgical garments, gloves and face masks. As a matter of fact, textile materials made of these antimicrobial agents are going to be good wears for Covid-19 front line workers and general populace to combat the current pandemic virus.

2.4 Synthetic dyes

These are developed in laboratory, from existing natural coloring materials to alleviate demerits fund in natural dyes/pigments. Some of the demerits of natural coloring matters are poor fastness properties, low yield, irreproducibility, lack of uniformity and high cost. Therefore, synthetic coloring matters are relatively cheap, reproducible and have uniform hue, very good to excellent fastness properties and high yield [1, 8]. Previous studies reported bio-based synthetic dyes to possess antibiotic, antiviral, antifungal, anti-oxidant, anticancer, anti-malarial, anti-inflammatory and ultraviolet protective properties [9].

These coloring matters are many and they are mostly applied onto textile fibers through exhaustion process. Pigments are mostly applied through coating or printing process [8].

It was reported by researchers that many synthetic dyes (especially Schiff base dyes) have antimicrobial property. In previous study, it was reported that synthesized novel 1, 3-bis[(furan-2-l) methylene]thiourea functional dyed wool and cotton fabrics have above 70% anti-fungi inhibition against *Fusarium oxysporum, Colletotrichum gloeosporioides* and *Cercospora zeaemaydis* fungi and moderate antibiotic property against *Staphylococcus aureus* (Gram-positive), *Pseudomonas aeruginosa* and *Xanthomonas axonopodis* (Gram-negative) bacteria. This study was jointly carried out at Textile and Polymer research laboratory, Department of Chemistry, Federal University of Technology, Akure, Nigeria and Industrial and Environmental Unit, Department of Chemical Sciences, College of Natural and Applied Sciences, Fountain University, Osogbo, Nigeria.

2.5 Natural dyes

Natural dyes are generally soluble in water and as a result, they are usually applied onto textile fibers through exhaustion technique [8]. Due to environmental awareness, there is interest restoration in the use of natural dyes for coloration of textiles. Besides this major reason of their application on textiles, many of them do impart antimicrobial property onto the textiles. The way and manner of their antimicrobial activity on textile fibers has not been clearly defined. Although, it was reported by Soares et al. [10] that antimicrobial active compounds in dyes are phenolics, terpenoids and anthraquinones. The cationic charge on these compounds interacts with anionic charged microbial cell wall to resist cell growth and facilitate microbial death.

Poor fastness properties that are associated with natural dye dyed fibers are usually overcome through the use of metal salts or compounds as mordant. If metal salt like silver nitrate is used as mordant prior to dyeing of textile material with natural dye, dual enrichment given to the dye on fiber matrix are formation of covalent bond between dye molecule-metal ions and fiber matrix and antimicrobial property enhancement of the natural dye dyed fibers [2]. If natural mordant like chitosan solution is used as mordant prior to dyeing process of the fiber, amino cationic charge on chitosan boosts antimicrobial property of the dye through degradation of proteineous microbial cell wall of microorganism. It equally enhances color strength of the dyed fibers [3]. Antimicrobial synergic effect of chitosan and natural dye called henna on jute fibers was studied against *Staphylococcus aureus* and *Klebsiella pneumoniae* by Bhuiyan et al. [3]. They reported that chitosan mordanted dyed fabrics have greater than 90% bacterial reduction. If chitosan is not crosslinked in the fiber matrix, dyed fiber mordanted with chitosan may no longer durable after the first laundry operation. Hence, natural dye dyed fibers mordanted chitosan can be used as wound management fibers in the hospital. Whereas, those mordanted with silver compounds are durable after several laundry operations due to their covalent bond with fibers. Therefore, they can be used in production of under wares, socks, hospital wares, face masks etc.

3. Application of antimicrobial agents onto textile fibers

The antimicrobial chemical reagents can be applied onto the textile fibers through (i) exhaustion and (ii) coating/dry pad techniques depending on interaction of fiber matrix with antimicrobial agents [10].

3.1 Antimicrobial exhaustion on textile fibers

Antimicrobial agents that chemically react with fibers are applied on fibers through this technique. Antimicrobial natural and synthetic dyes are usually applied through this technique [1].

Exhaustion technique involves solubilization of antimicrobial agent in a suitable solvent, immersion of the fibers in the solution and stirs the mixture for specific period of time at a particular temperature. At the end of reaction period, the treated fibers will be washed under running tap water to get rid of unattached antimicrobial molecules [11].

3.2 Coating/dry pad application of antimicrobial agent on textile fibers

Antimicrobial agents that cannot bind chemically with textile fibers are applied onto fibers through coating/dry pad technique. Chitosan is a good example of antimicrobial agent that can be applied onto fibers through this technique. A detail on this technique has been discussed in Section 2.3.

4. Fibers with inherent antimicrobial properties

The fibers reported for inherent antimicrobial properties are flax, bamboo, hemp and kapok cellulosic fibers with cationic functional group according to Soares et al. [10]. Others are wool and chitosan fibers with amine functional groups [2]. The cellulosic antimicrobial fibers function through interaction of their polycationic functional group with anionic components of microbs resulting into permeability of the microbial cell wall, which eventually leads into death of microorganisms.

The textile fibers with amine functional group (wool and chitosan) resist microorganisms through linkage of cationic amine group with thio group of microbial cellular enzymes, hindering growth and activities of microorganisms and of course results into their death [1]. Chitosan yarns prepared through wet spinning of dilute acetic acid solubilized chitosan fibers in appropriate alkaline coagulating bath are usually used for industrial scale production of wound bandage. Large scale production of fabric from chitosan fibers is yet to be reported due to low mechanical strength of chitosan yarn. In medical wound dressing and sutures, chitosan performs its wound healing ability through diffusion to the site of bacterial growth, reacts with anionic function group in bacteria and puncture the cell wall to kill the bacteria [2].

5. Conclusion

Treatment of textile fibers with antimicrobial agents is very essential because textile materials are potential media for microbial growth. Microbial growth is undesirable to textiles due to formation of offensive odor, discoloration and degradation. It also makes the textile consumers uncomfortable and occasionally it causes skin problems. Treatment of textiles with silver compounds/nanoparticles, quaternary ammonium compounds, chitosan, synthetic and natural dyes or combinations of these agents alleviates textile materials from aforementioned inherent microbial demerits. Antimicrobial functional textiles are good materials to be used in production of surgical gowns, gloves, socks and body masks for containing microbial borne diseases like ebola and pandemic Covid-19.

Acknowledgements

The author hereby acknowledge assistance rendered by Dr. MA Adebayo of Chemistry Department, Federal University of Technology Akure for the success of this research work.

Conflict of interest

The author declares no conflict of interest.

Author details

Jamiu Mosebolatan Jabar
Textile and Polymer Research Laboratory, Chemistry Department, The Federal University of Technology, Akure, Nigeria

*Address all correspondence to: jmjabar@futa.edu.ng

References

[1] Jabar JM, Alabi KA, Lawal AK. Mint: Synthesis, characterization and application of novel 1, 3-bis[(furan-2-l) methylene]thiourea functional dye on wool and cotton fabrics. SN Applied Sciences. 2020;2:1850. https://doi.org/10.1007/s42452-020-03653-7

[2] Mather RR and Wardman RH. The chemistry of textile fibres. 2nd ed. Royal Society Chemistry, UK; 2015

[3] Bhuiyan MAR, Islam R, Islam S, Hossain A and Nahar K. Mint: Improving dyeability and antibacterial activity of *Lawsonia inermis* L on jute fabrics by chitosan pretreatment. Textiles and Clothing Sustainability. 2017;3:1. hptts//doi.10.1186/s40689-016-0023-4

[4] Hong KH. Min: Effects of tannin mordanting on coloring and functionalities of wool fabrics dyed with spent coffee grounds. Fash Text, 2018;5:33 https://doi.org/10.1186/s40691-018-0151-3

[5] Botelho CM, Fernandes MM, Souza JM, Dias N, Teixeira JA, Fangueiro R, Zille A. Mint: New textile for personal protetive equipment–plasma chitosan/silver nanoparticles nylon fabric. Fibres. 2021;9:3. https://doi.org/10.13390/fib9010003

[6] Mondai MIH, Ahmed F, Roknuzzaman M, Huda MN, Habib MA. Min: Antimicrobial activity of chitosan and derivative exhausted cotton fabrics as ecofriendly antimicrobial agent. Journal of Textile Engineering and Fashion Technology. 2020;6(3):77-80. DOI.org/10.15406/jteft.2020. 06.000233

[7] Ferrero F, Periolatto M. Min: Antimicrobial finish of textiles by chitosan UV–curing. Journal of Nanoscience and Nanotechnology. 2012;12:4803-4810

[8] Jabar JM, Ogunmokun AI, Taleat TAA. Mint: Color and fastness properties of mordanted *Bridelia ferruginea* B dyed cellulosic fabric. Fash Text, 2020;7:1. https ://doi.org/10.1186/s4069 1-019-0195-z

[9] Abu-Dief AM, Mohamed IMA. Min: A review on versatile applications of transition metals incorporating schiff bases. Beni-Suef Uni J Basic Appli Sci. 2015;4:119-133. https ://doi.org/10.1016/j.bjbas.2015.05.004

[10] Soares G, Magalhaes A, Vasconcelos A, Pinto E, Santos O Min: Comfort and antimicrobial properties of developed bamboo, polyester and cotton knitted spacer fabrics. Cellulose Chem Technol 2018;(1-2):113-121

[11] Jabar JM Min: Effect of rate of dyeing on the fastness properties of indigo dyed cotton fabrics. Intl Refereed Multidisciplinary J Contemporary Research 2014;2(1):43-47

Chapter 11

Functional Textile for Active Wear Clothing

Ramratan Guru, Anupam Kumar and Rohit Kumar

Abstract

Moisture management property is an important aspect of any fabric meant for active wear fabric, which decides the comfort level of that fabric specially used as active wear garments. Regular physical activity is important to maintain consistency in human health. To achieve comfort and functional support during various activities such as walking, stretching, jogging etc., athletes and sports persons use active wear clothing. A fabric's moisture management performance is also influenced by its air and water vapour permeability. The moisture management finish (MMF) and Antimicrobial finish (AMF) have been used to increase moisture absorbency; improves wetting, wicking action and antimicrobial performance. In this study, influence of MMF and AMF finishes on the moisture management property of different knitted active wear fabrics had been carried out. For the study two different knit fabrics of 100% Polyester and 100% Nylon with three different GSM levels (100, 130 and 160) has been selected. Further two varieties of commercially available functional fabric finishes have been also taken for the study. The result shows that in case of finished fabric at certain concentration level, as the fabric GSM increases the value of Accumulative one-way transport index (OWTI) %, water vapour permeability but same time drying rate increases. The result shows that in case of finished fabric at certain concentration level, as the fabric GSM increases the value of accumulative one-way transport index (OWTI) %, water vapour permeability decreases but same time drying rate increases. The knitted fabrics of 100% Polyester and 100% Nylon composition follow the similar trend. Further with the increase of fabric finish concentration level, OWTI %, and water vapour permeability (WVP) factor decreases while the drying rate increases.

Keywords: Active wear knit fabrics, moisture management, dry rate performance, water vapour permeability

1. Introduction

Moisture management properties of knit fabrics are important factors for deciding not only the comfort but also the performance of functional clothing like active wear, inner wear and sportswear. Comfort refers to the way clothing interacts with the body, with respect to dissipation of heat and moisture generated by the metabolic processes [1, 2]. During normal activity, human body loses heat by conduction, convection as well as radiation processes. Under normal condition, body cools itself by insensible perspiration where water vapour is lost from the body. When heat generation is excessive, the body breaks into a sweat or liquid moisture, also known as sensible perspiration [3]. Those properties such as smoothness of the fabric surface, air permeability, heat transmittance, hydrophilicity, knit structure, and the presence of a bio-finish influence the comfort characteristics of the knitted fabric. Active sportswear is mostly made of polyester knitted fabrics. Polyester with a modified cross section like hexachannel in coolmax gives more comfort due to its rapid liquid transmission and drying [4].

Moisture management properties of fabrics are influenced by various constructional parameters of the fabric which give knit fabric a porous structure. Total porosity of a knit fabric comprises two types of porosity, viz. micro porosity caused by void spaces among the fibres in the yarns and the macro porosity, which is a consequence of void spaces among the yarns. The air permeability, UV transmission and screen printing depend on the macro porosity; absorption of liquids and capillary phenomenon depend on micro porosity; and thermal resistance and water vapour permeability of fabric depends on both micro- and macro porosity [5, 6]. Interaction of liquids with textile materials involve several physical phenomenon such as wetting of fibre surface, transport of liquid into assembly of fibres, adsorption on the surface or diffusion of liquid into the interiors of fibres, Evaporation of sweat during wear has the potential to cool the body besides restricting the additional weight of sweat being absorbed by the fabric [7].

Moisture is transported in textiles through capillary action or wicking. In textiles, the spaces between the fibres effectively form tubes, which act as capillaries, and transport the liquid away from the surface. The liquid moisture management performance of fabrics results from complex properties including their absorbent capacity, absorption rate, and evaporation [8, 9]. This observed that the water and moisture transmission process is controlled by the water vapour pressure gradient across the inner and outer faces of the fabric. The resistance to diffusion was governed by the fabric construction, i.e. the size and concentration of inter yarn pores and the fabric thickness. The efficiency of yarn wicking depends on the surface tension, i.e., wet ability of the fibre surfaces and on the size, volume and number of capillary spaces was determined by the choice of yarn and fabric construction [10]. The length of time for a fabric to dry depends mainly upon the amount of initial liquid water retained by the fabric per unit area for evaporation. Also, the drying process seems to be related to capillary penetration and porosity of the fabrics. The most significant influence of fibre properties was believed to be the manner in which fibre shape and surface reflect increased or decreased capillarity of the fabric, which in turn causes and enhanced or diminished water uptake on wetting and water retention on drying [11]. The noted finish on a fabric is the most important consideration when developing a dynamic fabric system, as the initial uptake of water depends on the presence of a hydrophilic finish on the fabric surface. This initial uptake is the rate-determining step of the wicking action and a hydrophilic surface finish enhances the moisture management capabilities of fabrics [12, 13].

Antimicrobial finish is manly important role play for the active or sportswear fabrics. The present time is more demand all textile products in better antimicrobial performances. Antimicrobial treatment apply on fabric surface are basically more reduces cross infection, microbial bacteria and skin infections like fungi and increases the performance of sports person infections [14].

Duration the sports activity is more generated sweat and temperature for this condition get for more growth bacteria. This bacteria and fungi cause loss for sports activity performance, ageing, staining, unpleasant odours and potential skin infections.

The basically is during the sports activity generated sweat and increases temperature. In this condition are increase bacteria. This bacteria and fungi cause loss for sports activity performance, ageing, staining, unpleasant odours and potential skin.

2. Fibre use for active or sportswear cloths

Recently some year now, in active sportswear clothing are used for basically fashionable with more comfort performance. Active sportswears are one of the most lucrative segments within knits apparel. Performance of the clothing helps to remain cool, comfort and dry through the moisture management, thermal performance and other techniques. Polyester based knit has come up as a favourite for the performance of the apparel and also it can be engineered to wick transport moistures away from the body for the users' comforts.

The polyester is most common fibre used in active or sportswear cloths. Other fibres are used for active wear cloth like cotton, cotton-polyester, nylon-spandex, polyester- spandex, polypropylene and wool blend. Fibre crossection mainly used in active or sportswear cloth like irregular cross section and hollow structures fibre used [15]. Now it is more use blend with natural fibre in case for active wear cloths because improved thermo physiological performances. The basically fibre use sportswear clothing are mention in following (**Table 1**).

Polyester	**Nylon (Polyamide)**	**Polypropylene**	**Lycra/Elastane**	**Cotton**
Strong	Strong	Strong	Medium strength	Strong
Non absorbent	Non absorbent	Non absorbent	Not very absorbent	absorbent
Crease resistant	Crease resistant	Crease resistant	Crease resistant	Crease easily
Durable hard wearing	Durable hard wearing	Durable hard wearing	Durable so hard wearing	Durable
Elastic so stretches	Elastic so stretches	Elastic so stretches	Very elastic So stretches well	Not Very elastic so does not stretch
Not very warm to wear	Not very warm to wear	Not very warm to wear	To make a stretchy & fitting fabric	Cool to wear

Table 1.
Basically fibre used in active sportswear cloths.

3. Design requirement for the active or sportswear cloths

The textile materials are basically used in all sports as active or sportswear, games like for athletic clothing, football-cricket clothing, jackets, pants, shirts, shorts, socks, sweatshirts, swimwear and tennis clothing.

Plan prerequisites of dynamic and execution athletic apparel have delivered architects with abilities and information in illustrations, materials and style to imagine tastefully satisfying and ergonomically practical reaches which exploit the most recent advances in utilitarian and 'shrewd' materials [16]. Driving style fashioners have rushed to understand that the presentation has really become the feel in athletic apparel. It is the fabrics and innovation that set the precedent. Fuse of microfibres, breathable boundary fabrics, inventive stretch materials, shrewd materials, intelligent materials, for example, stage change materials and shape-memory polymers, and wearable innovation as a piece of the useful plan framework in active apparel, will get standard in the item improvement measure.

The development of new materials and designs for active or sportswears cloths has produced an exceptionally aggressive market for sports cloths design. The desires of customers for active wear sport and sportswear are concert, protection and comfort associated. The basically are all activewear cloths need for light weight, more durable, fast absorbing performance, heat- liquid regulating materials mainly used for functional design sportswear (**Tables 2–7**).

Knitted fabric is commonly used as base layer for functionally active wear due to greater elasticity and stretch ability compared to woven cloth, which is very imperative for freedom of body movement in sports. The tactile sensations by clothing

Single jersey	Rib	Interlock
High extensible in length & width	Excellent width elasticity	Width wise stretch
The fabric has tendency to curl.	No tendency to curl	No tendency to curl
Use for sportswear & undergarment	Use for sportswear & collar, cuffs, socks	Shorts, tops & sports, technical textile

Table 2.
Commercially use knitted structure for active sportswear.

Requirements	Mechanisms	Role of material designing
• External climate • Thermo physiological requirement • Physical and mechanical requirements • Protection and maintenance	• Mechanism of thermo physiological comfort • Stretch ability and comfort • Functional finishes protection and maintenance	Designing of sportswear based on the knowledge of textile properties and construction, along with the characteristics of other materials (E.g. membrane, coatings etc.), in combination.

Table 3.
Designing process for active sportswear cloths.

Functional properties	Aesthetic properties	Other properties
Light weight, high tenacity, more stretch ability, thermal preservation, Antimicrobial- UV resistance, more cooling capacity, more sweat absorption, quickly drying performance, liquid-vapour permeability and moisture management performances basically are requirement.	Feeling of softness, surface texture, handle, shine, colour discrepancy, transparency and comfort in sports wear are essential factors.	**Protection:** From wind water and undesirable climate **Insulation:** Safety from cold **Vapour permeability:** To make certain that body vapour passes outward through all layers of the clothing system. **Stretch ability:** To offer the freedom of movement essential in sports.

Table 4.
Basic requirement for active sportswear cloths.

Properties	Yarn 1	Yarn 2
Type	Polyester filament yarn	Nylon filament yarn
Blend	100%	100%
Yarn fineness (Denier)	120 D	120 D
No. of filaments in cross-section	75	75

Table 5.
Characteristics of yarn polyester and nylon.

Parameters	Polyester	Nylon
Breaking force (gf)	623.2	646.3
Tenacity (g/tex)	23.14	24.36
Elongation (%)	13.85	14.70
Unevenness (%)	14.92	13.86

Table 6.
Yarn quality parameters.

Sample & blend ratio	Polyester (100%)			Nylon (100%)		
GSM (g/m^2)	100	130	160	100	130	160
Wales Per cm	22.04	17.32	22.04	22.44	22.04	21.25
Courses Per cm	16.53	20.47	16.53	17.32	16.92	16.53
Stitch density (Loop/cm^2)	364.32	354.54	364.32	388.66	372.91	351.26
Loop length (mm)	2.3	2.2	1.9	2.4	2.2	1.9
Count (Tex)	10	15	21	9	14	21

Table 7.
Fabric geometrical characteristics.

in direct contact with the wearer skin makes wearer more relaxed due to uneven surfaces provided by the knitted fabrics in comparison to smooth-surfaced woven cloth. In addition, the lesser number of contact points of fabric with skin results in reduced clinging sensation during sweat-wetted skin [17].

4. Basic mechanism in thermal and moisture transmission though active wear clothing

The basic process implicated in heat and vapour transport is essential aspect which effects dynamic comfort of active wear garments. The basic phenomena heat can be transferred within active wear in the shape of conduction, convection, radiation and concealed heat transfer by vapour - liquid transport. Conduction, convection and radiation are overwhelmed by the temperature distinction between skin surface and climate and are thusly assembled as dry heat transfer. Then again, dormant heat transfer is accomplished by moisture transmission identified with water vapour pressure between the skin surface and the climate [18–20].

4.1 Essentials of heat transfer through garments system

The active wear fabric layers can by heat transfer from conduction, convection, radiation and wind penetration mechanisms as shown in **Figure 1**.

4.2 Essentials of moisture transfer through garments system

The basic phenomena moisture form garment may be transfer in liquid- vapour form. In vapour structure extraordinary framework like diffusion, sorption, absorption, convection and condensation are included while if there should arise an occurrence of liquid structure wetting and wicking are two components which are for the most part happen as shown in **Figure 2**.

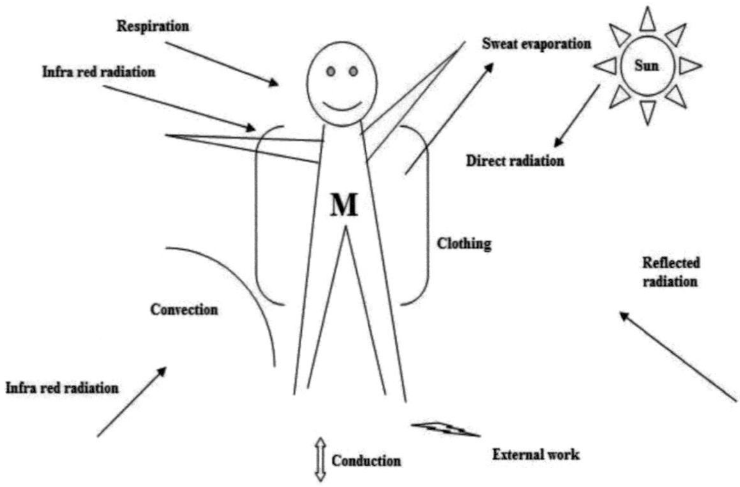

Figure 1.
The pathways for heat loss from the activities with human body.

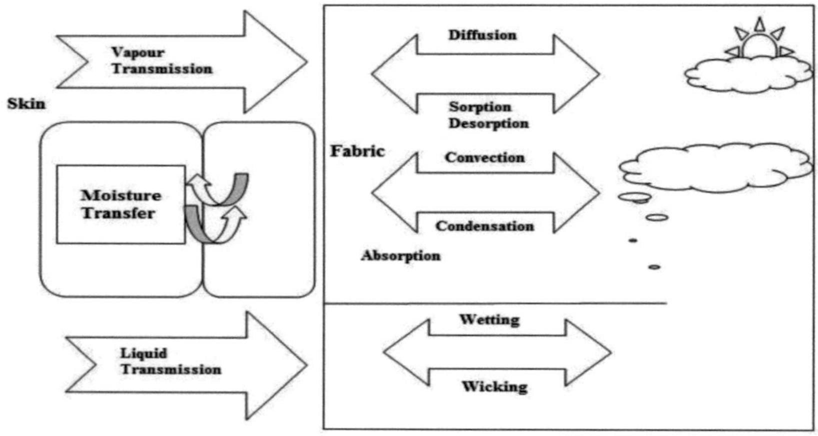

Figure 2.
The pathways for moisture loss from the activities with human body.

There are various finishes which are being applied nowadays on fabrics to improve its moisture management behaviour. So here in this research work various combinations of knit activewear fabrics (Polyester and Nylon) with varying moisture management finish and antimicrobial finish have been studied for its improvement in moisture management behaviours and antimicrobial activities for the activewear garments.

5. Material for active wear cloths

In this study work, polyester and nylon yarn count range has kept constant 120 denier, The mesh interlock knit activewear fabrics has prepared on circular knitting machine. Two different knitted fabrics of 100% Polyester and 100% Nylon were used for the study with three different GSM (100, 130 and 160). The fabrics used are scoured, bleached and ready for dyeing (RFD) fabrics (**Figures 3–5**).

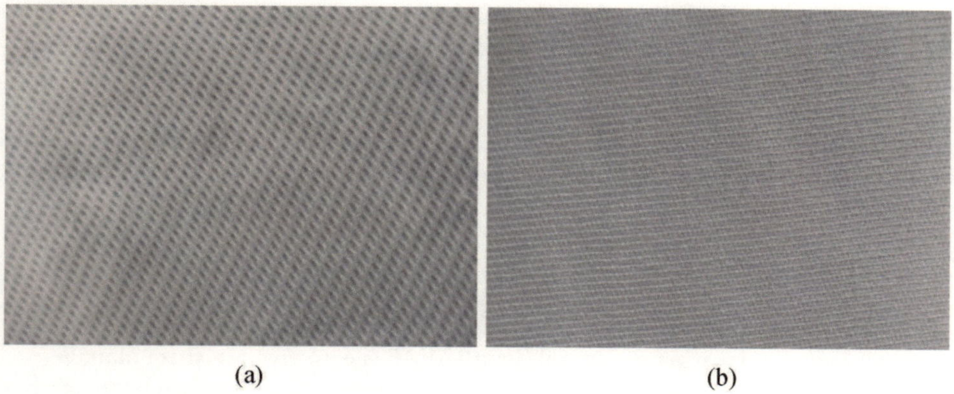

Figure 3.
(a) Polyester fabric. (b) Nylon fabric.

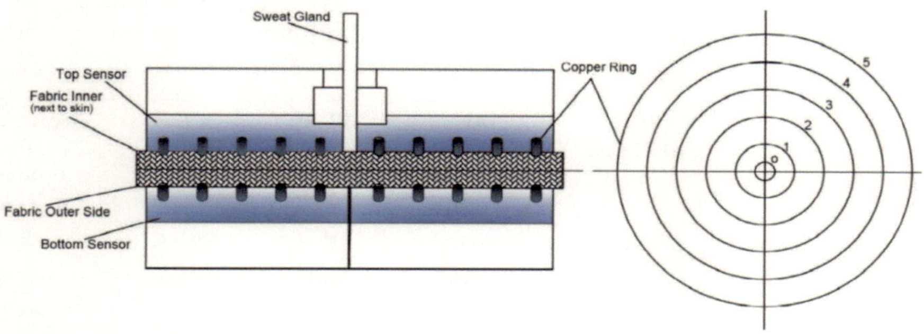

Figure 4.
Schematic diagram of MMT apparatus testing.

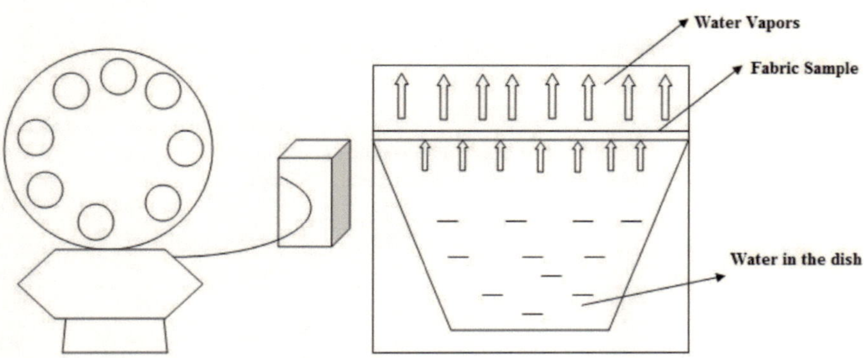

Figure 5.
Schematic diagram of water vapour permeability tester equipment (left) and testing procedure (right).

6. Moisture management (MMF) and antimicrobial finish (AMF)

The fabrics of different type and different GSM are finished with (i) Evo soft MMF finish, (ii) Evo AMF finish. In Evo soft MMF finish Silicone micro emulsion is done which increases the hydrophilic and moisture management characteristics of the fabric. Similarly AMF finish, antimicrobial cloth is used especially for activewear and leisure activities to feel clean and safe or to control malodour.

Anti-microbial finished textile lowers down the psychological discomfort associated with foul odour arising out of microbial growth and by fungi causing skin infections which is an important aspect as human body sweats during various sports activities and the temperature of human body also increases, favouring microbial growth. They also create a powerful barrier against the spread of antibiotic resistant bacteria, which are responsible for medical infections in hospitals other activities.

7. Application of finishes

Various finishes are applied on ready for dyeing fabrics, as per the following methodology. For treating the samples with MMF (i.e. to give moisture management finishing) solutions of 10 gpl and 20 gpl concentrations were prepared. For 10 gpl concentration, 10 gram of MMF was added to 1 gpl of acetic acid and 1 litter of water. Whereas, for 20 gpl concentration of finishing, 10 gram of MMF was added to 1 gpl of acetic acid and 1 litter of water. The same procedure was followed for preparing solution for other two finishes. Samples of dimension (25x 25) cm were prepared and treated with 100 ml of prepared solution by immersing it in

Fabric mass (g/m^2)	Material	Fabric thickness (mm)	Air Permeability ($Cm^3/Cm^2/Sec$)	WVP (g/m2/ day)	Rate of drying (mg/min^*inch^2)	Accumulative one-way transport index (%)
100	Polyester	0.41	56.22	321.52	0.0963	321.52
	Nylon	0.40	55.26	396.90	0.0863	396.90
130	Polyester	0.55	47.21	244.94	0.0757	244.94
	Nylon	0.47	45.87	293.41	0.0657	293.41
160	Polyester	0.56	43.65	199.01	0.0635	199.01
	Nylon	0.49	41.25	189.32	0.0535	189.32

Table 8.
Vapour and liquid moisture management properties of standard samples without finish.

Fabric mass (g/m^2)	Fabric	MMF finish conc.	Fabric thickness (mm)	Air Permeability (Cm3/Cm2/ Sec)	WVP (g/m2/ day)	Rate of drying (mg/ min*inch2)	Accumulative one-way transport index (%)
100	Polyester	10	0.43	57.65	1663.18	0.2381	241.094
		20	0.40	57.12	1108.79	0.2536	196.201
	Nylon	10	0.42	56.23	2217.58	0.2181	283.41
		20	0.41	56.10	1563.38	0.2526	196.13
130	Polyester	10	0.56	49.26	1663.18	0.292	181.094
		20	0.51	48.25	1523.12	0.363	162.667
	Nylon	10	0.48	47.25	1663.18	0.272	187.09
		20	0.46	46.45	1423.32	0.323	154.35
160	Polyester	10	0.57	44.56	1663.18	0.231	137.73
		20	0.53	43.89	1109.16	0.323	106.52
	Nylon	10	0.51	43.58	1563.18	0.241	132.73
		20	0.50	42.87	1109.16	0.321	97.52

Table 9.
Influence of variation in (MMF) finish on moisture management properties of fabrics.

Fabric mass (g/m^2)	Fabric	AMF finish conc.	Fabric thickness (mm)	Air Permeability (Cm^3/Cm^2/Sec)	WVP (g/m2/day)	Rate of drying (mg/min*inch^2)	Accumulative one-way transport index (%)
100	Polyester	10	0.42	57.15	2217.50	0.1794	176.39
		20	0.41	55.12	2014.50	0.1802	156.29
	Nylon	10	0.44	55.13	2117.38	0.2182	298.41
		20	0.41	53.10	1663.28	0.2226	186.23
130	Polyester	10	0.56	50.23	2217.58	0.1612	135.49
		20	0.52	48.15	1663.18	0.1708	98.87
	Nylon	10	0.48	49.35	1663.18	0.262	188.09
		20	0.47	46.45	1213.10	0.341	122.15
160	Polyester	10	0.55	44.56	2017.28	0.1371	96.29
		20	0.52	43.89	1562.12	0.1402	61.82
	Nylon	10	0.53	42.58	1663.18	0.243	132.73
		20	0.51	41.87	1215.20	0.308	96.52

Table 10.
Influence of variation in antimicrobial (AMF) finish on moisture management properties of fabrics.

the solution contained in a beaker for 10 minutes. Then the sample was taken out & sand with between two transparent sheets & was passed through the padding mangle to squeeze out the solution. The squeezed samples were dried at 150 ^{0}c for 1 minute in oven dryer. The same procedure was repeated for 2 samples for each level. The whole experimental work was carried out for 100, 130, 160 GSM 100% polyester and 100% Nylon knit fabrics. The variation in fabric geometrical characteristics after applying various finishes and their concentration (level) of finish is tabulated in the (**Tables 8–10**). Same processes applying antimicrobial finishes.

8. Moisture management tester

The knit fabric (Untreated and treated) samples were tested on SDL ATLAS M290 moisture management tester (MMT) according to AATCC test method 195–2009, 2011. The accumulative one-way transport index (OWTI) and the overall moisture management capacity (OMMC) measured by using the (Moisture management tester) MMT provide an insight about the liquid moisture transmission performance of fabrics. OWTC is the difference in the accumulative moisture content between the two surfaces of the fabric. OWTC reflects the one-way liquid transport capacity from the top (Inner next to the skin) to the bottom (Outer) surface of the fabric.

8.1 Drying rate testing

Dry rate testing was carried out using dry rate tester, which evaluates the weight of water evaporated in given time from the fabric. This device can be used independently to find a drying rate or in conjunction with the SDL Atlas Moisture Management Tester (MMT) in order to obtain a more complete understanding of the moisture management properties of a performance fabric. Sample size of 15 x 15 cm was used for the study, to which 2 ml water was added on its surface and allowed dry for required amount of time in the room conditions. The difference between initial and final weight gives the dry rate % of the fabric sample.

8.2 Water vapour permeability testing

Water vapour permeability testing is carried out to determine the resistance of textiles and textile composites (Particularly action wear fabrics) to water vapour penetration using testing standard BS 3424. It was carried out in the water vapour permeability tester which consists of 8 containers with water reservoirs, a standard permeable fabric cover, sample holder ring and precision drive system. The water vapour permeability (WVP) of the fabric was calculated in g/m^2/day is using the Equation (1).

$$WVP = \frac{24\,M}{A_t} \tag{1}$$

where, M- Loss of the assembly over the time period t (in g).

T- Time between successive weightings' of the assembly in hours.

A - Area of exposed test specimen (equal to the internal area of the test dish (in m^2) in this case. A = 0.0054113 m^2.

8.3 Scanning electron microscope

The surface of the coated fabrics was investigated using an SEM XL 30, Philips. According to SEM image confirm the impregnation of moisture management finish has used on the surface of the fabric. This can be also revealed from the SEM images of the moisture management finish shown as below **Figures 6** and 7. I have used coating on the polyester fibre with a particle size ranging 10 nm. The similar trend has also found for the nylon fibre.

This can be also perceived from **Figures 8** and **9** in SEM images at the uniform coating of the antimicrobial finishes on the polyester fabrics surface with a particle size ranging 10 nm. The similar trend has also found for the nylon.

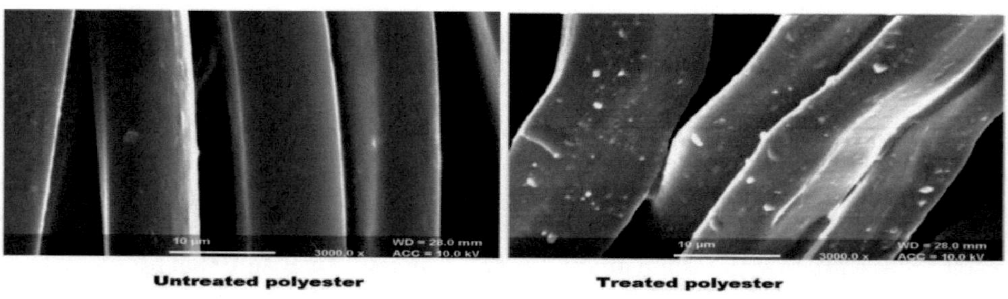

Figure 6.
SEM images of untreated and treated polyester fabric with moisture management finishes.

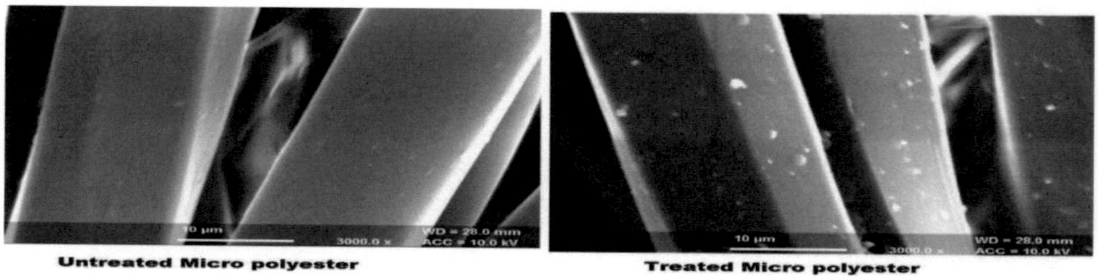

Figure 7.
SEM images of untreated and treated polyester fabric with moisture management finishes.

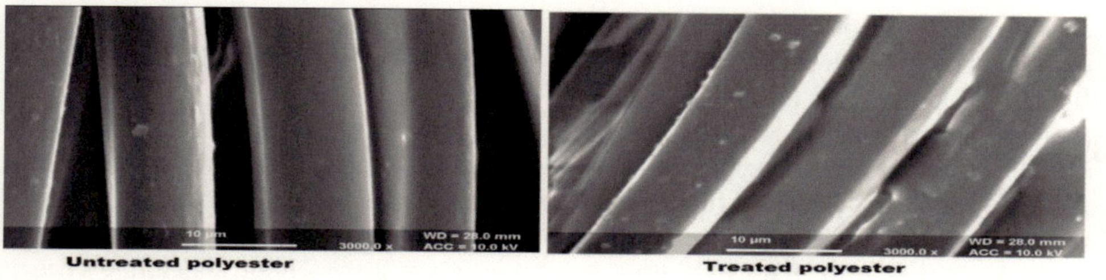

Figure 8.
SEM images of untreated and treated polyester with antimicrobial finish.

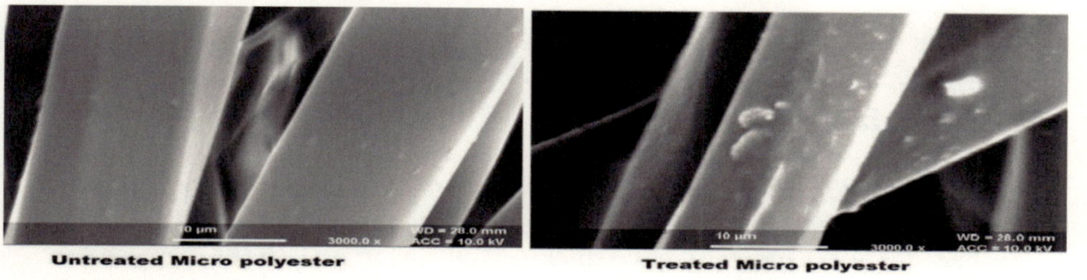

Figure 9.
SEM images of untreated and treated polyester fabric with antimicrobial finish.

9. Influence of moisture management finish (MMF) on fabric moisture management properties

In case of 100% polyester fabric it can be observed from the **Figure 10**. That as the fabric GSM increases from 100 to 160 grams, the value of accumulative one-way transport index (OWTI) % decreases. It is due to the increase in the thickness of the fabric with the increase in GSM as shown on the **Table 9**. The increased thickness

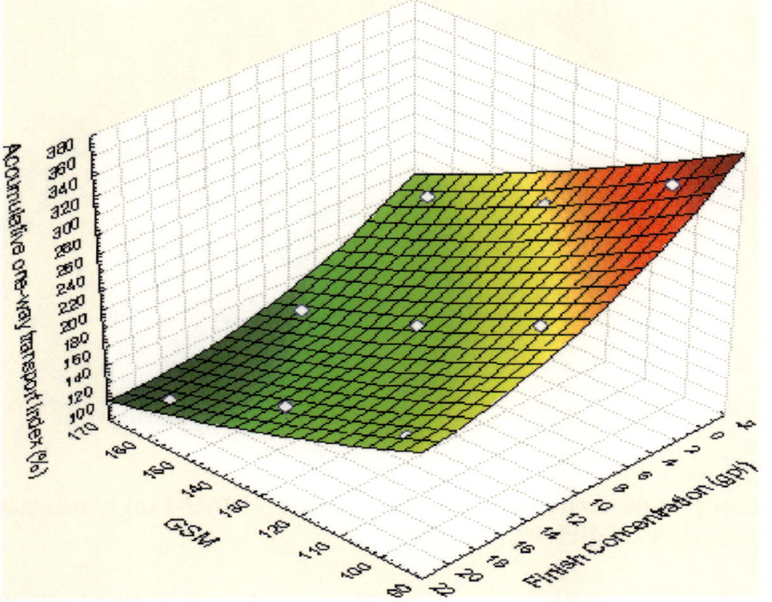

Figure 10.
Effect of MMF finish concentration and GSM on OWTI% in polyester fabric.

offers more restriction to the flow of moisture across the plane of fabric (reduced conductivity), which reduces the OWTI %. Also it was observed that the increased finishing concentration decreases the OWTI% of polyester fabric. It is due to the increased decreased pore size after finishing. HDS finish provides a surface finish on the fibre surface to increase its moisture management property. Since the finish is applied on the surface of the fibre, the fibre diameter increases and pore size decreases after finishing. The decreased pore size also decreases the air permeability of the fabric as shown in the **Table 9**. It can be seen that other fabric 100% Nylon also follows the similar trend but the rate of reduction in pore size and OWTI% was different for different fabrics. Basically the HDS softness to penetrate deeply into fibres with amorphous structure to create and increase core hydrophilicity and softness to the fabrics.

10. Influence of moisture management finish (MMF) on drying rate of different fabrics

In case of 100% polyester fabric it can be observed from the **Figure 11**. That as the fabric GSM increases from 100 to 160 grams, the value of drying rate increases. This is because of the increase in the thickness of the fabric with the increase in GSM. Increase in the thickness causes the water to spread in wider volume which causes the fabric to dry easily. Further with the increase of finish concentration level, drying rate increases. It is due to the blocking of pores of the fabric and so water remains on surface of the fabric not inside the pores and facilitating easy drying. It can be seen that other fabric 100% Nylon also follows the similar trend but the rate of increment is different due to its different physical properties than polyester.

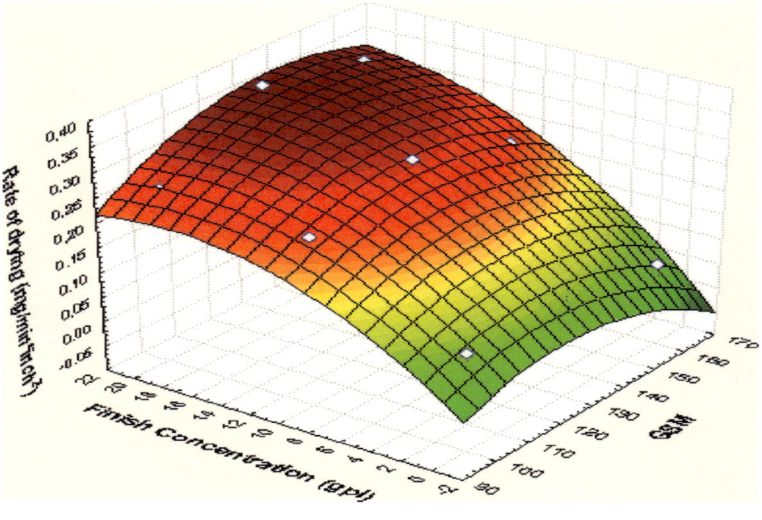

Figure 11.
Effect of MMF finishes concentration and GSM on rate of drying in polyester fabric.

11. Influence of moisture management finish (MMF) on water vapour permeability of different fabrics

In case of 100% polyester fabric it can be observed from the **Figure 12**. That as the fabric GSM increases from 100 to 160 grams, the value of water vapour permeability (WVP, gm/m^2/day) decreases. It may be due to the increase in the thickness of the fabric

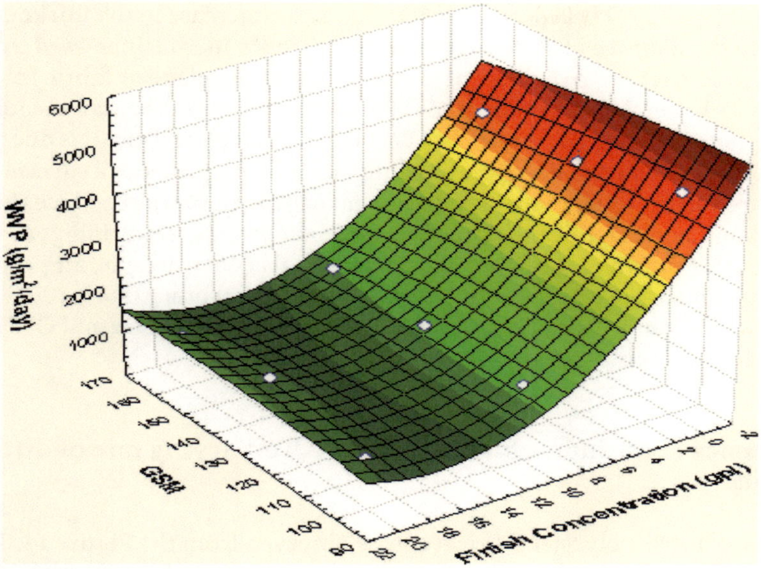

Figure 12.
Effect of MMF finish concentration, GSM on water vapour permeability in polyester fabric.

with the increase in GSM. Further with the increase of finish concentration level, WVP decreases. It is due to the increase in the fabric thickness after finishing, blinding of the fabric structural pores and reduction in fabric porosity with the increase of finish level. This may also be attributed that the reason of blocking of natural capillary action of the fibre/fabrics softener (HDS) [13]. It can be seen that other fabric 100% Nylon also follows the similar trend but the rate of reduction is different.

12. Influence antimicrobial (AMF) finish on fabric moisture management properties

In case of 100% polyester fabric it can be observed from the **Figure 13**. That as the fabric GSM increases from 100 to 160 grams, the value of accumulative one-way

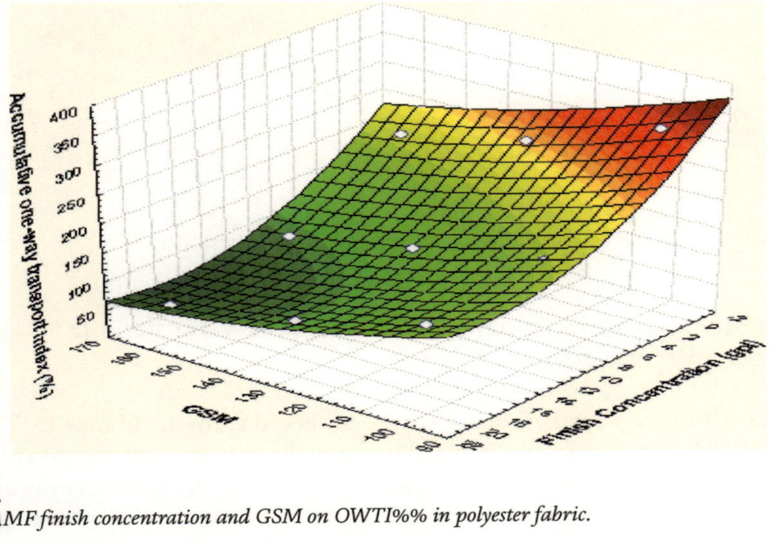

Figure 13.
Effect of AMF finish concentration and GSM on OWTI%% in polyester fabric.

transport index (OWTI) %decreases. It is due to the increase in the thickness of the fabric with the increase in GSM as shown on the **Table 10**. The increased thickness offers more restriction to the flow of moisture across the plane of fabric (reduced conductivity), which reduces the OWTI %. Also it was observed that the increased finishing concentration decreases the OWTI% of polyester fabric. It is due to the increased decreased pore size after finishing. PEH finish provides a surface finish on the fibre surface to increase its moisture management property. Since the finish is applied on the surface of the fibre, the fibre diameter increases and pore size decreases after finishing. The decreased pore size also decreases the air permeability of the fabric as shown in the **Table 10**. It can be seen that other fabric 100% Nylon also follows the similar trend but the rate of reduction in pore size and OWTI% was different for different fabrics.

13. Influence of antimicrobial (AMF) finish on drying rate of different fabrics

In case of 100% polyester fabric it can be observed from the **Figure 14**. That as the fabric GSM increases from 100 to 160 grams, the value of drying rate increases. This is because of the increase in the thickness of the fabric with the increase in GSM. Increase in the thickness causes the water to spread in wider volume which causes the fabric to dry easily. Further with the increase of finish concentration level, drying rate increases. It is due to the blocking of pores of the fabric and so water remains on surface of the fabric not inside the pores and facilitating easy drying. It can be seen that other fabric 100% Nylon also follows the similar trend but the rate of increment is different due to its different physical properties than polyester.

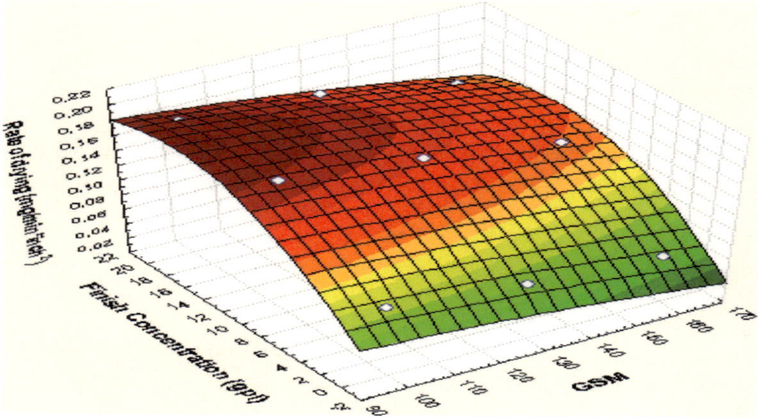

Figure 14.
Effect of AMF finish concentration and GSM on rate of drying in polyester fabric.

14. Influence of antimicrobial (AMF) finish on water vapour permeability of different fabrics

In case of 100% polyester fabric it can be observed from the **Figure 15**. That as the fabric GSM increases from 100 to 160 grams, the value of water vapour permeability ($g/m^2/day$) decreases. It may be due to the increase in the thickness of the fabric with the increase in GSM. Further with the increase of finish concentration

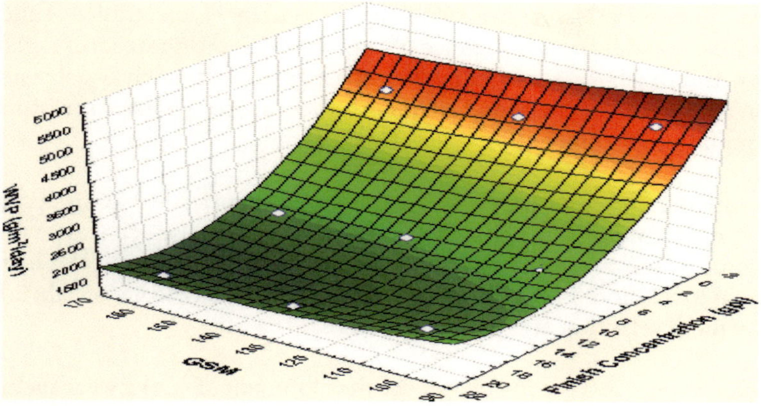

Figure 15.
Effect of AMF finish concentration, GSM on water vapour permeability in polyester fabric.

level, WVP decreases. It is due to the increase in the fabric thickness after finishing, blocking of pores of the fabric and reduction in fabric porosity with the increase of finish level [14]. It can be seen that other fabric 100% Nylon also follows the similar trend but the rate of reduction is different.

15. Conclusions

In this research an attempt has made to study the influence of MMF and AMF finishes on the moisture management behaviour, dry rate performance, water vapour permeability properties on different knit activewear fabrics. Therefore from the various combinations of fabrics, GSM, finishes and finish concentration level the following conclusions are drawn:

- The liquid and vaopur moisture management properties are much influenced by the GSM and finishing concentration. The moisture management property of the fabric was increased when the one way transport index (OWTI) %, dry rate performance and the rate of water vapour permeability (WVP) increases in the fabric. When the moisture management finishes (MMF) and antimicrobial finish are applied on the polyester and nylon fabrics, the OWTI% is much influenced by the GSM and finishing concentration. At higher GSM and finishing concentration, OWTI% reduced due to increased thickness and decreased pore size.

- The dry rate performance increases with increased GSM and finishing concentration which is due to more area for moisture spreading and evaporation from the fabrics. Water vapour permeability (WVP) performance of fabric reduces at increased GSM and finishing concentration; it is due to the smaller pores and reduced porosity at higher GSM and finishing concentration level. However in the different fibre type's fabrics, it was observed that to have less influence on the moisture management properties, it's because, these finishes applied were less penetrating into fibre and hence it's not reacting with fibre molecules of the yarns. Moisture transmission properties (Both vapour and liquid form) improve by moisture management, antimicrobial finish. Uniform coating of finishes is observed in polyester and then followed by the nylon fabrics, on the basis of SEM image.

- Active sportswear is a vast and challenging field in which required functionality can be designed by suitable choice of raw material, fabric structure, garment design and finishes. Due to suitable properties of fibres such as polyester, nylon and blends of fibres and filaments, their use in sportswear clothing is of paramount importance.

- Moisture management properties like sweat absorption sweat dissipation and faster drying are primary desirable functions of active sportswear, which affect the comfort sensation of players during sports, while ensuring the required thermal insulation. For performance apparel, the knowledge of fabric is useful for garment selection and design and development.

- To achieve required comfort level, the development of sportswear includes various modern approaches such as using special polymers, modifying the structure at fibre, yarn, and at fabric level techniques such as coating, laminating and finishing and other manufacturing technologies.

Author details

Ramratan Guru*, Anupam Kumar and Rohit Kumar
Department of Textile Engineering, Giani Zail Singh Campus College of Engineering and Technology, Maharaja Ranjit Singh Punjab Technical University, Punjab, India

*Address all correspondence to: ramratan333@gmail.com

References

[1] Uttam D. Active sportswear fabrics. International Journal of IT, Engineering and Applied Sciences Research, 2(1), 34-40, 2013.

[2] Ramratan and Choudhary A. K. Thermo-physiological study of active knitted sportswear: A critical review. Asian Textile Journal, 27(7), 53-64, 2018.

[3] Kothari, V. K. Fibers and Fabrics for Active Sportswear. Asian Textile Journal, 12 (3), 55-61, 2003.

[4] Manshahia M and Das A. High active sportswear a critical review. Indian Journal of Fiber and Textile Research, 39(2), 441-449, 2014a.

[5] Zupin Z, Hladnik A and Dimitrovski K. Prediction of one layer woven fabrics air permeability using porosity parameters. Textile Research Journal, 82(1), 117-128, 2011

[6] Moretz Herbert L and Brier Daniel L. Moisture-management garment and support pouch garment. 08/047,841. Accessed 30 Oct2012.

[7] Kissa, E. Wetting and wicking. Textile Research Journal, 66(1), 660-668, 1996

[8] Yoo S and Barker R. L. Moisture management properties of heat-Resistant work wear fabrics effects of hydrophilic finishes and hygroscopic fiber blends. Textile Research Joumal, 7491), 995-1000, 2004

[9] Manshahia, M and DAS, A. Comfort characteristics of knitted active sportswear heat and mass transfer. Research Journal of Textile and Apparel, 17(3), 50-60, 2013.

[10] Lomax, G.R. The designs of water proof, water vapour permeable fabrics. Journal of Coated Fabrics, 1(1), 15-18, 1985

[11] Cimilli, S, Nergis, B.U. Candan, C. and Ozdemir M. A Comparative study of some comfort related properties of socks of different fiber types. Textile Research Journal, 80(10), 948-957, 2010.

[12] A.C. Anand, D.A. Holmes and T. Rowe, In search of dynamic fabric for active wear. Textile Asia, 8(1), 77-80, 1996

[13] Ramratan and Choudhary A. K. Moisture management functionalization of sportswear fabrics with finishing on hydrophilic silicone softener. Asian Dyer Journal, 16 (3), 34-38, 2019 a.

[14] Ramratan and Choudhary A. K. Antimicrobial functionalization of activewear fabrics with quaternary ammonium compounds. Asian textile Journal 28(6), 71-75, 2019 b.

[15] Chakraborty J N, Deora D. Functional and interactive sportswear. Asian Textile Journal, 22(9), 69-74, 2013

[16] Oner, E., Okur, A. Thermo physiological comfort properties of selected knitted fabrics and design of T-shirts. Journal of The Textile Institute, 106, (12) 1403-1414, 2015

[17] Das A, Alagirusamy R. Sciences of clothing comfort. Woodhead Publishing India, New Delhi, 2010

[18] Bhatia D and Malhotra U. Thermo physiological wear comfort of clothing. Journal of Textile Science & Engineering, 6(2), 1-8, 2016.

[19] Manshahia, M and DAS, A. Moisture management of high active sportswear. Fibres & Polymers, 15(6), 1221-1229, 2014b.

[20] HU, Y. L, Li, Y. and Yeung, K.W. Liquid moisture transfer, Cambridge, Woodhead Publishing Limited, 2006.

Chapter 12

Textiles for Noise Control

Mallika Datta, Srijan Das and Devarun Nath

Abstract

This chapter includes the mechanism of sound absorption and the classes of sound absorbing material to control the noise. The basic phenomena related to the reduction of sound by allowing it to soak in and dissipate also were introduced first, which, can be realised by viscous effects, heat conduction effects, and internal molecular energy interchanges. Porous absorbers are materials where sound propagates through an interconnected pore network resulting in sound energy dissipation. They are only effective at the mid-to-high frequency range, which is most sensitive to the human ear. The applications of different textile fibres and their various forms were identified later in the chapter. Finally, specific discussions are given to sound parameters, noise absorption coefficient, and its measurement technique. The chapter also deals with various factors influencing sound absorption.

Keywords: noise absorption coefficient, noise control, nonwoven, porous absorber sound absorption

1. Introduction

"The International Committee of Standardization of Acoustical Terms" has defined noise as 'sound not desired by the recipient' i.e. 'unwanted sound'. Generally, the machines that have been developed for industrial purposes, for high speed transportation, or improved livelihood of human beings are accompanied by noise [1, 2]. A noise system is comprised of three component [1, 3].

Noise Source: The component which perturbs the air;

Noise Path: It is the medium that promotes the propagation of the acoustical energy from one point to other

Noise Receiver: The component which has the potential to adjudge the quantity or level of noise at a point of interest.

The unwanted noise can be reduced by [4, 5]: 1) Source Treatment 2) Transmission Path Treatment 3) Receiver Treatment. In general, there are four basic principles involved in noise control: isolation, absorption, vibration isolation, and vibration damping [6–8]. The noise control options are thus said to be: 1) Absorbers 2) Barriers 3) Composites 4) Enclosures 5) Lag Treatment.

2. Mechanism of sound absorption

Sound absorbers are soft, porous, open-celled materials such as Baffles or Quilted fibrous system blankets that reduce the reflection of sound waves by allowing them to soak in and dissipate also. The dissipation mechanism of sound absorption results in the conversion of acoustic energy to heat energy.

Attenuation or dissipation of acoustic energy as a sound wave moves through a medium attributed to three basic mechanisms [4, 6]: Viscous effects, Heat conduction effects, Internal molecular energy interchanges. Dissipation of acoustic energy due to fluid friction while moving through the medium is a thermodynamically irreversible propagation of sound responsible for viscous effects. In Heat conduction effects, heat transfer between high and low temperature regions in the wave results in non-adiabatic propagation of the sound. The sum of these two mechanisms, viscous and heat conduction, is called the classical attenuation, α_n is given by

$$\alpha_n(classical) = \frac{2\pi^r f^2}{\rho c_p{}^3}\mu\left(\frac{4}{3}+\frac{\varphi-1}{P_r}\right) \tag{1}$$

where,

μ is fluid viscosity, φ specific heat ratio (1.667 for monatomic gases and 1.400 for diatomic gases), Pr is Prandtl number and equals to $\mu c_p,/k_t$ where k_t = thermal conductivity c_p = specific heat at constant pressure ρ is the density of the fluid and f is frequency [4].

Attenuation of sound energy in fluid results from the finite time required converting translational kinetic energy into internal energies. This is associated with the rotation and vibration of the molecules. The attenuation coefficient can be written in terms of the sum of the individual contributions as follows [4]:

$$\alpha = \alpha_n(classical) + \sum \alpha_{ij} \tag{2}$$

where, the α_{ij} are the contributions of the various vibration energy relaxation effects.

3. Sound absorbing material

The sound absorption performance of the material is influenced by the amount of acoustic energy absorbed and reflected by the same. The concept of a perfect absorber can be understood by an open window that transfers all the incidence energy to the other side of the window, thereby results in 100% absorption (α= 1.0). An open window of 1 m^2 area gives 1 Sabine of absorption [5]. The maximum absorption of sound is found when the natural impedance of the material is equal to the characteristic impedance of the air (medium). Thus the sound absorbing capacity of a material is the function of its natural impedance (Zc). The sound absorption phenomenon of an acoustic material varies with the frequency and angle of incidence of the sound waves impinge upon the material.

There are mainly four types of sound absorption materials [2] available to achieve sound absorption namely porous absorber, Helmholtz resonator, membrane absorber and perforated panel absorber [7, 8] as discussed below which are

3.1 Helmholtz resonator

Helmholtz resonator can be used to estimate sound absorption at a lower frequency range [4, 8]. The quality factor (Q) indicates the quality of the resonator. The bandwidth in Hz is estimated from resonance frequency in the case of the Helmholtz resonator following the equation as described below [5].

$$Q = \frac{f_{res}}{\Delta f} \tag{3}$$

where Q is the quality factor, f_{res} is resonance frequency in Hz and Δf is bandwidth in Hz

A Helmholtz resonator is a cavity filled with air with a small opening or neck. The resonator is considered to be undamped when no porous fibres are found in the cavity. Equation (4) is used to determine the undamped resonance frequency of a Helmholtz resonator [9].

Under the condition of no porous fibres inside the cavity,

$$f_{res} = \frac{1}{2\pi\sqrt{BC}} \tag{4}$$

where,

B is inertance and C is an acoustical capacitance

$$B = \frac{\rho_0(L + 1.7R)}{\pi R^2} \tag{5}$$

where,

ρ_0 = Density of the air kg/m^3 (L+1.7R) = Effective length of the neck, m

πR^2 = Area of the opening, m^2

R = Radius of the hole in m

$$C = \frac{V}{\rho_0 c_0{}^2} \tag{6}$$

where,

C = Acoustical capacitance

V = Volume of the chamber, m^3.

c_0 = Speed of sound in air, m/sec.

ρ_0 = Density of air, kg/m^3

Alternatively, the approximate resonance frequency, f_{res} (for nonporous fibrous material) could be calculated using the following equation [9]

$$f_{res} = \frac{CD}{2\pi\sqrt{vV}} \tag{7}$$

where

c_0 = Speed of sound, m/s

D = Area of the neck, m^2

v = D (L + 1.7R) Effective volume of the neck, m^3

V = Volume of the chamber, m^3

The sound absorption performance of the Helmholtz resonator at its resonance frequency can be then estimated using the following equation

$$\alpha_B = 0.159\left(\frac{c_0}{f_{res}}\right)^2 \tag{8}$$

where,

α_B = Sound absorption, m^2 (Sabine)

c_0 = Speed of sound in air, m/sec

3.2 Membrane absorber

A Membrane absorber [5] or diaphragmatic absorber is used to absorb low frequencies sound energy. The membrane absorber offers resistance to rapid flexing and the surrounding enclosed air also shows resistance to compression during vibration at the low frequencies of sound and converts that to heat energy. A membrane type absorber is made of plywood or rubber stretched and attached to a rigid support/panel placed at some distance with respect to a solid wall. The stiffness of the panel and the method of fixing of membrane on the panel influence performance of the absorber as the panel itself tends to vibrate.

3.3 Perforated panel absorber

A perforated panel absorber [5] is the rigid thin perforated sheet with a circular opening/aperture. An air cavity is found behind the perforated panel absorber (PPA) as it is mounted at a distance from the wall. The performance of PPA is improved to tackle broader frequency range when the cavity is filled with porous fibre. The resonance frequency of the perforated panel can be estimated using the following equation [9].

$$f_{res} = \frac{c}{2\pi}\left\{\frac{P}{d(L+1.7R)}\right\} \tag{9}$$

where
c = Speed of sound in air, m/sec
P = Perforation ratio (hole area/plate area)
d = Distance of the perforated panel from the wall, m
L = Perforated panel thickness, m
$(L + 1.7R)$ = Effective length of the neck
R = Radius of the hole, m.

An air gap between the porous fibre and the wall increases the thickness of the perforated panel and also the depth of the air gap lowers its resonance frequency. Therefore, by varying the depth of the air space and the thickness of the perforated panel, the broader frequency range of sound absorption performance could be achieved. A perforation ratio of more than 20% with a small aperture does not affect the sound absorption of porous fibre. However, a smaller perforation ratio reduces the higher frequency sound absorption performance of porous fibre [7].

The thickness of the PPA increases with cavity depth or the air gap between rigid wall and absorber which in turn lowers its resonance frequency. Thereby, the change in cavity depth/depth of the air space and the thickness of the perforated panel make PPA more suitable for the broader frequency range of sound absorption performance. The small aperture/perforation with more than 20% perforation ratio has no effect on sound absorption by PPA. However, a smaller perforation ratio reduces the higher frequency sound absorption performance of porous fibre [7].

3.4 Porous absorber

In porous absorbers [10, 11], sound propagates through an interconnected pore [12] network resulting in sound energy dissipation. They are only effective at the mid-to-high frequency range, which is most sensitive to the human ear [12]. The porous absorbers are found their applications in noise control for industries, automobiles, building acoustics, and sound recording studios. The magnitude of undesirable sound/noise reflected from hard interior surfaces and reverberant noise level is reduced in presence of such absorbers [8]. The various classes of porous absorbers are i) cellular ii) granular, and iii) fibrous materials. The cellular absorbers are made of foam of polyurethane and sometimes metal like aluminum [8]. Sound absorbing foam with an open structure allows the propagation of air from one to the other face through the interconnected pores [12]. The polymer based foam is associated with fire hazard and the generation of combustible toxic gases while the metallic foam offers higher mechanical strength [13]. Some common examples of granular absorbers are panels made of wood chips, pervious road, and, porous concrete. The granular absorber can be used in the form of a consolidated structure made with the suitable binder(s) or in the loose form [13]. The used tires, rubber particles, and waste foam have been identified in making granular absorbers [12, 14]. Fibrous materials as reported compose of glass, mineral, or organic fibres in the form of nonwoven fabrics, boards, or preformed elements. The granular materials can achieve a broadband absorption limited to around 0.8, while in the case of fibrous absorbent, the value can rise to unity [12]. Fibrous and granular absorbers are often produced by bonding the fibres or granules with a binder. They are generally covered with a thin perforated sheet such as highly perforated panels of metal, wood or gypsum giving better aesthetic value protecting from damages and prevent the particles from polluting the air which may harm the eco-system [8].

4. Textile materials as porous absorber

4.1 Textile fibre

The textile fibre based porous absorber may be made of felt, glass wool, rock wool, polymer foams, waste cloth fillers [15–21]. The structure of the absorber consists of cavities that promote internal reflection of sound of waves, trap the sound energy and dampen the oscillation of the air particles by friction with absorber material [22]. They are, however, effective for the medium to high frequency range only [23]. The type of fibrous porous absorber is based on the raw material viz., metal, synthetic polymer, and natural. The fibrous forms of metal and its alloy viz., stainless steel, nickel, aluminum, etc., are also identified as suitable noise absorbers in harsh environments [19, 23, 24].

Presently, synthetic fibres have largely been used owing to some advantageous attributes viz., large specific surface area, good mechanical strength, and good permeability [23, 25, 26]. The various structural forms of such fibre as acoustic absorber include the felt, woven cloth, and fibre reinforced composites [23]. The various synthetic fibre options in noise control are glass wool, rock wool, basalt, carbon fibre [15, 17, 19, 20, 24, 27–30]. The crude based fibres like polyester [26, 31–38], polypropylene [39–41], nylon [42, 43] were also used for noise control study. Polyester microfibre felts showed improved noise absorption in the middle frequency, ranged from 1200 Hz to 4000 Hz [44]. The study on polyester and nylon microfibre fabric concluded that sound absorption increased with fabric density up to 0.14 g/cm [17, 45]. Synthetic fibres and their blend at different ratios towards the optimised acoustic absorption coefficient have been identified by several researchers [37, 38, 40, 46–50].

The use of metallic fibre in noise controls is limited due to their poor flexibility, heavyweight, poor formability [23], while the synthetic polymer fibres are non-biodegradable and cause serious health hazards during manufacturing, installation, and disposal [10]. The limitation of metallic and synthetic based fibrous porous absorbers wrenches the researcher in the exploration of suitable alternatives from a renewable resource.

Different works had been reported on exploration of using various natural fibres [51–53] viz., wool [54], cotton [50, 55], kapok [56], kenaf [57], hemp [58–60], ramie [61], flax [57, 61–63], banana [64], broom [65], coir [66], jute [51, 61, 67–70], tea leaf [71], combination of Luffa fibres with cotton layer [72] and agro residues viz., straw, oil palm [57] for producing sound absorbers. Most of the fibres showed encouraging sound absorption property.

The textile materials in various forms viz., woven [73–76], carpet [77, 78], non-woven [70, 76, 79–88], knitted [89], and composite [31, 48, 59, 68, 90–93] were used for control of noise.

The major disadvantage associated with fibrous absorbers is in tackling long wavelengths of low frequency sound energy. Different approaches were attempted by various researchers to improve the noise control performance of porous absorbers. Some works are available in the literature on the improvement of absorption performance of noise of low frequency range by adjusting the non-acoustical parameters viz., thickness, areal density, and bulk density of porous absorber [94].

4.2 Forms of textiles in noise control

4.2.1 Nonwoven fabric

The use of nonwoven or felt structures developed from different fibres and their blend or in composite form is gaining interest to use as a noise control solution [3, 13, 26, 31, 81, 84, 87, 95]. Effect of fibre fineness, fibre cross section, structure parameters of nonwoven viz., total surface area, the density of fibre packing, thickness and physical parameters including fibre mixing ratio (blend ratio), bulk density [51, 70, 96, 97] were investigated. The results revealed that total surface and fabric density determined sound absorption positively, and fibres with profiled cross-section shapes show a higher noise reduction coefficient [26]. The acoustic absorption coefficient increased with the increase of thickness of nonwoven/felt [98, 99]. Pore diameter [100], porosity, and the air gap behind the mounted non-woven samples influenced the sound absorption at low frequency [101]. Absorption behaviour changes with different fibre content in their blend [46, 47, 98, 99] and there was an optimum bulk density for noise reduction [98]. The use of fibre with a larger lumen in the mix of natural with synthetic fibres enhanced the noise control performance [98, 99]. The increase in hollow fibre percentage in nonwoven fabric was directly related to its thickness and sound absorption efficiency [85]. The use of nonwoven made from natural fibres was suitable for application in vibration control for the automobile industry due to their excellent strength and renewable properties [102] and offered excellent absorption in the mid-to-high frequency ranges [86]. The work revealed [40] that the orientation of fibres in nonwoven did not affect acoustic absorption whatever might be the pile orientation of 0°/90° and 45°/45° [40]. The smaller size of aerogel embedded in a nonwoven fibre matrix has positive effects on acoustic absorption [103]. Introduction of nano fibre in making nonwoven improve both sound absorption and sound transmission loss [104]. Acoustical nonwovens with activated carbon fibre on the surface exhibited improved acoustic properties [79–81].

4.2.2 Woven fabric

The use of woven structure in noise control was identified by various researchers [105–107]. The study on the effect of weave type, weft yarn linear density, thickness created by the layering of test fabrics, yarn spinning system, and depth of air space at the back of samples revealed that the sound absorption coefficient of woven fabrics is influenced by both density and porosity of fabrics. The weight and cover of woven cloth as an upshot thread density and thread count [107, 108] influenced transmission loss of sound transmitted through the structure. Higher thickness of woven fabric associated with improved noise reduction coefficient [106]. The plain weave structure offered higher sound absorption in comparison to other weave designs. Higher absorption woven structure was found with finer and low twisted weft yarns [109]. The woven structure made of fabrics rotor-spun yarns exhibited the highest absorption in comparison to structure made with ring-spun or compact yarns.

Pile carpet as noise control material was investigated [77] and it was found that pile thickness and weight of carpet have a minor influence on transmission loss of sound moving through the carpet. Effect of pile parameters namely fibre type, pile height, and carpet construction, pile density, air gap behind the mounted tufted carpets was studied which identified suitability of various factors in controlling sound at audible frequency range [78]. The air gap behind the carpet enhances the noise absorption capacity at low to medium frequencies.

4.2.3 Knitted fabric

NAC for plain knitted structures with the same thickness but different pore changes with the size of pore diameter. The smaller diameter pore under the influence of smaller stitch sizes with lower porosity offers a higher degree of sound absorption sizes [110]. For the same value of pore radius and porosity, the NAC changes with thickness in the case of knitted structure [111]. Introduction of spacer fabric inside the knitted structure improved the noise control performance of the knitted structure.

5. Factors influencing sound absorption

Studies on various parameters that influence the sound absorption properties of fibrous materials have been published widely in the literature [7, 10, 12, 94, 112, 113]. The factors in detail can be described as follows:

5.1 Textile factor

5.1.1 Fibre Size

The sound insulation behaviour of wool fibre based material sound absorption coefficient increase with a decreasing fibre diameter [54]. It is found that thin fibres can move more easily than thick fibres on sound waves. Moreover, with fine denier, more fibres are required for the same volume density which generates a more tortuous path and higher airflow resistance [94]. Studies of Tancan [82] revealed that the fine fibre increases sound absorption coefficient values due to an increase in airflow resistance through increased viscosity resulting from the vibration of the air.

5.1.2 Thickness

The thickness of the porous absorber directly influences the low frequency sound absorbing performance [94, 114]. The material with an apparent thickness (includes cavity depth from the rigid back) equal to the one-quarter wavelength at a resonant frequency gives peak absorption [94], however, the threshold thickness for effective absorption is one-tenth of wavelength. A study also showed that sound absorption increases with the increase in thickness of the material only in the case of low frequencies. Thickness becomes insignificant at higher frequency.

5.1.3 Density

The sound absorption performance of a material is a function of the bulk density of the material [10, 54]. It is to be kept in mind that the density of the acoustic material affects its cost. The sound absorption value of the absorber at the middle and higher frequency (> 500 Hz) increases with the density of the sample [115]. When the number of fibres increases per unit area, the apparent density (considering the entrapped air) is high. Conversion of sound energy to heat increases as the surface friction at the viscous boundary layer increases [12] and so the sound absorption coefficient, especially for nonwoven fibrous materials [23, 53, 94]. Open and light porous structure absorbs the sound of low frequencies (<500 Hz), while denser structure suitable for frequencies above 2000 Hz.

5.1.4 Porosity

The sound absorption mechanism of the porous absorber can be explained in the light of, pore parameters viz, number, size, and shape [100]. Dissipation of sound energy is owing to frictional resistance offered by the pores that allowing the propagation of sound through it. The porosity is generally thus defined as the ratio of the volume of the voids in the material to its total volume [12, 116]. Equation (9) defines porosity (∅) [53, 117].

$$\text{Porosity} = (\varnothing) = \frac{V_a}{V_m} \tag{10}$$

where:
Va = Volume of the air in the voids
Vm = Total volume of the sample of the acoustical material being tested

5.1.5 Tortuosity

Tortuosity is a measure of the crookedness of the passageway through the pores, compared to the thickness of the sample. Tortuosity enumerates the influence of the internal structure of a material on its acoustical properties. Con Wassilieff [118] describes it as a measure of the deviation of the pores from the normal, or meander about the material. The location of the quarter-wavelength peaks of sound energy is influenced by tortuosity, while the height and width of the peaks are persuaded by porosity and flow resistivity. The degree of crookedness/tortuosity determines the behaviour of absorbing porous materials at the high frequency level.

5.1.6 Compression

The porous fibrous textile structures are compressible in nature and experience compression on the application of load and thickness decreases. The factors like density, porosity, tortuosity, airflow resistivity, porosity, and density also vary with changes in thickness. The studies [119, 120] found that in the event of reduction of the thickness of a homogeneous layer of porous fibrous porosity and characteristic lengths (shape factor) [12, 94] decrease while the density and tortuosity or crookedness in the structure increase. The effect of compression of fibrous structure is found to be more profound in the case of automotive acoustics. The weight of the passenger causes cyclic compression and expansion of the seat padding that results in squeezing down the porous materials (fibrous or cellular) which in turn results in the variation of the above mentioned physical parameters [3].

5.1.7 Airflow resistance

One of the most important parameters that influence the sound absorbing characteristics of fibrous material is the specific flow resistance per unit thickness of the material [12, 32, 94, 116, 121]. The characteristic impedance and propagation constant, which describes the acoustical properties of porous materials, are governed to a great extent by the flow resistance of the material [94].

The presence of fibrous peg (due to interlocking of fibre at the time of needling) as frictional elements in the case of a needled nonwoven, provide resistance to acoustic wave motion. As the sound wave enters a fibrous nonwoven structure, its amplitude is decreased by friction while moving through the tortuous path, and sound energy is converted into heat [94]. The friction resistance of the material to the flow of air is called 'airflow resistivity' and is expressed as:

$$\sigma = \frac{\Delta P}{\Delta T} \times \frac{1}{u} \text{Pa.s/m}^2 \tag{11}$$

where,

σ = airflow resistivity Pa.s/m^2

u = Air velocity through sample m/sec

Δp = Sound pressure differential across the thickness of the sample measured in direction of particle velocity, N/m^2

ΔT = Incremental thickness [3, 94]

Based upon the airflow test following ASTM D737 [122], flow resistivity σ of the sample is obtained from the following equation:

$$\sigma = \frac{P}{ct} \tag{12}$$

where,

P = Static pressure differential between both faces of the sample, dyne/cm^2 (10^{-1} Pa)

c = Air velocity, cm/s

t = Thickness of sample, cm

The airflow resistance per unit thickness of a porous material is proportional to the coefficient of viscosity of the fluid (air) and inversely proportional to the square of the pore size of the material. For a fibrous material with a given porosity, the flow resistance per unit thickness is inversely proportional to the square of the fibre diameter

5.1.8 Surface impedance

For a given layer of thickness, the acoustic resistivity of an absorber is directly related to its ability to dissipate sound energy. Moreover, the surface impedance [123] of the layer increases with the degree of air resistance offered by the structure that results in a more reflection of sound from the surface layer. Thereby, sound absorption by the fibrous structure reduces. The mechanism of sound absorption is frequency dependent, thus at a lower frequency as the thickness of the layer increases resistivity decreases.

5.2 Other factor

5.2.1 Placement/position of sound absorptive materials

The placement/position of sound absorptive materials is known to affect the sound absorption of the material. It has been reported by Alton Everest [5], that if several types of absorbers are used, material applied to the lower portions of high walls can be as much as twice as effective as the same material placed elsewhere [124]. The porous structure behaves like a frequency dependent membrane of a certain mass under the influence of an air cavity behind a material. The presence of air inside the cavity has an analogy to a mechanical spring. The absorption property of the porous fibrous absorber enhances significantly with the air filled cavity between the absorber and the rigid back wall [125].

5.2.2 Temperature

The study [126] revealed that the sound absorption characteristics of mineral wool remained unaffected by the change in temperature in the range of 10–50 °C. Least square method was used to develop a theoretical relation between the noise reduction coefficient and the thermal conductivity at different temperature conditions.

5.2.3 Process parameters

Process parameters during absorbent material formation have an important impact on sound absorption due to their effects on the characteristics of the absorbent material. It reported that 'air laid' web based nonwovens offered higher sound absorption compared to carded ones irrespective of the fibre content. This might be due to higher flow resistivity of air laid nonwovens because of relatively random placement, and thus, higher tortuosity, the higher number of pores with smaller sizes, higher number of fibre to fibre contact points, and gradient in porosity due to gravity. Among web bonding methods [3], did not find a significant difference between needled and needled plus thermally bonded nonwovens. Thermal bonded and needle punched nonwoven (punching density of 28 cm^{-2}) from polypropylene and needle punched polyamide nonwoven offered maximum absorption of sound at a material density of 100 kg/m^3 over a frequency range of 63 to 8000 Hz [88]. The fibres of diameter 10 to 40 μm were used to manufacture the nonwovens with thickness varying from 3 to 20 mm. The study revealed that the needle punched nonwovens had more dependency on the frequency of sound energy as well as on the diameter of fibres compared to thermally bonded webs.

6. Testing and characterisation of sound absorbing material

6.1 Absorption coefficient

When a beam of sound wave (E_i) strikes against a barrier, gets divided into three parts: i) E_r is reflected part ii) E_a is the part absorbed in the barrier iii) E_t is transmitted part to the other side of the barrier as shown in **Figure 1**. These phenomena can be written as

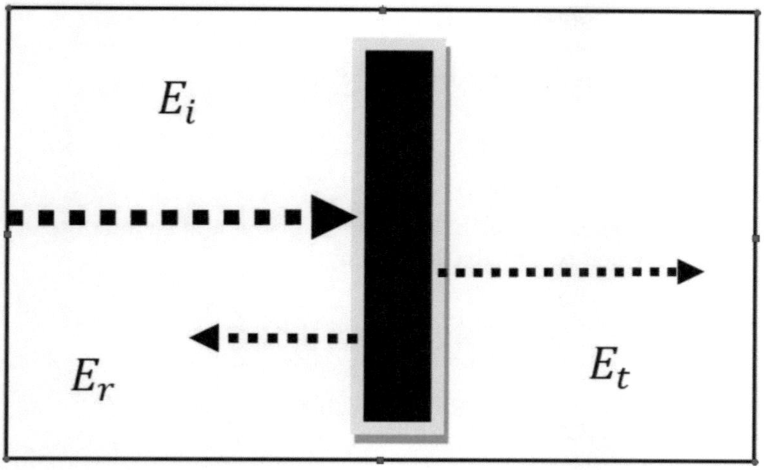

Figure 1.
Typical behaviour of sound wave striking a wall.

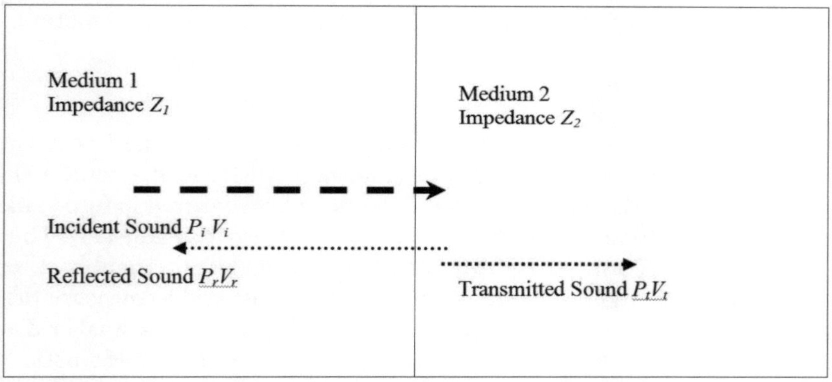

Figure 2.
Reflection and transmission at the boundary between two media.

$$E_i = E_r + E_a + E_t \tag{13}$$

Then the sound absorption coefficient, α is defined as follows [7, 93].

$$\propto = \frac{E_i - E_r}{E_i} = \frac{E_a + E_t}{E_i} \tag{14}$$

Equation (13) shows that all portion of the sound energy which is not reflected is considered to be absorbed.

When an infinitely large boundary plane presents between two media, the path of movement of the sound wave traveling from Medium 1 to 2 is normal to the boundary plane, as shown in **Figure 2**. The relation of reflected and transmitted sound can be expressed as:

$$P_i + P_r = P_t \tag{15}$$

$$V_i + V_r = V_t \tag{16}$$

where, P_i is incidence sound pressure, P_r is reflected sound pressure, $V = P/Z$ and Z is the inertance or impedance

$$Z_1 = \rho_1 c_1 \text{ and } Z_2 = \rho_2 c_2$$

The sound pressure reflection coefficient r_p is

$$r_p = \frac{P_r}{P_i} = \frac{Z_2 - Z_1}{Z_2 + Z_1} \tag{17}$$

The relationship between the absorption coefficient and sound pressure reflection coefficient between two media is given by:

$$\alpha_r = \frac{E_r}{E_i} = \left(\left|\frac{P_r}{P_i}\right|\right)^2 = \frac{(Z_2 - Z_1)^2}{(Z_2 + Z_1)^1} \tag{18}$$

$$\alpha = 1 - |r_p|^2 = 1 - \frac{E_r}{E_i} = 1 - \left(\left|\frac{P_r}{P_i}\right|\right)^2 = 1 - \frac{(Z_2 - Z_1)^2}{(Z_2 + Z_1)^1} \tag{19}$$

6.2 Principles of sound absorption measurement

Two standards and one relative measurement technique are used to measure the noise control property of absorbing material and describe in detail as follows.

6.2.1 Impedance tube method

This technique (Normal Incidence Technique) [4, 7, 127] is based on normal incident sound energy and requires a tube (**Figure 3**) where its diameter is smaller than the wavelength. The absorption coefficient can be measured in impedance tube as per ASTM E 1050-12 and ISO 10534-2: 1998 standards [128, 129]. The circular sample of φ 30 mm and φ 100 mm are tested against the rigid back wall using two tubes. The larger diameter tube (φ 100 mm) is used to measure the absorption coefficients in the frequency range, 63–1600 Hz, and a smaller diameter tube (φ 30 mm) for measuring in the frequency ranges from 1000 Hz–6300 Hz. The samples are tested five times to minimize the influence of variation of thickness and areal density.

6.2.2 Reverberant field method

Measurement of sound absorption is concerned with the performance of a material exposed to a randomly incident sound wave, which technically occurs when the material is in a diffusive field (Random Incidence Technique) [7, 130]. EN ISO 354 (2003) testing standard method [131] is employed to measure the

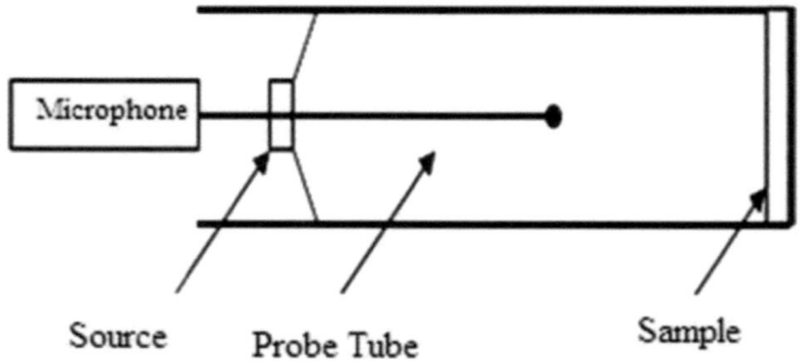

Figure 3.
Schematic diagram of impedance tube.

reverberation time (RT) to determine the absorption coefficient. This technique requires a reverberation chamber with a volume of 200 m^3. The diffused acoustic field is created inside the room to test the sample with an area between 10 m^2 and 12 m^2. The absorption coefficient α_r is calculated using the following formula from RT.

$$\alpha_r = \frac{KV}{s}\left[\frac{1}{T_m} - \frac{1}{T_0}\right] + \overline{\alpha} \tag{20}$$

Where,
V = Volume of reverberation chamber, m^3
S = Surface area of the material tested, m^2
T_m = time with tested material, sec.
T_0 = time empty, sec
$\overline{\alpha}$ = Average sound absorption coefficient of reverberation chamber
K = constant=0.16

6.2.3 Clemson-Boston differential sound insulation method

Clemson-Boston Differential Sound Insulation (CBDSI) Tester [95] is used to measure sound insulation of the material as shown in **Figure 4**.

CBDSI is comprised of a computer to process sound signal, an amplifier to amplify the sound signal, sound source, sound chamber, sample holder to mount the test sample, and a detector to detect the sound. The sound source generates white noise [7] in the frequency range of 73 Hz to 20,000 Hz for testing the sound insulation property of the material under investigation. The signal amplifier amplifies the signals generated by the computer which are then converted into sound waves via the sound source. The material under investigation mounted on the sample holder interferes with sound energy as it is in the path of sound. The sound moves through the sample and strikes the detector, which changes the received sound signals to electric signals that are then analyzed by the signal-processing computer. The test is also conducted with no sample in the sample holder to measure the insulation property of the material in reference to the background.

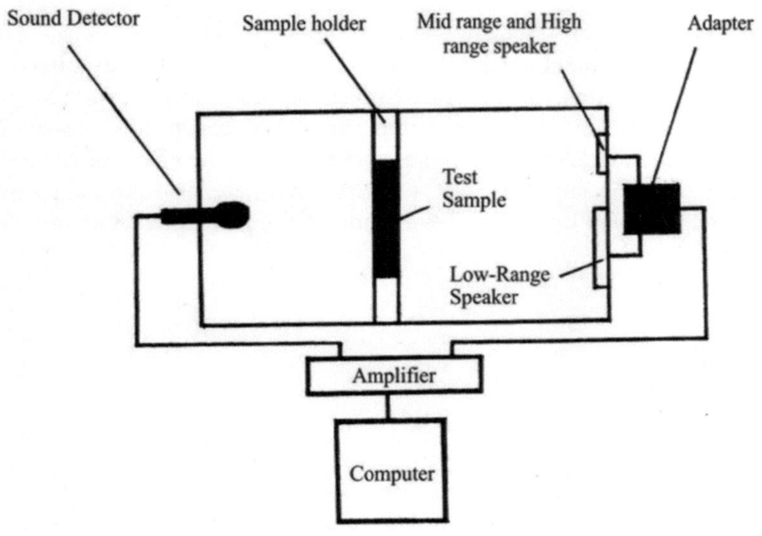

Figure 4.
Schematic diagram of Clemson-Boston differential sound insulation tester.

CBDSI tester provides a direct comparative analysis of various types of samples as they are tested under the same conditions. Sound insulation property in terms of 'Transfer Function Magnitude' in dB unit is evaluated by this instrument.

7. Commercial players and market overview

Currently, a wide range of synthetic fibres is taken for noise reduction applications. Various structural forms of synthetic fibre use in noise control are woven, nonwoven, knitted, fibre felts, and fibre reinforced composites. The different cross-sections of synthetic fibre such as hollow [35, 85], and triangular, trilobal are beneficial to improve acoustic absorption properties. It has been reported that hollow fibre has higher sound absorption and of lighter weight. Superior mechanical responses [23] of synthetic fibre and their various cross sectional in combination with natural fibre [22, 52, 53, 72] make them suitable for optimized applications in noise control. The use of recycled synthetic fibre [38] as a noise control solution addresses the concern related to the disposal of hard generated after the use of virgin nonbiodegradable synthetic fibre.

The forecasted growth size of the global noise control system market will be nearly 200% during 2017-2027 [132]. The noise control materials generally used in residential, industrial, and commercial applications in the form as follows

Acoustic panels
Acoustic tiles
Sound curtains
Acoustic surface
Sound insulating flooring
Sound barrier walls
Baffles
Sound blanket
Sound doors

8. Conclusion

In this chapter, the mechanism of sound absorption has been discussed first followed by a detailed description of the type of sound absorbers such as Helmholtz resonator, membrane absorber, perforated panel absorber, porous absorbers with emphasis on textile materials as sound absorbers. Then the effect of fibre size, thickness, density, porosity, tortuosity, compression, etc. on sound absorption has been discussed followed by a brief description of the principles of sound absorption measurement.

Conflict of interest

The authors declare no conflict of interest.

Author details

Mallika Datta*, Srijan Das and Devarun Nath
Government College of Engineering and Textile Technology, Serampore, Serampore, India

*Address all correspondence to: dattamallika8@gmail.com

References

[1] A. Nick, U. Becker, and W. Thoma, 'Improved acoustic behavior of interior parts of renewable', Journal of Polymers and the Environment, vol. 10, no. 3, pp. 115–118, Jul. 2002, DOI:10.1023/A:1021124214818.

[2] S. Amares, E. Sujatmika, T. W. Hong, R. Durairaj, and H. S. H. B. Hamid, 'A Review: Characteristics of Noise Absorption Material', J. Phys.: Conf. Ser., vol. 908, p. 012005, Oct. 2017, DOI:10.1088/1742-6596/908/1/012005.

[3] K. A. Jayaraman, 'Acoustical absorptive properties of nonwovens (M. S. Thesis).', North Carolina State University, USA., North Carolina, USA., 2005. Accessed: Oct. 17, 2020 [Online]. Available: https://fitfibers.com/wp-content/uploads/2018/06.0/4dg-acoustic-absorption.pdf

[4] F. B. Randall, Industrial noise control and acoustics. New York-Basel: Marcel Dekker Inc., 2001.

[5] F. A. Everest, The master handbook of acoustics, 4th ed. New York: McGraw-Hill, 2001.

[6] L. L. Faulkner, Handbook of Industrial Noise Control. New York: Industrial Press Inc., 1976.

[7] Z. Maekawa and P. Lord, Environmental and architectural acoustics. London; New York: E & FN Spon, 1994. Accessed: Jul. 26, 2020 [Online]. Available: http://www.myilibrary.com?id=3255

[8] H. Kuttruff, 'A simple iteration scheme for the computation of decay constants in enclosures with diffusely reflecting boundaries', The Journal of the Acoustical Society of America, vol. 98, no. 1, pp. 288–293, Jul. 1995, DOI: 10.1121/1.413727.

[9] M. N. Dimon and M. Ula, 'The study of normal incidence sound absorption coefficience (sound absorption) of wood circular perforated panel (CPP) using numerical modelling technique', p. 170.

[10] J. P. Arenas and M. J. Crocker, 'Recent trends in porous sound-absorbing materials', Sound and Vibration, no. MATERIALS REFERENCE ISSUE, pp. 12–17, Jul. 2010.

[11] X. Qiu, 'Principles of Sound Absorbers', in Acoustic Textiles, R. Padhye and R. Nayak, Eds. Singapore: Springer Singapore, 2016, pp. 43–72. DOI:10.1007/978-981-10-1476-5_3.

[12] T. J. Cox and P. D'Antonio, Acoustic absorbers and diffusers: theory, design and application, 2nd ed. New York-London: Taylor and Francis, 2004.

[13] N. D. Yilmaz, 'Acoustic properties of biodegradable nonwovens (PhD Thesis)', North Carolina State University, North Carolina, USA-Raleigh, 2009 [Online]. Available: https://repository.lib.ncsu.edu/bitstream/handle/1840.16/3248/etd.pdf?sequence=1&isAllowed=y

[14] J. Zhao, X.-M. Wang, J. M. Chang, Y. Yao, and Q. Cui, 'Sound insulation property of wood-waste tire rubber composite', Composites Science and Technology, vol. 70, pp. 2033–2038, 2010.

[15] Y. Yang, Z. Chen, Z. Chen, R. Fu, and Y. Li, 'Sound insulation properties of sandwich structures on glass fiber felts', Fibers Polym, vol. 16, no. 7, pp. 1568–1577, Jul. 2015, DOI:10.1007/s12221-015-5200-6.

[16] N. Lippitz, J. Rösler, and B. Hinze, 'Potential of metal fibre felts as passive absorbers in absorption silencers', Metals, vol. 3, no. 1, pp. 150–158, Mar. 2013, DOI:10.3390/met3010150.

[17] N. Lippitz, C. Blech, S. Langer, and J. Rösler, 'Identification of material parameters for the simulation of acoustic absorption of fouled sintered fiber felts', Materials, vol. 9, no. 8, p. 709, Aug. 2016, DOI:10.3390/ma9080709.

[18] N. Kino, 'A comparison of two acoustical methods for estimating parameters of glass fibre and melamine foam materials', Applied Acoustics, vol. 73, no. 6–7, pp. 590–603, Jun. 2012, DOI: 10.1016/j.apacoust.2011.12.007.

[19] E. Moretti, E. Belloni, and F. Agosti, 'Innovative mineral fiber insulation panels for buildings: Thermal and acoustic characterization', Applied Energy, vol. 169, pp. 421–432, May 2016, DOI:10.1016/j.apenergy.2016.02.048.

[20] C.-N. Wang and J.-H. Torng, 'Experimental study of the absorption characteristics of some porous fibrous materials', Applied Acoustics, vol. 62, no. 4, pp. 447–459, Apr. 2001, DOI: 10.1016/S0003-682X(00)00043-8.

[21] M. E. Delany and E. N. Bazley, 'Acoustical properties of fibrous absorbent materials', Applied Acoustics, vol. 3, no. 2, pp. 105–116, Apr. 1970, DOI:10.1016/0003-682X(70)90031-9.

[22] M. N. Yahya and D. D. V. Sheng Chin, 'A review on the potential of natural fibre for sound absorption application', IOP Conf. Ser.: Mater. Sci. Eng., vol. 226, p. 012014, Aug. 2017, DOI:10.1088/1757-899X/226/1/012014.

[23] X. Tang and X. Yan, 'Acoustic energy absorption properties of fibrous materials: A review', Composites Part A: Applied Science and Manufacturing, vol. 101, pp. 360–380, Oct. 2017, DOI: 10.1016/j.compositesa.2017.07.002.

[24] F. Sun, H. Chen, J. Wu, and K. Feng, 'Sound absorbing characteristics of fibrous metal materials at high temperatures', Applied Acoustics, vol. 71, no. 3, pp. 221–235, Mar. 2010, DOI: 10.1016/j.apacoust.2009.09.001.

[25] M. Tascan, E. A. Vaughn, K. A. Stevens, and P. J. Brown, 'Effects of total surface area and fabric density on the acoustical behavior of traditional thermal-bonded highloft nonwoven fabrics', Journal of the Textile Institute, vol. 102, no. 9, pp. 746–751, Sep. 2011, DOI:10.1080/00405000.2010.515731.

[26] M. Tascan and E. A. Vaughn, 'Effects of total surface area and fabric density on the acoustical behavior of needlepunched nonwoven fabrics', Textile Research Journal, vol. 78, no. 4, pp. 289–296, Apr. 2008, DOI:10.1177/0040517507084283.

[27] J. Zhu et al., 'Gradient-structural optimization of metal fiber porous materials for sound absorption', Powder Technology, vol. 301, pp. 1235–1241, Nov. 2016, DOI:10.1016/j.powtec.2016.08.006.

[28] Y.-J. Pan et al., 'Sound absorbent, flame retardant warp knitting spacer fabrics: Manufacturing techniques and characterization evaluations', Fibers Polym, vol. 16, no. 12, pp. 2682–2688, Dec. 2015, DOI:10.1007/s12221-015-5576-3.

[29] R. Wang, R. Yan, C.-W. Lou, and J.-H. Lin, 'Characterization of acoustic-absorbing inter/intra-ply hybrid laminated composites under dynamic loading', Fibers Polym, vol. 17, no. 3, pp. 439–452, Mar. 2016, DOI:10.1007/s12221-016-5322-5.

[30] B.-S. Kim, S.-J. Cho, D.-K. Min, and J. Park, 'Sound absorption structure in helical shapes made using fibrous paper', Composite Structures, vol. 134, pp. 90–94, Dec. 2015, DOI:10.1016/j.compstruct.2015.08.039.

[31] C.-W. Lou, J.-H. Lin, and K.-H. Su, 'Recycling polyester and polypropylene nonwoven selvages to produce functional sound absorption

composites', Textile Research Journal, vol. 75, no. 5, pp. 390–394, May 2005, DOI:10.1177/0040517505054178.

[32] M. Garai and F. Pompoli, 'A simple empirical model of polyester fibre materials for acoustical applications', Applied Acoustics, vol. 66, no. 12, pp. 1383–1398, Dec. 2005, DOI:10.1016/j.apacoust.2005.04.008.

[33] Y. Na and G. Cho, 'Sound absorption and viscoelastic property of acoustical automotive nonwovens and their plasma treatment', Fibers Polym, vol. 11, no. 5, pp. 782–789, Aug. 2010, DOI:10.1007/s12221-010-0782-5.

[34] N. Kino and T. Ueno, 'Evaluation of acoustical and non-acoustical properties of sound absorbing materials made of polyester fibres of various cross-sectional shapes', Applied Acoustics, vol. 69, no. 7, pp. 575–582, Jul. 2008, DOI:10.1016/j.apacoust.2007.02.003.

[35] S. Jiang, Y. Xu, H. Zhang, C. B. White, and X. Yan, 'Seven-hole hollow polyester fibers as reinforcement in sound absorption chlorinated polyethylene composites', Applied Acoustics, vol. 73, no. 3, pp. 243–247, Mar. 2012, DOI:10.1016/j.apacoust.2011.09.006.

[36] P. Soltani and M. Zarrebini, 'Acoustic performance of woven fabrics in relation to structural parameters and air permeability', Journal of The Textile Institute, vol. 104, no. 9, pp. 1011–1016, Sep. 2013, DOI:10.1080/00405000.2013.771427.

[37] C.-H. Huang, J.-H. Lin, and Y.-C. Chuang, 'Manufacturing process and property evaluation of sound-absorbing and thermal-insulating polyester fiber/polypropylene/thermoplastic polyurethane composite board', Journal of Industrial Textiles, vol. 43, no. 4, pp. 627–640, Apr. 2014, DOI:10.1177/1528083712471697.

[38] A. Patnaik, M. Mvubu, S. Muniyasamy, A. Botha, and R. D. Anandjiwala, 'Thermal and sound insulation materials from waste wool and recycled polyester fibers and their biodegradation studies', Energy and Buildings, vol. 92, pp. 161–169, Apr. 2015, DOI:10.1016/j.enbuild.2015.01.056.

[39] C. Zhao, P. Wang, L. Wang, and D. Liu, 'Reducing Railway Noise with Porous Sound-Absorbing Concrete Slabs', Advances in Materials Science and Engineering, vol. 2014, pp. 1–11, 2014, DOI:10.1155/2014/206549.

[40] J.-H. Lin, T.-T. Li, and C.-W. Lou, 'Puncture-resisting, sound-absorbing and thermal-insulating properties of polypropylene-selvages reinforced composite nonwovens', Journal of Industrial Textiles, vol. 45, no. 6, pp. 1477–1489, May 2016, DOI:10.1177/1528083714562088.

[41] G. Pan, Y. Zhao, H. Xu, B. Ma, and Y. Yang, 'Acoustical and mechanical properties of thermoplastic composites from discarded carpets', Composites Part B: Engineering, vol. 99, pp. 98–105, Aug. 2016, DOI:10.1016/j.compositesb.2016.06.018.

[42] A. Khan, M. Mohamed, N. Al Halo, and H. Benkreira, 'Acoustical properties of novel sound absorbers made from recycled granulates', Applied Acoustics, vol. 127, pp. 80–88, Dec. 2017, DOI:10.1016/j.apacoust.2017.05.035.

[43] F. Suvari, Y. Ulcay, B. Maze, and B. Pourdeyhimi, 'Acoustical absorptive properties of spunbonded nonwovens made from islands-in-the-sea bicomponent filaments', Journal of The Textile Institute, vol. 104, no. 4, pp. 438–445, Apr. 2013, DOI:10.1080/00405000.2012.740330.

[44] A. Nick, U. Becker, and W. Thoma, 'Improved acoustic behavior of interior parts of renewable resources in the

automotive industry', Journal of Polymers and the Environment, vol. 10, no. 3, pp. 115–118, 2002, DOI:10.1023/A:1021124214818.

[45] Y. Na, J. Lancaster, J. Casali, and G. Cho, 'Sound absorption coefficients of micro-fiber fabrics by reverberation room method', Textile Research Journal, vol. 77, no. 5, pp. 330–335, May 2007, DOI:10.1177/0040517507078743.

[46] G. Thilagavathi, E. Pradeep, T. Kannaian, and L. Sasikala, 'Development of natural fiber nonwovens for application as car interiors for noise control', Journal of Industrial Textiles, vol. 39, no. 3, pp. 267–278, Jan. 2010, DOI:10.1177/1528083709347124.

[47] A. Veerakumar and N. Selvakumar, 'A preliminary investigation on kapok/polypropylene nonwoven composite for sound absorption', Indian Journal of Fibre and Textile Research, vol. 37, no. 4, pp. 385–388, 2012.

[48] E. Markiewicz, S. Borysiak, and D. Paukszta, 'Polypropylene-lignocellulosic material composites as promising sound absorbing materials', Polimery, vol. 54, no. 06, pp. 430–435, Jun. 2009, DOI: 10.14314/polimery.2009.430.

[49] J. Merotte, A. Le Duigou, A. Bourmaud, K. Behlouli, and C. Baley, 'Mechanical and acoustic behaviour of porosity controlled randomly dispersed flax/PP biocomposite', Polymer Testing, vol. 51, pp. 174–180, May 2016, DOI: 10.1016/j.polymertesting.2016.03.002.

[50] S. Santhanam, B. M., S. Temesgen, D. Atalie, and G. Ashagre, 'Recycling of cotton and polyester fibers to produce nonwoven fabric for functional sound absorption material', Journal of Natural Fibers, vol. 16, no. 2, pp. 300–306, Feb. 2019, DOI:10.1080/15440478.2017.1418472.

[51] S. Sengupta, G. Basu, M. Datta, S. Debnath, and D. Nath, 'Noise control material using jute (Corchorus olitorius): effect of bulk density and thickness', The Journal of The Textile Institute, vol. 112, no. 1, pp. 56–63, Jan. 2021, DOI:10.1080/00405000.2020.1744222.

[52] T. Yang et al., 'Sound Absorption Properties of Natural Fibers: A Review', Sustainability, vol. 12, no. 20, p. 8477, Oct. 2020, DOI:10.3390/su12208477.

[53] S. Gokulkumar, P. Thyla, L. Prabhu, and S. Sathish, 'Measuring Methods of Acoustic Properties and Influence of Physical Parameters on Natural Fibers: A Review', Journal of Natural Fibers, pp. 1–20, Apr. 2019, DOI:10.1080/15440478.2019.1598913.

[54] K. O. Ballagh, 'Acoustical properties of wool', Applied Acoustics, vol. 48, no. 2, pp. 101–120, Jun. 1996, DOI:10.1016/0003-682X(95)00042-8.

[55] N. N. M. Hassan and A. Z. M. Rus, 'Influences of Thickness and Fabric for Sound Absorption of Biopolymer Composite', AMM, vol. 393, pp. 102–107, Sep. 2013, DOI:10.4028/www.scientific.net/AMM.393.102.

[56] H. Xiang, D. Wang, H. Liua, N. Zhao, and J. Xu, 'Investigation on sound absorption properties of kapok fibers', Chin J Polym Sci, vol. 31, no. 3, pp. 521–529, Mar. 2013, DOI:10.1007/s10118-013-1241-8.

[57] M. N. Yahya, M. Sambu, H. A. Latif, and T. M. Junaid, 'A study of Acoustics Performance on Natural Fibre Composite', IOP Conf. Ser.: Mater. Sci. Eng., vol. 226, p. 012013, Aug. 2017, DOI:10.1088/1757-899X/226/1/012013.

[58] U. Berardi and G. Iannace, 'Acoustic characterization of natural fibers for sound absorption applications', Building and Environment, vol. 94, pp. 840–852, Dec. 2015, DOI:10.1016/j.buildenv.2015.05.029.

[59] P. Glé, E. Gourdon, and L. Arnaud, 'Modelling of the acoustical properties of hemp particles', Construction and Building Materials, vol. 37, pp. 801–811, Dec. 2012, DOI:10.1016/j.conbuildmat.2012.06.008.

[60] P. Glé, E. Gourdon, L. Arnaud, K.-V. Horoshenkov, and A. Khan, 'The effect of particle shape and size distribution on the acoustical properties of mixtures of hemp particles', The Journal of the Acoustical Society of America, vol. 134, no. 6, pp. 4698–4709, Dec. 2013, DOI:10.1121/1.4824931.

[61] W. Yang and Y. Li, 'Sound absorption performance of natural fibers and their composites', Sci. China Technol. Sci., vol. 55, no. 8, pp. 2278–2283, Aug. 2012, DOI:10.1007/s11431-012-4943-1.

[62] F. Asdrubali, S. Schiavoni, and K. V. Horoshenkov, 'A Review of Sustainable Materials for Acoustic Applications', Building Acoustics, vol. 19, no. 4, pp. 283–311, Dec. 2012, DOI:10.1260/1351-010X.19.4.283.

[63] H. P. Lee, B. M. P. Ng, A. V. Rammohan, and L. Q. N. Tran, 'An Investigation of the Sound Absorption Properties of Flax/Epoxy Composites Compared with Glass/Epoxy Composites', Journal of Natural Fibers, vol. 14, no. 1, pp. 71–77, Jan. 2017, DOI: 10.1080/15440478.2016.1146643.

[64] M. K. bin Bakri, E. Jayamani, S. K. Heng, S. Hamdan, and A. Kakar, 'An Experimental and Simulation Studies on Sound Absorption Coefficients of Banana Fibers and their Reinforced Composites', NHC, vol. 12, pp. 9–20, Nov. 2016, DOI:10.4028/www.scientific.net/NHC.12.9.

[65] U. Berardi, G. Iannace, and M. Di Gabriele, 'The Acoustic Characterization of Broom Fibers', Journal of Natural Fibers, vol. 14, no. 6, pp. 858–863, Nov. 2017, DOI:10.1080/15440478.2017.1279995.

[66] Zulkifli, 'Noise Control Using Coconut Coir Fiber Sound Absorber with Porous Layer Backing and Perforated Panel', American Journal of Applied Sciences, vol. 7, no. 2, pp. 260–264, Feb. 2010, DOI:10.3844/ajassp.2010.260.264.

[67] D. J. Oldham, C. A. Egan, and R. D. Cookson, 'Sustainable acoustic absorbers from the biomass', Applied Acoustics, vol. 72, no. 6, pp. 350–363, May 2011, DOI:10.1016/j.apacoust.2010.12.009.

[68] S. Fatima and A. R. Mohanty, 'Acoustical and fire-retardant properties of jute composite materials', Applied Acoustics, vol. 72, no. 2–3, pp. 108–114, Feb. 2011, DOI:10.1016/j.apacoust.2010.10.005.

[69] P. V. Bansod, T. Mittal, and A. R. Mohanty, 'Study on the Acoustical Properties of Natural Jute Material by Theoretical and Experimental Methods for Building Acoustics Applications', Acoust Aust, vol. 44, no. 3, pp. 457–472, Dec. 2016, DOI:10.1007/s40857-016-0073-4.

[70] S. Sengupta, 'Sound reduction by needle-punched nonwoven fabrics', Indian Journal of Fibre and Textile Research, vol. 35, pp. 237–242, 2010.

[71] S. Ersoy and H. Küçük, 'Investigation of industrial tea-leaf-fibre waste material for its sound absorption properties', Applied Acoustics, vol. 70, no. 1, pp. 215–220, Jan. 2009, DOI: 10.1016/j.apacoust.2007.12.005.

[72] G. Thilagavathi, S. Neela Krishnan, N. Muthukumar, and S. Krishnan, 'Investigations on Sound Absorption Properties of Luffa Fibrous Mats', Journal of Natural Fibers, vol. 15, no. 3, pp. 445–451, May 2018, DOI:10.1080/15440478.2017.1349016.

[73] A. Putra, Y. Abdullah, H. Efendy, W. M. Farid, M. R. Ayob, and M. S. Py, 'Utilizing Sugarcane Wasted Fibers as a Sustainable Acoustic Absorber', Procedia Engineering, vol. 53, pp. 632–638, 2013, DOI:10.1016/j.proeng.2013.02.081.

[74] X. Tang, X. Liu, and X. Yan, 'Investigation on the Sound Absorption Properties of Waste Green Tea Residues Covered by Woven Fabric', Journal of Natural Fibers, pp. 1–10, May 2020, DOI:10.1080/15440478.2020.1764455.

[75] H. Li, N. Zhang, X. Fan, J. Gong, J. Zhang, and X. Zhao, 'Investigation of effective factors of woven structure fabrics for acoustic absorption, Applied Acoustics, vol. 161, p. 107081, Apr. 2020, DOI:10.1016/j.apacoust.2019.107081.

[76] P. Segura-Alcaraz, J. Segura-Alcaraz, I. Montava, and M. Bonet-Aracil, 'The effect of the combination of multiple woven fabric and nonwoven on acoustic absorption', Journal of Industrial Textiles, vol. 50, no. 8, pp. 1262–1280, Mar. 2021, DOI:10.1177/1528083719858771.

[77] M. E. Nute and K. Slater, '62—A STUDY OF SOME FACTORS AFFECTING SOUND REDUCTION BY CARPETING', The Journal of The Textile Institute, vol. 64, no. 11, pp. 645–651, Nov. 1973, DOI:10.1080/00405007308630314.

[78] Y. Z. Shoshani and M. A. Wilding, 'Effect of pile parameters on the noise absorption capacity of tufted carpets', Textile Research Journal, vol. 61, no. 12, pp. 736–742, Dec. 1991, DOI:10.1177/004051759106101207.

[79] Y. Chen and N. Jiang, 'Carbonized and activated non-wovens as high-performance acoustic materials: part II noise absorption', Textile Research Journal, vol. 79, no. 3, pp. 213–218, Feb. 2009, DOI:10.1177/0040517508093593.

[80] Y. Chen and N. Jiang, 'Carbonized and activated non-wovens as high-performance acoustic materials: part I noise absorption', Textile Research Journal, vol. 77, no. 10, pp. 785–791, Oct. 2007, DOI:10.1177/0040517507080691.

[81] N. Jiang, J. Y. Chen, and D. V. Parikh, 'Acoustical evaluation of carbonized and activated cotton nonwovens', Bioresource Technology, vol. 100, no. 24, pp. 6533–6536, Dec. 2009, DOI:10.1016/j.biortech.2008.10.062.

[82] M. Tascan and E. A. Vaughn, 'Effects of Fiber Denier, Fiber Cross-Sectional Shape and Fabric Density on Acoustical Behavior of Vertically Lapped Nonwoven Fabrics', Journal of Engineered Fibers and Fabrics, vol. 3, no. 2, p. 155892500800300, Jun. 2008, DOI:10.1177/155892500800300206.

[83] Y. Shoshani and Y. Yakubov, 'A Model for Calculating the Noise Absorption Capacity of Nonwoven Fiber Webs', Textile Research Journal, vol. 69, no. 7, pp. 519–526, Jul. 1999, DOI:10.1177/004051759906900709.

[84] D. V. Parikh, Y. Chen, and L. Sun, 'Reducing Automotive Interior Noise with Natural Fiber Nonwoven Floor Covering Systems', Textile Research Journal, vol. 76, no. 11, pp. 813–820, Nov. 2006, DOI:10.1177/0040517506063393.

[85] A. Mahmoud, G. El-Shenawy, and E. Ramadan, 'Using Nonwoven Hollow Fibers to Improve Cars Interior Acoustic Properties', Textile and Apparel, vol. 16, no. 3, pp. 49–56, Aug. 2012, DOI:10.1108/RJTA-16-03-2012-B005.

[86] M. Küçük and Y. Korkmaz, 'The effect of physical parameters on sound absorption properties of natural fiber mixed nonwoven composites', Textile Research Journal, vol. 82, no. 20, pp. 2043–2053, Dec. 2012, DOI:10.1177/0040517512441987.

[87] Y. Shoshani and Y. Yakubov, 'Use of Nonwovens of Variable Porosity as Noise Control Elements', International Nonwovens Journal, vol. os-10, no. 4, pp. 1558925001os–10, Dec. 2001, DOI: 10.1177/1558925001os-1000407.

[88] A. V. Genis, E. Yu. Kostyleva, L. N. Andrianova, and V. A. Martem'yanov, 'Comparative evaluation of acoustical properties of heat-bonded and needle-punched fibrous materials prepared from polymer melts', Fibre Chem, vol. 21, no. 6, pp. 479–482, 1990, DOI: 10.1007/BF00544434.

[89] Mozhdeh Ghahremani Honarvar, Ali Asghar Asgharian Jeddi, and M. Amani Tehran, 'Noise Absorption Modeling of Rib Knitted Fabrics', Textile Research Journal, vol. 80, no. 14, pp. 1392–1404, Sep. 2010, DOI:10.1177/0040517509357649.

[90] R. Brencis, J. Skujans, U. Iljins, and I. Preikss, 'Research of hemp fibrous reinforcement effect to bending strenght and sound absorption of foam gypsum', in 3rd International Conference Civil Engineering`11, Jan. 2011, pp. 32–37.

[91] H. Mamtaz, 'Modelling of acoustic wave propagation through the natural fiber composites', presented at the 27th Symposium of Malaysian Chemical Engineers (SOMChE 2014), Nov. 2014.

[92] L. Lyu et al., 'Sound absorption properties of multi-layer structural composite materials based on waste corn husk fibers', Journal of Engineered Fibers and Fabrics, vol. 15, p. 155892502091086, Jan. 2020, DOI: 10.1177/1558925020910861.

[93] E. Jayamani and S. Hamdan, 'Sound Absorption Coefficients Natural Fibre Reinforced Composites', AMR, vol. 701, pp. 53–58, May 2013, DOI:10.4028/www.scientific.net/AMR.701.53.

[94] H. S. Seddeq, 'Factors influencing acoustic performance of sound absorptive materials', Australian Journal of Basic and Applied Sciences, vol. 3, no. 4, pp. 4610–4617, 2009.

[95] M. Tascan and K. L. Gaffney, 'Effect of Glass-Beads on Sound Insulation Properties of Nonwoven Fabrics', Journal of Engineered Fibers and Fabrics, vol. 7, no. 1, p. 155892501200700, Mar. 2012, DOI: 10.1177/155892501200700111.

[96] M. Datta, 'Engineering of jute based acoustic insulator and its structural analysis', University of Calcutta, Kolkata. India, 2015 [Online]. Available: http://hdl.handle.net/10603/163760

[97] P. Ray, M. Datta, and B. Chatterjee, 'Evaluation of the Acoustic and Non-Acoustic Properties of Sound Absorbing Jute Nonwoven Fabric', presented at the International Seminar On Strengthening for Jute, Kenaf And Allied Fibres Research and Development, International Jute Study Group, Jun. 2011.

[98] X. Liu, X. Yan, L. Li, and H. Zhang, 'Sound-Absorption Properties of Kapok Fiber Nonwoven Fabrics at Low Frequency', Journal of Natural Fibers, vol. 12, no. 4, pp. 311–322, Jul. 2015, DOI:10.1080/15440478.2014.919891.

[99] P. Ganesan and T. Karthik, 'Development of acoustic nonwoven materials from kapok and milkweed fibres', The Journal of The Textile Institute, vol. 107, no. 4, pp. 477–482, Apr. 2016, DOI:10.1080/00405000.2015.1045251.

[100] G. Basu, M. Datta, S. Sengupta, D. Nath, and S. Debnath, 'Jute Felt for Noise Reduction: Understanding Effect of Pore Size Distribution', Journal of Natural Fibers, pp. 1–15, May 2021, DOI: 10.1080/15440478.2021.1921663.

[101] X. Liu, X. Yan, and H. Zhang, 'Effects of pore structure on sound absorption of kapok-based fiber nonwoven fabrics at low frequency',

Textile Research Journal, vol. 86, no. 7, pp. 755–764, May 2016, DOI:10.1177/0040517515599742.

[102] Y. Chen, L. Sun, O. Chiparus, I. Negulescu, V. Yachmenev, and M. Warnock, 'Kenaf/Ramie Composite for Automotive Headliner', J Polym Environ, vol. 13, no. 2, pp. 107–114, Apr. 2005, DOI:10.1007/s10924-005-2942-z.

[103] K. W. Oh, D. K. Kim, and S. H. Kim, 'Ultra-porous flexible PET/Aerogel blanket for sound absorption and thermal insulation', Fibers Polym, vol. 10, no. 5, pp. 731–737, Oct. 2009, DOI:10.1007/s12221-010-0731-3.

[104] A. Rabbi, H. Bahrambeygi, A. M. Shoushtari, and K. Nasouri, 'Incorporation of Nanofiber Layers in Nonwoven Materials for Improving Their Acoustic Properties', Journal of Engineered Fibers and Fabrics, vol. 8, no. 4, p. 155892501300800, Dec. 2013, DOI:10.1177/155892501300800412.

[105] M. Barburski, J. R. Blaszczak, and Z. Pawliczak, 'Influence of designs of weaves on acoustic attenuation of fabrics', Journal of Industrial Textiles, vol. 49, no. 1, pp. 33–45, Jul. 2019, DOI:10.1177/1528083718769945.

[106] M. Datta, D. Nath, S. Nandi, S. Bhakat, and G. Basu, 'Study on Noise Control Behaviour of Jute Plain Woven Fabric', Kolkata. India, Jan. 2017, pp. 289–292.

[107] Y. Shoshani and G. Rosenhouse, 'Noise absorption by woven fabrics', Applied Acoustics, vol. 30, no. 4, pp. 321–333, 1990, DOI:10.1016/0003-682X(90)90081-5.

[108] I. Prasetyo, B. Yuliarto, K. Anwar, and D. R. Andhika, 'Characterization of woven fabrics for development of micro perforated panel absorbers', p. 8.

[109] P. Soltani and M. Zerrebini, 'The analysis of acoustical characteristics and sound absorption coefficient of woven fabrics', Textile Research Journal, vol. 82, no. 9, pp. 875–882, Jun. 2012, DOI:10.1177/0040517511402121.

[110] T. Dias and R. Monaragala, 'Sound absorbtion in knitted structures for interior noise reduction in automobiles', Meas. Sci. Technol., vol. 17, no. 9, pp. 2499–2505, Sep. 2006, DOI:10.1088/0957-0233/17/9/018.

[111] R. M. Monaragala, 'Knitted structure for sound absorption', in Advances in knitting technology, K. F. Au, Ed. Oxford: Woodhead Publ, 2011, pp. 2662–2285.

[112] N. N. Voronina, 'Influence of the structure of fibrous sound absorbing materials on their acoustical properties', Soviet Physics-Acoustics, vol. 29, no. 5, p. 355-357, 1983.

[113] N. Voronina, 'Improved empirical model of sound propagation through a fibrous material', Applied Acoustics, vol. 48, no. 2, pp. 121–132, Jun. 1996, DOI:10.1016/0003-682X(95)00055-E.

[114] M. A. Ibrahim and R. W. Melik, 'Physical Parameters Affecting Acoustic Absorption Characteristics of Fibrous Materials', in Proceedings of the Mathematical and Physical Society of Egypt, Egypt, 1978, vol. 46, pp. 125–130.

[115] T. Koizumi, N. Tsujiuchi, and A. Adachi, 'The development of sound absorbing materials using natural bamboo fibers', in High Performance Structures and Composites, vol. 59, C. A. Brebbia and W. P. de Wilde, Eds. Southampton; Boston: WIT Press, 2002, pp. 157–166.

[116] J. Y. Chen, 'Reducing Noise in Automotives Interiors', in Textile advances in the automotive industry, 1st ed., R. Shishoo, Ed. Cambridge, England: Woodhead Publishing, 2008, pp. 198–228.

[117] X. Sagartzazu, L. Hervella-Nieto, and J. M. Pagalday, 'Review in Sound

Absorbing Materials', Arch Computat Methods Eng, vol. 15, no. 3, pp. 311–342, Sep. 2008, DOI:10.1007/s11831-008-9022-1.

[118] C. Wassilieff, 'Sound absorption of wood-based materials', Applied Acoustics, vol. 48, no. 4, pp. 339–356, Aug. 1996, DOI:10.1016/0003-682X(96)00013-8.

[119] C.-N. Wang, Y.-M. Kuo, and S.-K. Chen, 'Effects of compression on the sound absorption of porous materials with an elastic frame', Applied Acoustics, vol. 69, no. 1, pp. 31–39, Jan. 2008, DOI:10.1016/j.apacoust.2006.08.006.

[120] B. Castagnède, A. Aknine, B. Brouard, and V. Tarnow, 'Effects of compression on the sound absorption of fibrous materials', Applied Acoustics, vol. 61, no. 2, pp. 173–182, Oct. 2000, DOI:10.1016/S0003-682X(00)00003-7.

[121] L. L. Beranek, 'Acoustic Impedance of Porous Materials', The Journal of the Acoustical Society of America, vol. 13, no. 3, pp. 248–260, Jan. 1942, DOI: 10.1121/1.1916172.

[122] ASTM D737 – 04, "Standard Test Method for Air Permeability of Textile Fabrics, 7.01 vols. ASTM International, 2008.

[123] E. Brandão, E. Tijs, A. Lenzi, and H.-E. de Bree, 'A Comparison of Three Methods to Calculate the Surface Impedance and Absorption Coefficient from Measurements Under Free Field or in situ Conditions', Acta Acustica united with Acustica, vol. 97, no. 6, pp. 1025–1033, Nov. 2011, DOI:10.3813/AAA.918482.

[124] E. F. Alton, The master handbook of acoustics, 4th ed. USA-New York: McGraw-Hill, 2001.

[125] M. Hakamada, T. Kuromura, Y. Chen, H. Kusuda, and M. Mabuchi, 'Sound absorption characteristics of porous aluminum fabricated by spacer method', J. Appl. Phys., vol. 100, no. 11, p. 114908, 2006, DOI:10.1063/1.2390543.

[126] R. K. Srivastava, R. L. Dhabal, B. M. Suman, A. Saini, and P. Panchal, 'An estimation of correlation on thermo-acoustic properties of mineral wool', Journal of Scientific and Industrial Research, vol. 65, pp. 231–236, Mar. 2006.

[127] F. Forouharmajd and Z. Mohammadi, 'Assessment of normal incidence absorption performance of sound absorbing materials', Int J Env Health Eng, vol. 5, no. 1, p. 10, 2016, DOI:10.4103/2277-9183.184224.

[128] ASTM E 1050-12. Standard test method for impedance and absorption of acoustical materials using a tube, two microphones and a digital frequency analysis system. New York, USA: ASTM International, 2012.

[129] ISO 10534-2: 2001 Acoustics-Determination of Sound absorption coefficient and impedance in impedance tubes—Part 2: Transfer-function method. Geneva, Switzerland: International Organization for Standardization, 2001.

[130] H. Kuttruff, Room Acoustics, 6th ed. Sixth edition. | Boca Raton: CRC Press [2017]: CRC Press, 2016. DOI: 10.1201/9781315372150.

[131] ISO 354:2003 (E) Acoustics—Measurement of sound absorption in a reverberation room, 2nd ed. International Organization for Standardization, 2003.

[132] 'Global Noise Control System Market Growth, Share, Size, Trends and Forecast (2021–2027)', Rn518767179, Apr. 2021 [Online]. Available: https://www.reanin.com/report-store/global-noise-control-system-market

Chapter 13

Sportswear: Acumen of Raw Materials, Designing, Innovative and Sustainable Concepts

Yamini Jhanji

Abstract

Sportswear constitutes an integral part of technical textiles and encases great potential as far as technological and design innovations are concerned. The sports textiles have witnessed tremendous evolution and that too at a much faster pace compared to ready to wear segment. The sports clothing is no longer restricted to sportsperson involved in performance sports or strenuous physical activities. However, there has been a surge for sports apparels and accessories among health conscious, fitness freak and gym enthusiasts. Accordingly, the sportswear industry has witnessed revolutionary advancements in development of different sportswear categories like active wear, leisurewear and athleisure to fulfill the requirements of sportsperson as well as health conscious millennials. The basic and functional requirements of comfort, breathability, light weight, anti-static and anti-odor properties can be engineered into sportswear by optimum selection of fibers, yarns, fabrics and garments' designing aspects. The chapter will provide an insight on the classification, requirements, design aspects, raw material procurement, innovative and sustainable concepts employed in sportswear to enhance the functionality and comfort characteristics of sportswear. Furthermore, the role of technology and fashion in sportswear transformation is also covered in the last sections of the chapter.

Keywords: Accessories, Active wear, Comfort, Sportswear, Knitted fabrics, Leisure Wear

1. Introduction

Sportswear has emerged as one of the most promising and technologically driven textile segment with massive innovations and advancements involved right from raw material procurement to design and development of sports specific clothing. The field is promising and innovative with several avenues as far as research and development, pioneering new technologies and trailblazing concepts are concerned.

The basic requirements of thermo physiological as well psychological comfort, dexterity, agility to wearer, breathability, moisture management, light weight, antimicrobial and anti-odor properties can be incorporated into sportswear by correct selection of fibers, yarns and fabric variables for sportswear. The sports clothing is no longer restricted to sportsperson involved in performance sports or strenuous physical activities. However, there has been a surge for sports apparels and accessories among health conscious, fitness freak and gym enthusiasts. Accordingly, the sportswear industry has witnessed revolutionary advancements in development of different sportswear categories like active wear, leisurewear and athleisure to fulfill the requirements of sportsperson as well as health-conscious millennials. Apart from functional requirements, a lot of emphasis is being laid on esthetic aspects as well considering increasing number of females involved in yoga, gyming and other sporting activities who give precedence to silhouette, colors and other design details of sportswear. Accordingly, the technological as well as ergonomic advancements in sportswear design and development have opened new avenues for researchers to explore the field further.

2. Sportswear categorization

The sportswear can be categorized based on a number of factors such as:

- Sportsperson's level of physical activity and fatigue
- Stress involved during strenuous activity
- Duration for which the sportsperson doffs the clothing
- Ambient conditions.

2.1 Categorization based on level of physical activity

The sportswear can be classified into active and leisure wear based on the sportsperson's level of physical activity.

2.1.1 Active wear

Active wear also referred to as professional sportswear encompasses the sportswear attire that are usually worn by sportspersons for short time duration when indulging in rigorous, high level of physical activities such as skiing, long jump, high jump and other such adventure sports etc. All such sports demand active, stressful and maximum physical performance thereby resulting in profuse sweating (sensible perspiration) experienced by the sportsperson. The designing of active wear is not as challenging a task as the design considerations for leisure wear because sportsperson during the entire duration of active sport is exposed to constant ambient conditions within the boundaries of the playing ground irrespective of indoor or outdoor conditions Moreover, the factors like sportsperson's age, gender and frequency of doffing the clothing is predetermined which can serve as quick guide for designers while designing active wear.

2.1.2 Leisure wear

Leisure wear comprises sportswear worn during sports activities like cricket, hockey and golf. The aforesaid sports activity demands intermittent performance with alternating active and rest phases by sportsperson and with prolonged exposure to varying ambient conditions. Leisure sportswear are worn by players belonging to varying age and gender groups and those indulging in low to moderate physical activity. Moreover, the duration and frequency of wearing and ambient conditions are all variable during the course of the activity. Consequently, the designing of leisure wear is a challenging task for designers as they need to consider the varying ambient conditions and extended durations on the field to which wearer would be subjected. The wearer is expected to don the clothing the entire day or several hours at stretch in changing environmental conditions. Thus, the designing of leisurewear needs special consideration of wearer's physiological requirements and changing environmental conditions to which sportsperson will be exposed while indulging in sports activity. Furthermore, casual and exercise wear, parkas, hoodies, pants, and crew neck fleece sweaters that provide a combination of esthetic, style, comfort, and functionality in a less competitive mode can also be included in category of leisurewear.

2.2 Categorization based on sport specific requirements

The sportswear can also be classified based on specific requirement of sports. Different sports involve different level of physical exertion and are performed in varying ambient conditions. Consequently, the clothing worn for a particular sport like cycling may not be suitable for another sport such as under water sports, mountaineering etc. performed in contrastingly different environment. The sportswear can be classified into dry, damp and wet fast action sportswear based on sports specific requirements.

2.2.1 Dry fast action sportswear

Dry fast action sportswear are worn during sports activities such as football, rugby, tennis and track games that demand optimum moisture management properties enabling quick sweat absorption and dissipation thereby providing cooling effect to wearer.

2.2.2 Damp-fast action sportswear

Damp-fast action sportswear is suitable for sports where rapid sweat evaporation from the skin surface is a prerequisite. Apart from rapid transfer of liquid perspiration, the sportswear should ensure good water vapor permeability, water proofing and protection from cold along with high degree of stretch ability.

2.2.3 Wet-fast action sportswear

Wet-fast action sportswear is specially designed for sporting activities like swimming and other under water sports activities which require a high degree of stretch and form fitting. The clothing plays a vital role in enhancing the athlete's performance by reducing drag and fatigue.

2.3 Categorization based on weather conditions

The weather conditions to which sportsperson are exposed during the activity also dictates the classification of sportswear. Accordingly, the sportswear may be classified as cold, moderate and hot weather sportswear. The material selection and design aspects for three categories will vary drastically owing to different set of properties required for each clothing type.

2.3.1 Cold weather sportswear

Cold weather sportswear is generally worn during ice skating, mountaineering and any such winter outdoor activity where the wearer is at risk of heat loss and thus hypothermia. The cold weather sportswear should be able to trap the body heat and provide protection against cold and humid conditions. Consequently, the clothing is designed in such a manner that it exhibits high thermal insulation for entrapment of body heat and breathable for moisture vapor sweat to easily escape out but prevent the ingress of liquid from external sources through clothing.

2.3.2 Moderate weather sportswear

Moderate weather sportswear is preferred by sports enthusiasts when ambient conditions are conducive with moderate temperature and humidity. Accordingly, sportswear worn in moderate climate should be breathable, permeable to air and heat passage to ensure dry and comfortable feel to wearer.

2.3.3 Hot weather sportswear

Hot weather sportswear are generally preferred when the ambient temperature is high and the wearer may be at risk of hyperthermia as he experience profuse sweating (sensible perspiration) and elevated body temperature as a synergistic effect of his own metabolic heat generation and the hot weather. The clothing should thus be light weight, quick drying, and wick able to push the liquid moisture away without sweat absorption in next to skin layer and should exhibit high thermal conductivity for rapid heat dissipation thereby ensuring dry and cool feel next to skin [1–3].

The requirements and key design aspects of sportswear will be discussed elaborately in the next sections of the chapter.

3. Requirements of sportswear

The categorization of sportswear discussed in previous section highlighted that sportswear are categorized based on level of wearer's physical activity, specific sports and ambient conditions. The requirements for each sportswear category will be drastically different as the clothing is worn in altogether different ambient conditions, for varying durations and frequency. Sportswear designed as active wear for outdoor applications should provide protection to wearer against external elements and environmental extremities such as wind, sunlight, rain and snow. Moreover, the clothing should possess optimum thermal and moisture management properties in order to maintain the heat balance between the metabolic heat produced as a result of physical activity and the outside environment. Perspiration both in vapor (insensible) and liquid (sensible) form should be readily dissipated to the outside environment to provide dry microclimate next to skin for the wearer. This requirement can be met by designing the sportswear that exhibit low resistance to heat transfer and evaporative heat loss. Sportswear should ensure rapid liquid transfer by means of wicking and should have good drying ability to prevent condensation of liquid sweat near skin. However, a high level of thermal insulation is prerequisite for cold weather sports clothing so as to prevent body heat to escape to outside environment. Contrastingly, low thermal insulation is desirable for sportswear intended for warmer climates. The concept of "Onion-skin" principle encompassing clothing system with several layers and consisting of several clothing items is applied in sportswear to achieve variable thermal insulation as per the capricious ambient conditions. The clothing can thus be adapted to the changing environment by donning or doffing individual clothing items for effective protection against the external elements [1–4].

Furthermore, the requirements for sportswear can be as categorized into functional and esthetic requirements, both of which play a crucial role in determining the performance and consumers acceptability for the clothing. Functional attributes of sportswear pertain to light weight, low fluid resistance, high tenacity, strechablility, thermal regulation, UV protection, vapor permeability, and sweat absorption and release while esthetics requirements entail softness, surface texture, handle, luster and color of the sportswear.

In general, the most common characteristics sought in sportswear can be enlisted as follows:

- Optimum thermal and moisture regulation
- Good air and water vapor permeability
- Rapid moisture absorption and wicking property
- Absence of dampness & dry feel next to skin
- Rapid Drying ability
- Low water absorption of next to skin layer clothing
- Dimensionally stable even when wet
- Durable, easy care and lightweight
- Soft and pleasant touch
- Effective protection against external elements such as extreme cold, sunlight, wind, rain etc.
- Stretch ability, form fitting and shape retention
- Antimicrobial & antistatic properties.

4. Fiber, yarn and fabric interplay for sportswear design and development

The type of sport, ambient conditions and level of physical activity as discussed in the previous section dictates the functional requirements and performance characteristics of sportswear.

Sportsperson involved in high active sports such as tennis and soccer usually experience heat stress owing to high amount of metabolic heat generation and profuse sweating. Therefore, the thermo-physiological comfort aspect of sportswear is of utmost importance for such sports to ensure well-being of sports person without any hindrance to their performance and efficiency. Dry microclimate for wearer involved in intensive physical activity and in hot, humid conditions is ensued by engineering fabrics exhibiting effective moisture vapor and liquid moisture transmission through fabric. The effective heat and moisture dissipation through fabrics intended for active wear requires special consideration of geometry, packing density and structure of the constituent fibers in yarn and fabric construction.

Thermo-physiological comfort properties of sportswear are influenced by multitude of fiber, yarn and fabric variables that influence inter yarn spaces, capillary geometry and in turn the moisture vapor and liquid moisture transmission through textile structures.

Sportswear engineered with specialized fibers, yarns and fabric structures exhibit excellent moisture management properties. Accordingly, sportswear designers experiment with variable fiber cross-sectional shape, shape factor and specific surface area of fiber, yarn variables like twist, linear density, structure and packing coefficient and fabric variables like loop length and porosity, varying knit structures like plated, elastene fabrics and those designed with bio mimic concepts to design sportswear intended for performance sports to keep the wearer comfortable with dry sensation next to skin.

Undoubtedly, the role of fibers, yarns and fabric structure in engineering textile structures suitable for sportswear cannot be undermined. The following section will discuss the role of fibers, yarns and fabric variables and their selection criteria for sportswear design and development.

4.1 Fiber variables and their influence on thermo-physiological comfort aspects

A combination of natural and synthetic fibers is an optimal solution when designing clothing for next to skin and sportswear applications. However no single fiber or different fiber blends can ensure ideal clothing suitable for varied applications. The right type of fiber needs to be in the right place according to the fabric's end use. Any wrong selection of fiber combinations may lead to thermal and wetness discomfort to the wearer if water absorption and liquid transfer properties of the selected fibers are not according to level of sweat generated.

The primary requirements of effective liquid transmission, better wick ability and faster drying in sportswear can be achieved by incorporation of varying fiber profiles like tetra channel, hexa channel, five-leaf, trilobal and triangular cross-sections that offer enlarged surface area for transmission of liquid sweat compared to their circular counterparts (**Figure 1a** and **b**).

Coolmax is modified polyester fiber developed by Dupont. The fiber resembles double scallop with four channels having 20% more surface area than conventional polyester fiber therefore offering better wicking, moisture vapor permeability and water spreading over greater area in fabric.

4 DG fiber is speciality fiber with eight-legged cross section made of polyester and other polymers and large surface area/volume and bulk compared to round fibers. The fiber is capable of moving, storing and trapping the fluids owing to the unique grooved shape. Accordingly, fibers of varying cross sections are finding applications in sportswear owing to their effectiveness in heat, moisture and liquid transmission through fabrics.

Incorporation of non-circular fiber profile are characterized by increase in fiber's shape factor which influences the fiber capillary spaces, inter yarn pore spaces, packing density, specific surface area and in turn the thermo-physiological properties of fabrics.

Fibers with greater specific surface area possess good moisture absorption and release properties. The micro grooves present on fiber surface enhance capillary absorbency, cause siphoning of moisture which can thus be dissipated by spreading over fiber surface. **Figure 1b** shows the different fiber cross-sections generally used in sportswear.

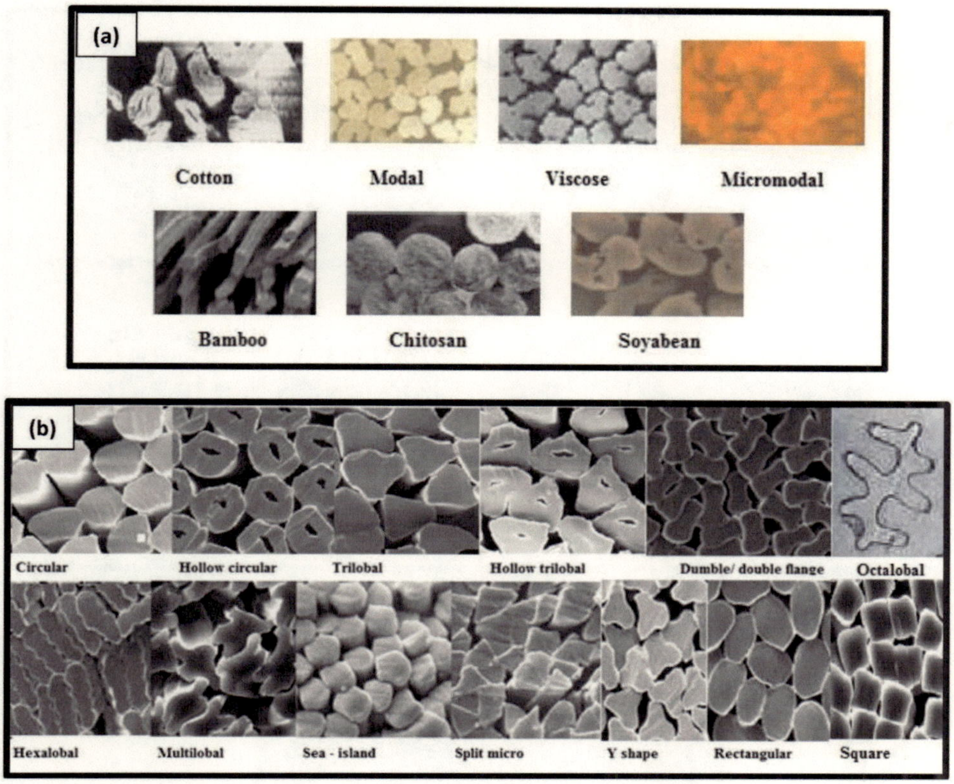

Figure 1.
(a) Different fibers for sportswear, (b) fibers of varying cross sections.

Moisture transmission properties of individual components can be drastically improved by blending two or more fibers into single yarns. Polyester and cotton fibers in blended form are increasingly being used for specialized yarn production to achieve good wicking and low absorbency.

Wicking and thermal resistance can further be improved by creation of hollow and microporous yarn core by combination of different fibers such as cotton and PVA fibers (**Figure 2a**).

Welkey is fiber with hollow core and body of fiber has proliferation of small holes. Thermal resistance increases as a result of increased number of air spaces inside fibers. Wicking of sweat next to skin is possible by capillary action caused by small holes forming proliferations in fiber body. The fiber can thus be effectively utilized for designing winter wear sportswear to obtain efficient moisture management along with rapid sweat dissipation.

Bicomponent fiber is classified based on fiber cross-section into side- by side, sheath core, islands in the sea, segmented pie cross-section. Matrix of one polymer contains another polymer and micro denier fibers can be generated by this type of bi component structure. Polyester, polypropylene, nylon forms the island in the structure.

Bi-component filament yarn, *Naiva* developed by combination of 55% Naiva (Eval/ Nylon) yarn and 45% nylon microfiber is used for development of light weight, soft and moisture absorbing Naiva fabric suitable for mountaineering and other active sportswear. The extraordinary thermal and moisture management properties of Naiva fabric may be attributed to micro loops on the surface of Naiva fabric (**Figure 2b**) as a result of high thermal shrinkage property of yarn.

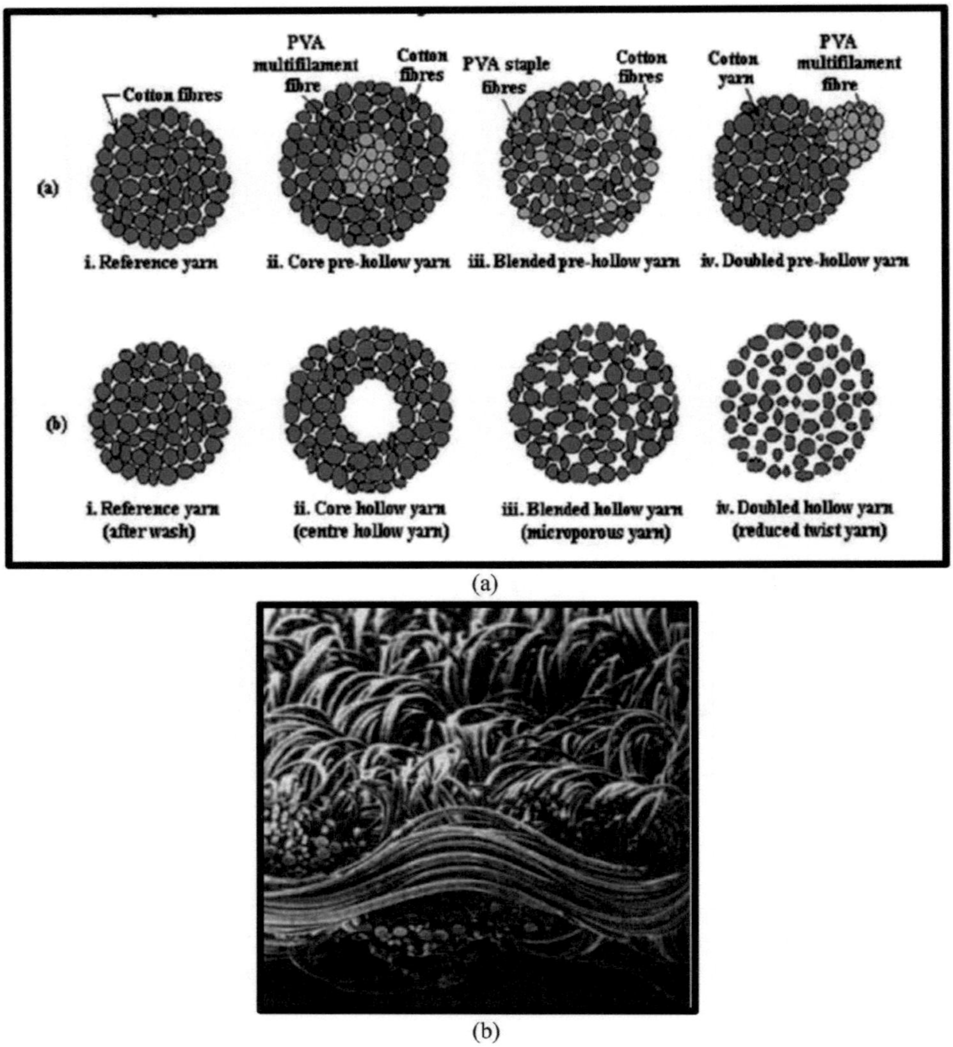

Figure 2.
(a) Core hollow yarn composed of cotton & PVA fibers. (b) Micro loops on surface of Naiva Fabric.

Eval, one of the components of bi component yarn is the copolymer resin of ethylene vinyl-alcohol [1–3].

Several researchers have explored the possibilities of combining different commodity and speciality fibers to engineer a textile structure suitable for sportswear with desirable thermal and moisture management properties.

Gurudatt et al. [5] studied the absorption and drying behavior of textile using cotton, polyester of regular cross section, polyethylene glycol modified polyester and scalloped oval cross-section fiber. It was suggested that absorption capacity of polyester enhances by cross-section and polymer modification. Knitted fabrics using scalloped oval cross section had higher absorption rate compared to regular polyester fiber.

Das et al. [6] studied the effect of fiber cross-sectional shape on moisture transmission properties of the fabrics and suggested that wicking rate through fabrics increased while water vapor permeability reduced as the fiber shape factor increased.

Matsudaira and Kondo [7] reported that more water could be absorbed by polyester fibers by making grooved or non-grooved hollow in fiber due to increase in space ratio and surface area of fiber in their studies on moisture transport properties of fabrics having different ratio of space to polymer in fiber cross-section.

Su et al. [8] developed composite knitted fabrics by blending profiled polyester fibers and cotton fibers. Fabrics with decreasing cotton content showed higher diffusion rate and drying rate. Worst water absorption ability was shown by fabrics made of profiled polyester alone. They suggested that moisture absorption and release of fabrics could be improved by making fabrics from core and cover yarns of polyester profile filament, profile polyester spun and cotton in different blend ratios.

Troynikov and Wardiningsih [9] suggested that blending wool fiber with polyester and regenerated bamboo fiber, produced fabrics with better moisture management properties than fabrics without blending.

Fangueiro et al. [10] studied the wicking and drying ability of knitted fabrics produced from blends of wool- coolmax and wool- fine cool. It was reported that fabrics with coolmax fibers could transport perspiration quickly from the skin to environment and showed the best capillarity performance, fine cool fabrics had higher drying rates whereas wool fiber-based fabrics showed low water absorption but good drying rate.

Oner et al. [11] observed higher overall moisture management capacity values for polyester fabrics compared to cellulose based fabrics and suggested that cotton fabrics caused wetness to be felt more than other fabrics.

Long [12] stated that liquid water transfer from the back to the face layer depends upon the water absorption of the fiber materials of the two layers and to a greater extent their difference.

Adams and Rebenfeld [13] observed that polyester fabrics showed better liquid water diffusion due to fast capillary action as the contact angle of polyester and water is small compared to wool. Highly hygroscopic fibers like wool took longer to reach equilibrium during process of water diffusion compared to less hygroscopic fibers like polyester.

Supuren et al. [14] investigated the moisture management properties of the double face fabrics and suggested that polypropylene (back) and cotton (face) fabric had better moisture management property.

Mehrtens & Mcalister [15] reported low wick ability for nylon fabrics when compared to cotton and orlon fabrics and suggested combination of lower fabric weight and thickness led to better comfort in their studies on knitted sport shirts for hot and humid conditions.

Ozturk et al. [16] studied the influence of fiber type on wicking properties of cotton- acrylic yarns and fabrics and suggested that wicking ability of yarns and fabrics increased with the increase in acrylic content in the blends.

The exhaustive reported research emphasizes that the fiber types owing to difference in their chemical nature and surface geometry have strong influence on heat, moisture, liquid transfer and moisture management properties of textiles.

4.2 Yarn variables and their influence on thermo-physiological comfort aspects

The yarn variables namely twist level, linear density, spinning system and yarn types play a crucial role in influencing the moisture vapor and liquid moisture transmission and in turn the thermo-physiological comfort aspects of sports textiles. The variation in any of these yarn variables influence the yarn structure which in turn depends on fiber geometry. Distribution of fibers in yarn dictates thermal as well as moisture transfer properties of fabrics.

Yarn structure is not rigid and capillary flow may produce lateral stress, which affects capillary sizes during liquid rise. Disruption of the continuity, length and orientation of the capillaries occurs due to changing packing density throughout yarn structure. Heterogeneity of pore size, shape and orientation affects the penetration of liquid into the yarn structure and hence its liquid retention properties. Likewise, number of filaments, yarn tension and twist significantly affect the yarn wicking performance by influencing the way in which individual filaments can pack in the yarn thus determining the amount of void spaces between filaments.

Moisture transfer is affected by degree of yarn twist, higher twist yarns improve capillary effect in moisture transfer as they are compact and provide less air volume. Lower twist generally results in reduced water transport through fabrics due to reduction in number and continuity of inter fiber capillaries. Twist in the yarn also affects the size of capillaries due to helical path of fibers in the yarn. More liquid on surface of twisted yarn is retained due to rough surface profile of these yarns compared to filament yarns.

Awadesh Kumar and Ramratan [17] studied the moisture management properties of different knit structures composed of micro polyester, texturized polyester and polyester –spandex blend and concluded that micro polyester fiber fabrics exhibited better liquid transmission properties compared to their counterparts owing to more capillary channels.

Linear density of constituent yarns affects the radial spread of water in fabrics. Fast liquid flow through inter yarn spaces in fine yarns is possible due to reduced capillary radius and low water retention of finer count yarns.

The yarns produced on different spinning system play a crucial role in dictating thermo-physiological properties of textiles intended for varied applications. The difference in the yarn structure and packing density of yarns produced on different spinning systems account for different thermal, moisture and liquid transfer properties of fabrics made from these yarns. Physical features of yarns and fabrics produced from these yarns are influenced by the type of yarn production (ring-spun, compact, open end) and in turn affect the performance properties of fabric.

A variety of yarns like ring, rotor, friction, vortex and compact spun yarns are used for varied applications in sports textiles. Dimensions and structure of inter yarn and intra yarn pores, pore size and their distribution along fabrics are influenced by density and structure of yarn.

Ring and rotor spun yarns vary widely in their structure which contributes to the entirely different properties of the two yarns. Ring-spun yarn has an ideal cylindrical helical structure with same number of turns per unit length in each helix, uniform specific volume and maximum packing density in the outermost zone of the yarn cross-section. Rotor spun yarn has a bipartite structure with an inner core which forms the bulk of the yarn and an outer zone of wrapper fibers occurring irregularly along the core length. Rotor yarn shows maximum packing density in first zone from core. Core part of rotor yarn is relatively dense structure; sheath part is less dense structure with belly-bands (**Figure 3**).

Yarn types can significantly influence the performance properties of textiles by affecting the fabric's bulk properties. Yarn hairiness and roughness can bring about changes in thermal properties of fabrics by entrapment of still air layer. Likewise, the moisture and liquid transfer properties of textiles are significantly affected by yarn types owing to difference in yarn roughness and arrangement of fibers in yarns. Increase in yarn roughness results in reduced rate of water transport through fabrics due to increase in effective advancing contact angle of water on yarn. Yarns with more random fiber arrangement can retard the liquid transfer by wicking as a result of disruption in continuity of capillaries formed by fibers. Wicking of yarns and fabrics is affected by difference in yarn surface roughness. Rough yarns are

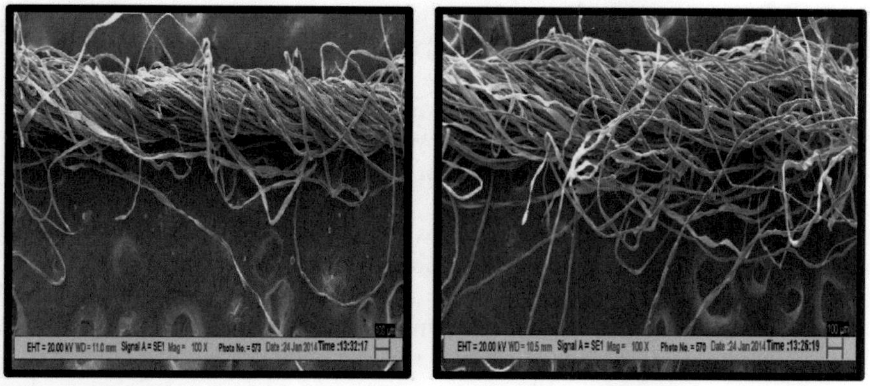

Figure 3.
SEM images of ring & rotor spun yarns.

formed by wool fibers with high apparent contact angle owing to random distribution of fibers in the yarns and the natural crimp. Yarns made of synthetic fibers have smooth surfaces and are well aligned.

Water transfer by capillary process is thus affected by two factors:

- Increase in yarn roughness causes an increase in effective advancing contact angle of water on yarn

- Random fiber arrangement decreases the continuity of capillaries formed by fibers in yarn

The following section reviews the studies undertaken and reported to determine the effect of various yarn variables on thermo-physiological aspects of textiles.

Y Jhanji et al. [18] studied the moisture management properties of polyester-cotton plated fabrics of ring vis a vis rotor yarns. They observed that ring yarn fabrics exhibited higher moisture vapor transmission rate, trans planar wicking, lower wetting time and higher one-way transport capacity as compared to rotor yarn fabrics, making the former suitable where body needs to dissipate sweat both in vapor and liquid forms, with respect to fabrics using combination of rotor-spun cotton yarns, which show higher absorbent capacity and would be slow drying with poor one way transport capacity. They concluded that yarn spinning system plays an important role in influencing moisture management properties of fabrics intended for next to skin applications.

Ansary [19] studied the influence of number of filaments on air permeability of polyester woven fabrics and reported a decrease in air permeability with increase in the number of filaments in the cross section of filling yarns.

Li and Joo [20] compared nano-scale filament, micro filament and normal filament knitted fabrics for their liquid transfer properties and concluded that nano-scale filament fabrics showed low porosity, high aerial density and increased absorption capacity and absorption rate. Better water absorption ability of nano scale filament fabrics compared to micro filament fabrics was attributed to smaller pore size of nano scale filaments compared to micro filaments.

Das et al. [21] varied the denier per filament for polypropylene knitted fabric to assess its influence on thermo-physiological comfort properties and observed that water uptake and wicking increases with increase in the number of filaments.

Behera et al. [22] compared the comfort properties of ring, rotor and friction spun yarn fabrics and suggested that ring and rotor spun yarns were comparable in thermal comfort aspects, friction spun yarn being the most suitable. They pointed out that in the normal wear conditions and in the absence of perspiration, rotor spun yarn would be superior to ring-spun yarns.

Kumar et al. [23] compared ring, rotor and vortex yarn knitted fabrics and observed that ring yarn knitted fabrics showed good knitting performance and smooth feel, however abrasion resistance of rotor and vortex spun yarn fabrics were higher than ring spun yarn fabrics.

Erdumlu and Saricam [24] studied the wicking and drying properties of vortex spun yarns and knitted fabrics in comparison with ring-spun yarns and fabrics. They observed that yarn type significantly affected the yarn wicking, fabric wicking and water absorbency. Vortex spun yarn owing to crimped yarn axis and tight wrappings along yarn length had lower yarn and fabric wicking values than ring-spun yarn fabrics. Fabrics knitted from ring-spun yarns wicked and absorbed water more evenly than fabrics knitted from vortex spun yarns.

Singh and Nigam [25] compared carded, combed and compact spun yarn woven fabrics for their comfort performance and reported that carded weft yarn-based fabric samples showed higher resistance against air drag than combed and compact weft filled fabric samples. Compact weft yarn fabrics showed high water vapor permeability and were reported to be suitable for summer wear shirting. Carded yarn woven fabrics showed high thermal insulation and were.

Sengupta and Murthy [26] reported that open- end spun yarns showed lesser wicking time for any given vertical weight compared to ring- spun yarn fabrics. They observed that owing to dense core and less dense skin of open-end yarns it showed differential dyeing behavior in core and skin with dye wicking to greater height in the core than in surrounding sheath fibers.

Chattopadhyay and Chauhan [27] compared ring and compact yarns for their wicking performance and suggested that ring yarns showed faster wicking compared to compact yarns as evident from higher equilibrium heights for ring yarns. They explained the lower wicking of compact yarn due to less average capillary size of compact yarn compared to ring yarn owing to higher packing coefficient of compact yarn.

4.3 Fabric variables and their influence on thermo-physiological comfort aspects

The thermo-physiological properties of textile materials particularly sportswear depend on constructional variables and bulk properties of fabrics. Fabric structure, thickness, cover factor, aerial density, bulk density, fabric porosity and finishing treatments affect the thermal and moisture management properties and hence determine the comfort properties of fabrics.

Woven and knitted fabrics are generally used for varied applications like inner wears, outerwear, work wear and sportswear. Knitted fabrics owing to lower cover factor have more pores in their structure and the porous structure ensures good air, moisture and heat transfer properties and show better liquid transmission properties than woven fabrics. The difference in basic structures of textile materials account for variation in amount of water absorbed by different fabric constructions. The structural differences are related to fiber arrangement in yarn thereby affecting yarn roughness factor Cos θ and size and continuity of capillaries. Random fiber arrangement leads to high contact angle; while lower contact angle associated with faster movement of water in yarns and fabrics is attributed to high degree of fiber alignment.

The different fabric structures used for sportswear vary in their bulk properties such as fabric tightness, porosity, aerial density and thickness that in turn dictate the heat, moisture and liquid transfer through the fabrics. Availability of inter yarn spaces for heat transmission, passage of air and moisture diffusion depend on the fabric's tightness factor. Thus, the bulk properties of fabric structures are crucial for optimum air, heat and moisture transmission through sportswear.

Several researchers have attempted to engineer different knit structures and compared the structures in terms of their comfort and performance properties intended for sportswear and other functional textiles. Innovative knit structures like plated fabrics, moisture management fabrics with different combinations of yarns in alternating courses, multilayered fabrics and fabrics mimicking the biometrics of plant structure have been developed for providing effective thermal and moisture management properties and sense of well-being to the wearer.

Structured or engineered fabrics are used in application areas relevant to commercial interest. Class of structured fabrics is moisture management fabrics; utilizing two or more fiber types in layered structures rendering two sides of fabrics distinctly different in character. Each side of fabric has the ability to exhibit different performance characteristics and thermo-physiological properties. Light weight two sided fabrics finding applications in varied areas are produced by plated knitted technique.

Both hydrophobic and hydrophilic yarns can be fed to single set of knitting needles and two separate yarns thus pass through each single needle of the set appearing distinctly on face and back sides of fabrics. Careful control of feed and positioning of two yarns is important to position distinct yarns in the two layers.

Plated knit structure is a double layered construction characterized by distinct face and back layers. The two layers are composed of different materials and accordingly serve different roles in providing wearer comfort.

One layer of plated fabric is the inner layer which is in direct contact with skin and serves the role of quick removal and transportation of sweat from body in vapor and liquid form. This layer serves as a separation layer and is composed of conductive and diffusive yarns generally characterized by low water absorption properties.

Another layer of plated fabric is the outer layer which is not in direct contact with the skin and prevents humidity build up near skin and vaporizes it to environment. This layer serves as absorptive layer and is composed of hydrophilic fibers and governs the liquid spreading and drying ability of fabrics. **Figure 4** shows the schematics of face and back layers of plated fabric.

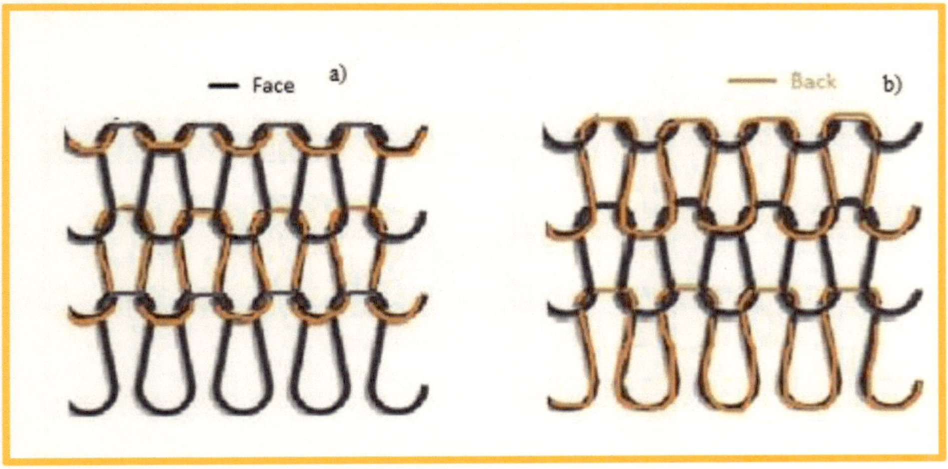

Figure 4.
Schematics of plated fabric (a) face and (b) back layer.

Selection of fiber and yarn combinations in the two layers can have a great bearing on the comfort properties, performance, esthetic appeal and end use of the knit structures.

Fibers of different chemical nature and thus different water absorbing properties can be used in different combinations to appear in face and back layers of plated fabrics.

Double layered knitted fabrics can be divided into following four types based on different fiber combinations and difference in water absorption properties of different fibers used in the two layers.

4.3.1 Double layered fabrics with hydrophobic fibers in face and back layer

The fabric has hydrophobic fiber in both face and back layers as shown in **Figure 5a**.

Liquid sweat next to skin cannot be absorbed by inner layer owing to its hydrophobicity and the only means by which sweat can be removed from skin is water vapor diffusion through pores within fabric. The diffused water vapor will evaporate slowly from the face layer in turn causing thermal and wetness discomfort to the wearer.

4.3.2 Double layered fabrics with hydrophilic fiber in back and hydrophobic fiber in face layer

The fabric has hydrophilic fiber in back/next to skin layer and hydrophobic fiber in face layer as shown in **Figure 5b**.

Liquid sweat next to skin can be absorbed by the back hydrophilic layer but the transfer of sweat to the face layer is restricted owing to hydrophobicity of the face layer. Thermal insulation of fabric decreases and fabric gives sensation of wetness and coolness as the pores in the inner layer are filled with water, removing the static air from the pores.

4.3.3 Double layered fabric with hydrophilic fibers in face and back layers

Figure 5c shows the fabric with hydrophilic fiber in face as well as back layers.

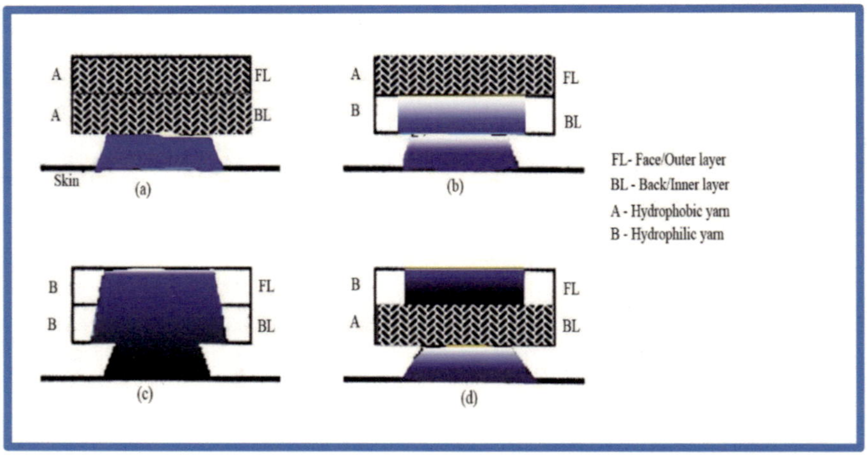

Figure 5.
Water transfer from skin to different fabric layers. (a) Hydophobic yarns in inner & outer layer, (b) Hydrophillic yarn in inner & hydrophobic yarn in outer layer, (c) Hydrophilic yarn in inner & outer layers, (d) Hydrophobic yarn in inner & hydrophilic yarn in outer layer.

Sweat from skin is picked up by hydrophilic fibers of back layer resulting in moisture accumulation and poor transfer to face layer. Water remains in the back layer and evaporation rate will be small owing to smaller wet area. The fabric will feel cool and wet to the wearer.

4.3.4 Double layered fabric with hydrophobic fiber in back and hydrophilic fiber in the face layer

Figure 5d shows the fabric with hydrophobic fiber in the back and hydrophilic fiber in the face layer. The back hydrophobic layer without absorbing the sweat itself transfers it to the face layer by means of capillary wicking. Face layer owing to hydrophilic fibers has good water absorption property and hence enables quick evaporation of sweat to environment by providing larger wet area.

Based on classification of double layered fabrics, Lord [28] indicated that that structure (d) with hydrophobic fiber in the back and hydrophilic fiber in the face layer would be most effective in maintaining dry skin micro climate by rapid liquid transfer to face layer. Additionally, several other researchers have unanimously recommended the use of hydrophobic fibers in next to skin and hydrophilic fibers in the face layer to achieve desirable moisture management and comfort properties in plated fabrics.

Plated fabrics designed with contrastingly different fiber and yarns exhibit the push- pull effect. Layer of hydrophobic fibers repel the perspiration next to skin and pushes or wicks it into outer layer of hydrophilic fibers which absorb or pulls away the moisture. Structured arrangement of hydrophobic and hydrophilic fibers in the two layers of plated fabrics and large difference in humidity between inner layer and ambient environment causes moisture movement from skin to outer atmosphere thus making the structures preferred choice for sportswear.

The structures are increasingly gaining popularity in apparels, next to skin applications, active wear and leisure sportswear owing to freedom in selection of contrastingly different constituents in the two layers. Therefore, the functional clothing intended for such applications are often specially engineered or structured such that the fabrics are normally two sided and are produced from a minimum of two yarns of different fiber content or characteristics.

Toda developed multi layered knitted structures composed of non-hygroscopic fibers. The structure was characterized by smaller inter fiber spaces in the face layer than in back layer by careful selection of fiber fineness, knitted structure and yarn type in face and back layers.

Yamini Jhanji et al. [29] investigated the effect of fiber type and yarn linear density on the thermal properties such as thermal resistance, thermal conductivity and thermal absorptivity along with air permeability and moisture vapor transmission rate of single jersey plated fabrics. They suggested that plated fabrics with nylon in the next to skin layer seemed suitable choice for warm conditions as these fabrics would feel cooler on initial skin contact owing to high thermal absorptivity and were permeable to passage of air and moisture vapor. Fabrics knitted with yarns of high linear density were found to be unsuitable in warm conditions owing to higher value of thermal resistance and lower values of air permeability and moisture vapor transmission rate.

Jhanji et al. [30] compared the moisture management properties of plated fabrics with altering hydrophilic and hydrophobic fibers in top and bottom layers and different types of hydrophobic fibers in top layers. They concluded that fabrics knitted with hydrophobic fibers (polypropylene, polyester) in top layers were suitable for next-to-skin applications as they were classified as moisture management fabrics owing to high values of accumulative one-way transport index and bottom spreading speed. It was further suggested that fabric knitted with nylon in top layer was classified as water penetration fabric due to poor liquid transfer properties. Fabrics knitted with cotton in top layer irrespective of the hydrophobic fiber in bottom layer were poor in moisture management properties.

Ghosh and Kaur [31] studied the effect of tightness factor on liquid transport properties of plain knitted fabrics and observed that with increase in tightness factor, fabrics showed higher wicking and lower water absorbency. They suggested that higher tightness factor resulted in less tortuosity thus providing less complicated path for liquid flow and offering less resistance to fluid flow compared to fabrics knitted with lower tightness factor.

Suganthi and Senthilkumar [32] studied moisture management properties of double layered fabrics varying the fiber types in inner and outer layers and observed that bi layered fabrics with micro fiber polyester in inner and modal in outer layer was the preferred choice for active sportswear owing to fabric's better moisture management properties.

The published literature suggests that fabric structures engineered by strategic combination of hydrophilic and hydrophobic fibers, speciality fibers and yarns exhibit variations in their bulk, physical and comfort characteristics thereby influencing thermal and mass transport properties of textiles. The fabric structure and in turn the fabric properties determine the suitability of textiles for sportswear applications. Having discussed, the significance of fiber, yarn and fabric variables on functional aspects of sportswear in the previous section, it becomes necessary to highlight the key trends and innovations in sportswear which serve to enhance the performance as well as esthetic attributes of the sportswear. The designing aspects and innovative approaches employed to render smart functionality to sportswear will be covered in details in the following sections of the chapter.

5. Key trends in sportswear design and development

Key trends in sportswear design and development encompasses performance and esthetic evolution of sportswear from next to skin to exterior or outer wear.

The inception of new functional and high-performance fibers and waterproof and breathable materials like polypropylene, polyester, polyamide in micro fine denier and Goretex respectively led to innovations in first layer sportswear such as performance underwear. The functional properties like wicking, fast drying, anti-odor and UV blocking have been considerably enhanced by inclusion of new, innovative fibers. However, the raw material selection has not brought about radical changes in design aspects of the first layer.

5.1 Designing sportswear as first layer garments with enhanced functionality and unconventional styling

The first layer garments have undergone a major transformation with more emphasis on design and development of all-in-one suits in competition swimming and running, winter sport wear and athletics.

Furthermore, sportswear manufacturers are exploring the avenues for creating garments offering multiple functionalities in a single layer as per specific requirements of wearer's body parts.

The first layer sportswear is particularly popular among runners and top level athletes who seek comfort, unhindered bodily movement, light weight, fast drying and stretch ability in their attires. Apart from functional aspects, first layer sportswear have witnessed huge esthetic transformation with emergence of racier styles featuring attractive and variable designs, funky colors, quirky prints, patterns and strategic placement of trimming as means of surface ornamentation. The sportsperson and fitness freaks who once merely considered the performance aspects of their clothing, no longer follow a taciturn approach to doff a stylish, funky sportswear that can render psychological well-being to wearer and visual delight to viewers.

Accordingly, designers are fostered to include innovative design concepts such as elaborate patchwork, asymmetrical styling and unconventional placement of trimmings, notions and labels in their sportswear design collections with due consideration to the changing preferences of sportspersons and consumers.

The functional aspects of performance under wears are enhanced by incorporation of innovative technologies like application of moisture management, UV protective, bacteriostatic finishes, controlled release of chemicals and other auxiliaries via microencapsulation. Accordingly, the underwear exhibit exceptionally superior moisture management properties, thermal and UV protection, antimicrobial and antistatic properties. Apart from functional attributes, the performance underwear have evolved significantly with vibrant fabric colors, contrasting trimming and off-center patterns widely used in their designing.

Introduction of asymmetrical design concepts like placing the closures along the side seam serve both esthetic and functional aspects by rendering unorthodox fashion appeal, layering and enhanced wearer agility. The trendy styles are thus becoming asset for youth oriented sportswear.

The first and second skin sportswear segment once considered a dowdy category, has emerged as top notch sportswear segment bringing new dynamics to sportswear market with all the innovative design concepts enjoying consumer acceptance.

5.2 All in one suits

The classic example of all in one suit is the body-covering Speedo swimwear intended for competition swimming introduced during Olympics.

The swimwear design fostered the concept of bio mimetics in sportswear designed later as the former closely mimicked the sharkskin as far as design orientation was concerned.

The success story of all-in-one swim suits paved the way for designing athletic sportswear, speed skating and cross country skiing suits. Nike, a popular sportswear brand was trailblazer in designing an elaborate, paneled speed skating suits comprising of seven different fabric types for cyclists. The novel suit with patchwork was designed to enhance the cyclist's performance, protection level and comfort in spite of the unfavorable ambient environment and excruciating conditions which cyclists generally encounter. The high tech suits are the state of the art suits offering multiple functionalities such as elasticity, compression, thermal insulation, protection against external elements and aerodynamics. The patch work design unique to high-end cycling sportswear has been adopted in second skin and first layer garment design as well.

5.3 Designing smartly via incorporation of sensors and electronic components

Another design perspective in sportswear segment envisages the incorporation of smart features via sensors and other electronic components that are comparable to high tech trimmings. A microphone with its associated embroidered control buttons on a garment sleeve or collar renders graphic yet functional embellishment to the clothing. The elimination of wind and rain flaps by inclusion of water tight zippers for medium level performance outerwear, switching to leaner and pared styles of trims and notions like printed and embroidered labels and motifs, drawstrings, velcro, snap closures and mesh lining for pockets to offer storage and ventilation both are some approaches to enhance the functionality and esthetic appeal without adding any additional bulk to the sportswear.

The sportswear designers are thus fascinated by concept of stealth design that implies less detailing, fewer accessories yet not at tradeoff with functional and smart features.

5.4 Design approaches to render breathability and waterproofing to sportswear

Waterproofing and breathability becomes all the more crucial while designing sportswear intended for outdoor sports where sports person is doomed to be exposed to humid, rainy conditions.

The technologies generally employed for development of waterproof breathable sportswear include:

- Development of high density fabric
- Application of polymeric coating
- Film lamination.

Development of High density fabric - The densely woven fabrics consisting of cotton or synthetic microfilament yarns with individual filament diameter of less than 10 micron and produced with high cover factor exhibit water proofing and breathability. The high cover factor of fabrics reduces the inter yarn spaces thereby preventing liquid penetration through fabric structure. The fabric on exposure to liquid causes cotton fibers to swell transversely reducing the pore size in the fabric structure. The dimensions of pore are smaller than water droplet thus effectively preventing water penetration however the pores allow the transmission of water vapor molecules (insensible perspiration) on account of smaller size of vapor molecules compared to water droplets thereby rendering breathability to the fabrics. The classic example of water proof breathable fabric is VENTILE, a high density oxford woven cotton fabric that effectively prevents the penetration of fluid but is permeable to passage of water vapor through the clothing.

Coated fabrics - Fabrics intended for sportswear can be imparted water proofing and breathability by application of polymeric coating either on one or both fabric surfaces. Polyurethane is the most commonly used coating for imparting water proofing to textiles. Micro-porous and hydrophilic membranes can be used for development of coated textiles. The micro-porous membrane features a coating containing very fine inter connected channels of the dimensions smaller than the finest raindrop. However, the size of channels is larger than that of water vapor molecules enabling water vapor passage through the air-permeable channels. Although, the hydrophilic membrane exhibit similar structure as that of micro-porous membrane, however, the mechanism of water vapor transmission in former is via adsorption-diffusion and de-sorption in contrast to passage of water vapor molecules through the air-permeable channels in the latter.

Lamination involves bonding a waterproof and breathable film to textile substrate. Thin polymeric membranes of maximum thickness up to 10 micron when bonded with base fabrics offer water proofing and breathability to textile substrate. Micro-porous membrane of poly-tetra fluoro ethylene (PTFE), poly-vinyldene fluoride PVDF and hydrophilic membrane composed of poly ethylene oxide are utilized for development of laminated water proof textiles for sports applications.

The ingress of water through seams in a water proof garment needs to be prevented through seam sealing. Apart from waterproofing, the laminated garments should be lightweight, flexible and comfortable to wearer. Thus, thinner strips, elasticized tapes and improved glues are increasingly being used for designing bulk free laminated sportswear. The traditional three ply composite construction comprising of fabric, film and mesh lining have undergone major transformation by elimination of mesh linings and addition of silicone touch finish to films imparting cleaner finishing and convenient doffing of the clothing. The overall freedom of wearer movement is thus ensured as a result of reduced friction within garment layers.

The sportswear designers prefer to do away with seams as they are a major source of friction, added fabric layers and bulk. Thus, designers prefer seamless knitting or heat sealing for reduction and elimination of seams to achieve a clean, compact performance wear.

The three layer sportswear are generally preferred for outdoor activities like hiking and cycling owing to their ability to provide protection against external elements (extreme cold or humidity) along with basic sportswear requirement of being lightweight, breathable and comfortable.

Each layer of a three layered assembly is designed to serve a specific function. The first, next to skin layer is designed with hydrophobic fibers to wick away sweat from skin to the outer layers, thereby rendering dry feel next to skin. Additionally, the innermost layer offers thermal protection to wearer in cold ambient conditions.

Second layer garments generally composed of fleece, assist in keeping the wearer warm and dry by drawing sweat from skin to the outer layer. The modifications in second layer are targeted to achieve high warmth to weight ratio without compromising the thermal insulation of clothing. However, the traditional three layer protective clothing assemblies are being rapidly replaced by advanced composite textile structures referred to as soft shell clothing designed by bonding multiple knits and fleece layers together.

The latter offers agility to wearer, protection against adverse environmental conditions with an additional advantage of being light weight and compact.

The second layer is further improvised to impart multiple functionalities such as warmth retention and insulation, water resistance, elasticity and wind protection. Therefore, sportswear has been witnessing a transition from complete water proofing by hard shell to water resistance by soft shell.

There are three approaches to design soft shell with augmented thermal insulation and wind protection. The first approach involves the utilization of windproof shell as a separate clothing entity while the second involves bonding fleece to wind blocking membrane. The membrane laminated sportswear thus offer thermal insulation along with water proofing and breathability. A new range of laminates designed with wind defender type membranes namely Gore – Tex Windstopper, Symptex Windmaster underscore protection against wind over water proofing, are being specifically developed for windy climatic conditions (**Figure 6**).

Adequate warmth and wind protection can also be achieved by third approach wherein fleece is bonded to tightly woven fabric or knitted structure.

Moreover, the comfort level, warmth and protection to wearer can further be provided by four layer system comprising of four garments - first layer, fleece, soft shell and hard shell.

Soft shells comprising of fleece and treated with water repellant surface finish are ideal candidates for outdoor activities as they primarily focus on enhanced thermal insulation, elasticity and abrasion resistance. The jackets have evolved radically

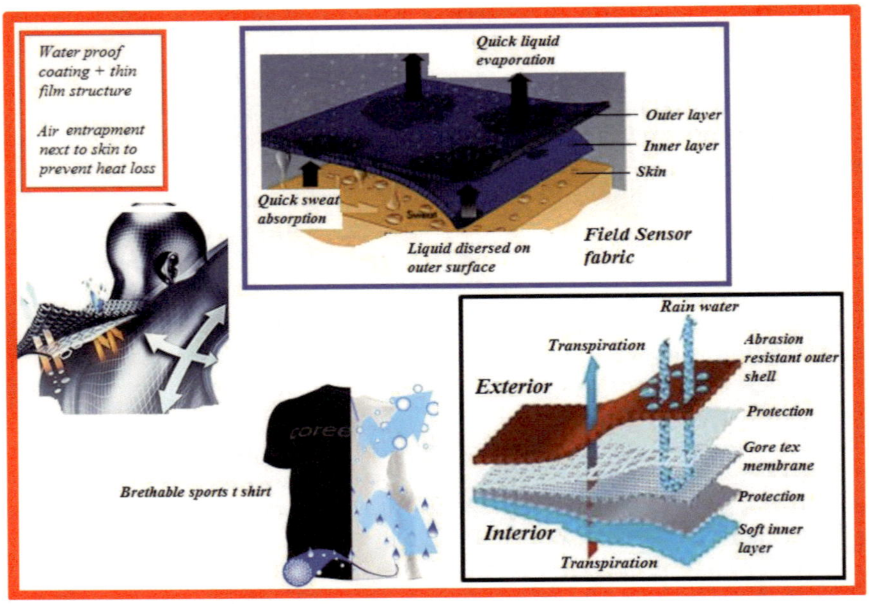

Figure 6.
Water proof & breathable sports wear.

as far as design and style elements are concerned and are increasingly being designed devoid of multiple drawstrings, elasticized hems or double storm flaps thereby eliminating cumbersome and bulky garment features. A closer-fitting, bulk free silhouette for better mobility, warmth retention and comfort to wearer has thus become synonymous to performance outerwear. The designing of hard shell jacket is also not aloof of the close fitting approach and thus designers have been striving to design leaner, fitted hard shell attires taking design inspirations from soft shells.

5.5 Design approaches for enhanced thermal insulation of sportswear

Other approaches for designing outdoor, winter sportswear are based on fundamental concept of exploiting the good insulation properties of still air layer and thus engineering textile structures with an ability to trap large volumes of still air. The entrapped air layer being good thermal insulator can provide enhanced thermal insulation to clothing incorporating hollow fibers, three dimensional spacer fabrics and alveolar or nodular raised knit structures. Hollow fibers on account of their light weight and improved thermal regulation outshines conventional fibers and are considered ideal for all such applications where high thermal resistance is sought for.

Accordingly, knitted structures like pique, honeycomb or ribbed raised textures are generally used for designing sportswear intended for winters. The honeycomb knits in conjunction with raised fabric in next to skin layer offers effective thermal insulation and are suitable for cold weather clothing and sportswear. Likewise, the structure of fleece can be modified for enhanced warmth by trimming the piles and creation of three dimensional grid thereby increasing the air entrapment next to skin.

The high performance thermo regulation along with light weight and wearer comfort can be engineered into sports apparels and accessories via three dimensional knit structures, spacer fabrics.

5.6 Nature as source of inspiration for sportswear designers

Nature is a big source of inspiration for human beings and capturing nature's beauty and functionality by biomimetic is a concept frequently explored in functional clothing particularly sportswear and protective clothing.

Speedo, a sportswear manufacturing company developed one of its own kinds of Fast skin biomimetic swimsuit taking inspiration from shark skin (**Figure 7**). The denticles of shark's skin were imitated on the fabric to impart super stretch property and thus the performance of swimmer donning the swimsuit could be considerably enhanced by shape retention, muscle compression and reduced drag coefficient.

Inotek® fabric based on "Pine cone effect" is quite popular among sportswear designers and manufacturers owing to exceptionally excellent thermal regulation and moisture management properties exhibited by the fabric. Pine cones comprises of two layers of stiff fibers that are oriented in different directions (**Figure 7**). The cones tend to close as the humidity increases to prevent moisture from getting in while they open up releasing their seeds and falling to the ground as a response to decreasing humidity. Likewise, the Pine Cone Effect based on reaction of plants to humidity is explored for designing fabrics that can respond to changing humidity conditions. The textiles based on biomimetic concept are composed of layer of thin wool spikes that open up on encountering increased humidity as a result of sweating by wearer. However, as the sweat evaporates and humidity drops down, the spikes on the fabric closes again in response to changing humidity.

The designers of long jump suit named SKYNFEEL exclusively designed for professional athletes might have been enticed by the salient characteristics of fauna to exhibit unhindered flights with their wings. The suit is designed with laterally

Figure 7.
Biomimetic principles for sportswear designing.

positioned flaps much like the wings of dragonfly. The dragon fly wing inspired flaps feature geometric lased cut panels that enhance the athlete's elevation during jumping via closing and opening up as per athlete's movement. The flaps remain closed during run-up however they open up as the wearer is preparing for jump. The opening of panels leads to creation of air pockets thereby resulting in aerodynamic effect. Consequently, the jumper's performance is enhanced due to his ability to suspend in air for longer duration and gaining distance while jumping.

Stomatex®, another smart fabric designed using the biomimetic principles finds application in compression athletic wear for enhanced performance and recovery. The salient feature of fabric is dome and pore mimicking the stomata (tiny pores) on the plant's leaves responsible for respiration and gaseous exchange in plants (**Figure** 7). The phenomenon of opening of stomata in daylight and closure at night is attempted to be recreated by the way of opening and closing of pores present on domes embossed in the outer knitted layer of fabric. The sportswear utilizing aforesaid fabric is generally designed in close-fitting silhouette to be able to react to wearer's bodily movements. During static conditions, the energy consumption by sports person is reduced and thus wearer comfort is ensured by release of excess heat and moisture rising into the domes and ultimately released via the pore. As the wearer is actively involved in some physical activity, the flexing and movement of domes (and pores) enables passage of cooler air into clothing and escape of heat and moisture to outer environment.

6. Innovative approaches for sportswear design and development

The sportswear industry is taken by storm by path breaking innovations as far as procurement of raw materials like high performance and specialty fibers, yarns and engineering of fabric structures like double layered, elastane and breathable fabrics are concerned. Furthermore, smart functionalities like antimicrobial, antistatic, anti-odor properties, monitoring sportsperson's physiological parameters, incorporation of smart materials like phase change materials and shape memory polymers, wearable sensors, tracking performance record of sportsperson and incorporation of smart technologies like smart coatings, nano technology and wearable electronics are engineered into sportswear (**Figures 8** and **9**).

6.1 Innovative raw materials for sportswear

The fibers suitable for sportswear have already been discussed in the previous section. However, the innovations in sportswear cannot be conscripted without the mention of high performance fibers and their role in improving the moisture transmission properties of sportswear. The utilization of high performance fibers such as Coolmax®, Thermolite®, Thermocool® in performance and active wear results in increased surface area, better wicking and moisture management and in turn dry, cooler microclimate to wearer. Thermolite® is particularly suitable for cold weather sportswear due to fabric's exceptionally high thermal insulation and moisture transmission properties. Winter sportswear comprising hollow core fibers possess the ability to trap higher volumes of static air and thus provide enhanced warmth and wearer comfort without any additional weight or bulk unlike conventional fleece fabrics. The clothing is thus gaining popularity among sports persons indulging in outdoor, winter sports like ice skating, mountaineering etc.

Dryarn, an innovative sustainable fiber from Aquafil is recyclable polypropylene microfiber and a preferred choice for developing sportswear fabrics that are sustainable, soft, anti-bacterial, light weight, quick drying, comfortable and exhibits high thermoregulatory capacity.

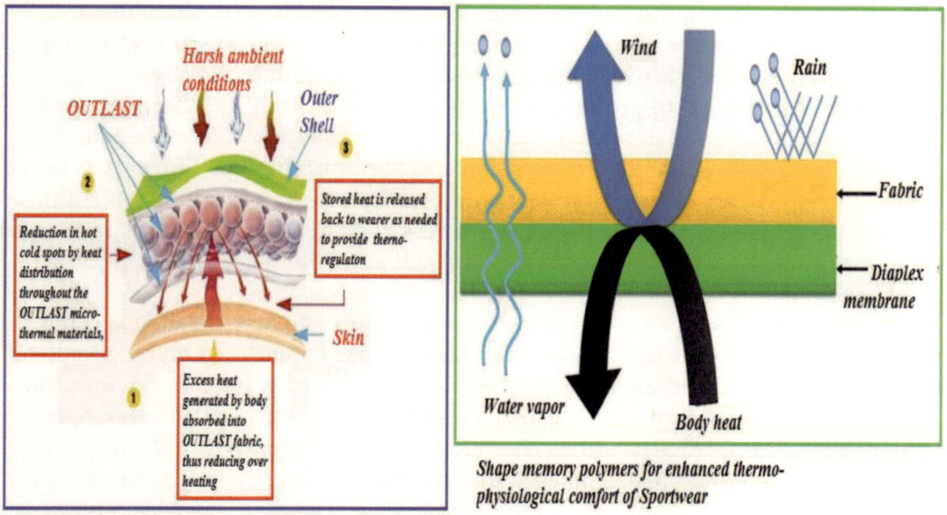

Figure 8.
Phase change materials & shape memory polymers for sportswear.

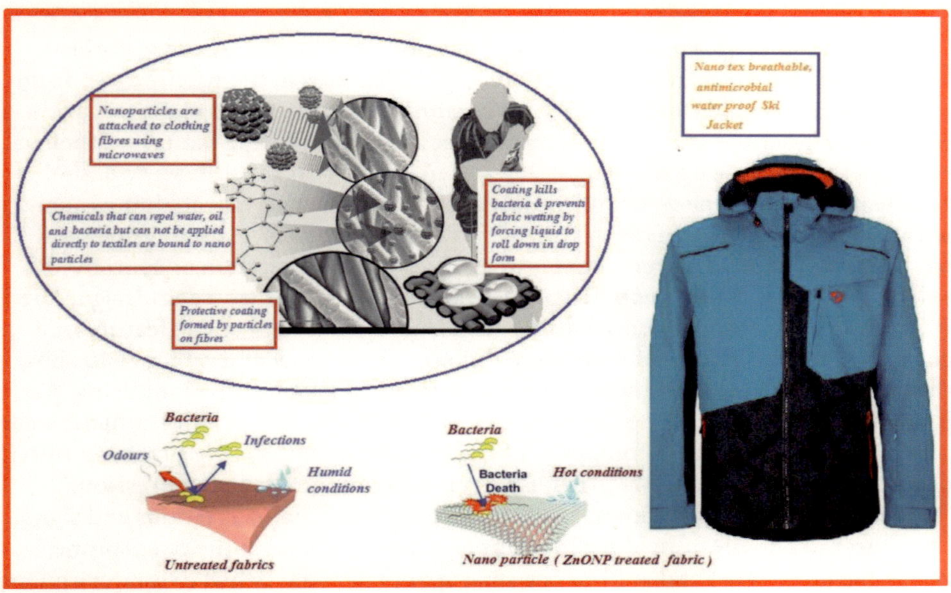

Figure 9.
Nano technology for antimicrobial sportswear.

Sportwool®, a two layered moisture management fabric featuring wool on the inner side and synthetic fiber on the outer side and Field sensor TM® with brushed inner side are other options for winter sportswear.

Field Sensor, high performance fabric from Toray is a multilayered structure suitable for varied sports applications. The excellent moisture management properties and wick ability for rapid liquid transmission from next to skin to outer layer can be attributed to the fabric's specially engineered structure with distinct inner and outer layers composed of coarser denier yarn and fine denier hydrophobic polyester yarn in a mesh construction respectively.

Additionally, thermo regulating materials like phase change or latent heat storage materials capable of sensing varying ambient conditions and responding by changing their phase are increasingly finding application in sportswear where sportsperson is exposed to prolonged, drastic environmental conditions. Outlast technology is involved in development of microencapsulated PCM coated fabric intended for sportswear and other smart textile applications. The sportswear developed with PCM treated fabrics provides thermal balance and maintain constant body temperature to wearer by absorbing excess body heat at elevated temperatures due to metabolic heat production and releasing it as the temperature drops down during cooling.

The potential of shape memory polymers to obtain effective thermal and moisture management properties was first explored in sailor suit designed for Swedish sailors. The suit based on membrane technology employed waterproof, windproof and breathable Diaplex membrane. The smart membrane can sense the changing ambient conditions and respond by changing its shape, memorizing the original shape and returning to the orginal, memorized shape accordingly. The membrane undergoes Micro-Brownian motion as it senses elevated temperature thereby creating micro-pores for heat and moisture transmission through the membrane (**Figure 8**).

6.2 Innovative wearable sensors and AI based technologies for sportswear

The concept of sportswear design and development has been drastically changing with sportsperson anticipating technological features in their attire apart from basic requirements of functionality and comfort. It thus becomes mandate for sportswear designers and manufacturers to conform to the expectations of their consumers and come up with technology laced sportswear that can serve best of both worlds by offering comfort, protection and other functional attributes along with serving as a personal trainer, activity tracker and can monitor physiological parameters of the sports persons (**Figure 10**). The new generation of sportswear exhibits such smart features which would be considered fantasy a decade ago. The myriad of innovations in sportswear as far as incorporation of smart features are concerned will be discussed in detail in this section of the chapter. **Figure 10** show the technology laced smart sportswear for performance & health monitoring. An innovation in athletic wear is development of Skin® 400 compression athletic wear series. The athletic wear is composed of elastane warp knitted fabric that can foster oxygen delivery to athlete's active muscles via dynamic gradient compression.

Smart sportswear like fitness pants feature built-in haptic vibrations and signal the wearer to be agile or hold position as per pulse generated at the stress prone zones like hips, knees and ankles. The smart pants can be synched to wearer's phone via bluetooth and provides additional feedback through the companion app.

Ambiotex's smart shirt intended for athletes is designed with integrated sensors and clip-on box for recording athlete's data pertaining to heart rate variability, threshold, fitness and stress levels.

Sabine Seymour's smart bra is designed with an integrated invisible biometric sensors and can be effectively worn to track heart rate and workout routines during sporting, gyming activities.

Compression sleeve by Komodo technologies is beneficial for sportspersons and heart patients as the sleeve is capable of monitoring heart rate activity via electrocardiogram (ECG) technology. The sensors integrated in the sleeve can monitor wearer's body temperature, sleeping patterns, workout intensity, air quality and UV rays.

Figure 10.
Technology laced smart sportswear for performance & health monitoring.

A smart shirt laced with blue tooth smart sensor can be paired with fitness apps like Map My Run, Run Keeper and Strava to capture the real time data and monitor physical activity and physiological parameters of sportspersons such as intensity and recovery; consumed calories fatigue level and sleep patterns.

Athos deals in range of training clothes intended for gym goers. The ensemble is designed with micro-EMG sensors that can detect muscle movement and in turn transfer the workout data such as sports person's muscle effort, heart rate and breathing to a smartphone. The app serves as a personal trainer for fitness freaks by providing insights on correct exercise protocol and injury avoidance.

Hexo skin smart sportswear are designed with embedded textile sensors for monitoring cardiac, respiratory phenomenon, physical training, sleep patterns, and mundane activities of an individual involved in rigorous sports. The visualization, reporting and analysis of data becomes very convenient with Hexo skin as the smart clothing is equipped with an accelerometer to quantify body movements, track heart rate to be viewed in real time. Furthermore, it prevents the sports person from over training by determining lung capacity for each activities performed and measurement of stress and training fatigue.

Boltt, a sports tech-brand pioneer in design and development of consumer-centric health and fitness clothing and smart shoes laced with stride sensors and activity tracker. The real-time audio feedback and customized workout suggestions generated by the brand's advanced artificial intelligence (AI) ecosystem can provide customized health and fitness coaching to wearers.

An on demand inflatable- deflate able textile tubing network in a jacket developed by Sympatex can provide the extra warmth to wearer when inflated while the jacket can be deflated to release the air held inside the tubes into the environment as it senses an elevation in body temperature ow wearer's activity level.

A Cyberia survival suit intended for arctic environment serves as a personal GPS and monitors the wearer's physical conditions. The suit derives its sensing and monitoring capabilities from an array of sensors and connecting electrodes embroidered onto textile substrate.

Qardio Core, an ECG monitor is a hardware fitness tracker capable of providing continuous medical grade data by incorporated sensors in the tracker. The smart tracker is designed for people indulging in active lifestyles but with a family history of chronic diseases. The doctors can analyze the obtained data from sensors to monitor the health record and can act instantly in event of any abnormal data.

Wearable X, pioneer in bringing design and technology together, launched smart yoga wear incorporating haptic feedback. Posture monitoring and vibrational reaction by smart garment assists in guided yoga.

Vitali smart bra is another state-of-the-art smart sport wear designed for fitness freak females. The bra is equipped with sensors to track heart and breathing rates. The stress levels of women can be monitored via data collected from sensors thereby sending reminder to wearer to take deep breath on detection of high stress levels.

Exercise routine for elderly people can be managed by smart knitted cardigan designed by Dutch designer Pauline van Dongen. The ordinary looking cardigan is equipped with four stretch sensors comprising conductive yarns and can transmit the information to an app for generating feedback. The obtained feedback serves as a guide for physiotherapist to suggest the best exercising options as per age and physical stamina of the wearer.

7. Conclusions

Sportswear constitutes an integral part of technical textiles and encases great potential as far as technological and design innovations are concerned. The sports textiles have witnessed tremendous evolution and that too at a much faster pace compared to ready to wear segment. The sports clothing is no longer restricted to sportsperson involved in performance sports or strenuous physical activities. However, there has been a surge for sports apparels and accessories among health conscious, fitness freak and gym enthusiasts. Accordingly, the sportswear industry has witnessed revolutionary advancements in development of different sportswear categories like active wear, leisurewear and athleisure to fulfill the requirements of sportsperson as well as health-conscious millennials. The basic and functional requirements of comfort, breathability, light weight, anti-static and anti-odor properties can be engineered into sportswear by optimum selection of fibers, yarns, fabrics and garments' designing aspects.

Sportswear has emerged as one of the most promising and technologically driven segments of technical textiles with massive innovations and advancements involved in design and development of sport specific attires.

The basic requirements of sportswear vary as per the sportsperson's level of physical activity, the specific nature of sport and the ambient conditions to which the sportsperson is exposed. The technological and ergonomic aspects for designing under water diving suit will be contrastingly different from those incorporated in designing clothing for a golfer. Thus, a lot of brain storming and research is involved in designing a clothing that meets the specific requirement of sports along with providing comfort, dexterity, agility to wearer, breathability, moisture management, light weight, antimicrobial and anti-odor properties. The correct selection of fibers, yarns and fabric variables for sportswear is of paramount importance to engineer the desired properties in the sport clothing.

Apart from functional requirements, a lot of emphasis is being laid on esthetic aspects as well considering increasing number of females involved in yoga, gyming and other sporting activities who give precedence to silhouette, colors and other design details of sportswear. Accordingly, the technological as well as ergonomic advancements in sportswear design and development have opened new avenues for researchers to explore the field further.

8. Future scope in design and development of sportswear

The field is promising and innovative with several avenues as far as research and development, pioneering new technologies and trailblazing concepts are concerned. Furthermore, the role of technology, bio mimics, fashion and mediation of interdisciplinary fields like wearable electronics, bio medical avenues have brought about a major transformation in design and development of smart sportswear thereby enhancing functionality and esthetics of sportswear. Apart from functional and esthetic appreciation of sportswear, the need of the hour is switching over to sustainable practices in sportswear supply chain. Consequently, the sportswear designers and manufacturers understanding their social and economic responsibilities should commit to sustainable practices for design and development of sportswear.

It can thus be recapitulated that the future belongs to smart sportwear spanning from ethical, to technology laced wearable electronics to camouflage clothing to convertible, modular sports ensembles which not merely serves as clothing for wearer but can be multifunctional entities with an ability to be transformed into a travel bag or a sleeping bag as per the sportsperson's convenience and requirements.

Author details

Yamini Jhanji[1,2]

1 The Technological Institute of Textile and Sciences, India

2 Indian Institute of Technology, Delhi, India

*Address all correspondence to: yjhanji@gmail.com

References

[1] Chaudhari S.S, Chitnis R.S & Ramakrishnan R (2008), Water proof breathable active sports wear fabrics,http://www.sasmira.org/sportswear.pdf.

[2] Das A & Manshahia M, High active sportswear, International Conference on Emerging Trends in Traditional & Technical Textiles ISBN: 978-93-5156-700-4 11th - 12th April 2014, NIT Jalandhar.

[3] Patnaik A., Rengasamy R. S., Kothari V. K. and Ghosh A., 2006, Wetting & wicking in fibrous materials, Textile Progress, 38, pp.1-10.

[4] Li Y., 2001, The science of clothing comfort, Textile progress, 31(1), pp. 1-135.

[5] Gurudatt K., Nadkarni V. M. and Khilar K. C., 2010, A study on drying of textile substrates and a new concept for the enhancement of drying rate, Journal of The Textile Institute, 101(7), pp. 635-644.

[6] Das B., Das A., Kothari V. K., Fanguiero R. and Araujo M., 2008, Effect of fiber diameter and cross-sectional shape on moisture transmission through fabrics, Fiber Polym, 9(2), pp. 225-231.

[7] Matsudaira M. and Kondo Y., 1996, The effect of a grooved hollow in a fiber on fabric moisture and heat transport properties, J Text Inst, 1 (3), pp. 409-416.

[8] Su C., Fang J. X., Chen X. H. and Wu W. Y., 2007, Moisture absorption and release of profiled polyester and cotton composite knitted fabrics, Text Res J, 77, pp. 764-769.

[9] Troynikov O. and Wardiningsih W., 2011, Moisture management properties of wool/polyester and wool/bamboo knitted fabrics for the sportswear base layer, Text Res J, 81, pp. 621-631.

[10] Fangueiro R., Goncalves P., Soutlnho F. and Freitas C., 2009, Moisture management performance of functional yarns based on wool fibers, Indian J Fiber Text Res, 34 pp. 315-320.

[11] Oner E., Atasagun H. G., Okur A., Beden A. R. and Durur G, 2013, Evaluation of moisture management properties on knitted fabrics, J Text Inst, 104, pp. 699-707.

[12] Long H. R., 1999, Water transfer properties of two-layered weft knitted fabrics, International Journal of Clothing Science and Technology, 11, pp. 198-205.

[13] Adams K. L. and Rebenfeld L., 1987, In-plane flow of fluids in fabrics: Structure/flow characterization, Text Res J, 57(11), pp. 647-654.

[14] Supuren G., Oglakcioglu N., Ozdil N., and Marmarali A, 2011, Moisture management and thermal absorptivity properties of double-face knitted fabrics, Text Res J, 81, pp. 1320-1330.

[15] Mehrtens D. G. and Mcalister K. C. 1962, Fiber properties responsible for garment comfort, Text Res J, 32, pp. 658-665.

[16] Ozturk M. J., Nergis B. and Candan C., 2011, A study of wicking properties of cotton-acrylic yarns and knitted fabrics, Text Res J, 81, pp. 324-328.

[17] Awadesh Kumar Chaudhary and Ramratan, The influence of yarn and knit structure on moisture management properties of sportswear fabrics, 101, 77-90 (2020).

[18] Jhanji Y, Deepti Gupta & V K Kothari, Moisture management properties of ring vis a vis rotor yarn plated knit structures, Indian Journal of Fiber & Textile Research 46, March 2021, pp. 48-51.

[19] Ansary R., 2012, The influence of number of filaments on physical and mechanical characteristics of polyester woven fabrics, Life Science Journal, 9, pp. 79-83.

[20] Li Y. and Joo C. W., 2013, Effect of filament size on the water transport of weft and warp knitted fabrics, Fiber Polym, 14(12), pp. 2169-2175.

[21] Das B., Bhattacharjee D., Kumar K. and Srivastava A., 2013, Thermo-physiological comfort characteristics of fine denier polypropylene fabrics, Res J Text Apparel, 17, pp. 133-140.

[22] Behera B. K., Singh M. K., 2013, Role of filament cross-section in properties of PET multifilament yarn and fabric. Part II: effect of fiber cross-sectional shapes on fabric hand, Journal of the Textile Institute, http://dx.doi.org/10.1080/00405000.2013.774132, pp. 1-12.

[23] Kumar R. C., Kumar A. P., Senthilnathan P., Jeevith R. and Anbumani N., 2008, Comparative studies on ring, rotor and vortex yarn knitted fabrics, Autex Res J, 8, pp. 100-105.

[24] Erdumlu N. and Saricam C., 2013, Wicking and drying properties of conventional ring and vortex spun cotton yarns and fabrics, J Text Inst, 104(12), pp. 1284-1291.

[25] Singh M. K. and Nigam A., 2013, Effect of various ring yarns on fabric comfort, Journal of Industrial Engineering, http://dx.doi.org/10.1155/2013/206240, pp. 1-7.

[26] Sengupta A. K. and Murthy H. V. S., 1985, Wicking in ring spun vis a vis rotor spun yarns, Indian J Text Res, 10(4), pp.155-157.

[27] Chattopadhyay R. and Chauhan A., Wicking behavior of compact and ring spun yarns and fabrics in One day seminar on Comfort in Textiles, Department of Textile Technology, IIT Delhi, New Delhi, India, 2004, October 16, pp.20.

[28] Lord P. R., 1974, A comparison of the performance of open end and ring spun yarns in terry toweling, Text Res J, 44, pp. 516-522.

[29] Yamini Jhanji, Deepti Gupta & V K Kothari, Comfort properties of plated knitted fabrics with varying fiber types, Indian Journal of Fiber & Textile Research, 40 (3) 2015, pp. 11-18.

[30] Y Jhanji, D Gupta & V K Kothari, Moisture management properties of plated knit structures with varying fiber types, The Journal of The Textile Institute, 2014, DOI:10.1080/00405000.2014.934044

[31] Ghosh S. and Kaur A., 2014, Wicking and absorbency behavior of plain knitted fabrics, paper presented at the International Conference on emerging trends in traditional and technical textiles, NIT Jalandhar.

[32] Suganthi T & Senthilkumar P, Moisture-management properties of bi-layer knitted fabrics for sportswear, Journal of Industrial Textiles, 47 (7), 1447-1463 (2018).